Archaeology and Ethnoarchaeology of Mobility

UNIVERSITY PRESS OF FLORIDA

Florida A&M University, Tallahassee
Florida Atlantic University, Boca Raton
Florida Gulf Coast University, Ft. Myers
Florida International University, Miami
Florida State University, Tallahassee
New College of Florida, Sarasota
University of Central Florida, Orlando
University of Florida, Gainesville
University of North Florida, Jacksonville
University of South Florida, Tampa
University of West Florida, Pensacola

Archaeology and Ethnoarchaeology of Mobility

Edited by Frédéric Sellet, Russell D. Greaves, and Pei-Lin Yu

University Press of Florida
Gainesville/Tallahassee/Tampa/Boca Raton
Pensacola/Orlando/Miami/Jacksonville/Ft. Myers/Sarasota

First cloth printing, 2006
First paperback printing, 2015

A record of cataloging-in-publication data is available from the Library
of Congress.
ISBN 978-0-8130-2956-2 (cloth)
ISBN 978-0-8130-6140-5 (pbk.)

The University Press of Florida is the scholarly publishing agency for the State
University System of Florida, comprising Florida A&M University, Florida
Atlantic University, Florida Gulf Coast University, Florida International
University, Florida State University, New College of Florida, University of
Central Florida, University of Florida, University of North Florida, University
of South Florida, and University of West Florida.

University Press of Florida
15 Northwest 15th Street
Gainesville, FL 32611-2079
http://www.upf.com

Contents

Figures

Tables

Preface

The following collection of essays investigates human mobility across space and time. In these studies, anthropologists examine the reasons for and effects of movements in modern and past populations and task groups. Our purpose is to present new research that raises questions about human adaptations and to offer fresh methodological and analytical frameworks to extract information about mobility from the ethnographic and archaeological record. Rather than provide the reader with an allegedly comprehensive definition that narrows the research focus to only a few of the many facets of human mobility, this volume embraces a diversity of approaches. As a result, we intentionally sought contributions that addressed not only archaeological perspectives but also human behavioral ecology, ethnoarchaeology, and biological anthropology. These multiple research trajectories offer opportunities to examine how human mobility is practiced in the ethnographic world and the possible ways in which those activities might be represented in the archaeological record.

Mobility is a critical aspect of human adaptation to environments. In small-scale societies, mobility is a strategy for organizing individuals, labor groups, and consumers to cope with the variation in spatial and temporal distributions of critical resources. As important as subsistence strategies are in dictating human movement, factors other than the search for food also structure population mobility and its archaeological effects. Information gathering, raw material collection, social networking, trade, mate search, and population fission present mobility needs that contrast with those of daily food searches. Although usually considered most pertinent to hunter-gatherer studies, mobility strategies also are significant components of more sedentary adaptations. Residential mobility, hunting, wild plant resource gathering, gardening, pastoral nomadism, labor exchange, trade, and a range of social and demographic movement are vital tactics among food-producing social systems of all sizes.

In the simplest terms, decisions regarding mobility affect many aspects of social organization, population distribution, and subsistence strategies. This in itself makes it an important area of anthropological research. But the wealth of potential lines of inquiry also creates numerous analytical and methodological hurdles. Due to the ever-changing nature of the scientific problem and the dense web of interactions between causal factors and mobility-related decisions, it would be naive to narrow the study of human movements to a simple equation of external stimuli with specific outcomes. This complexity is compounded by the many ways that mobility can be measured. The number of moves, distance traveled, frequency of movement, and even who is moving are all relevant to the study of human mobility and are legitimate units of comparison. Realistically, our knowledge of how these few descriptive variables are expressed among modern peoples in contrasting environments, with a range of subsistence and social organization, is still underdeveloped. Understanding how environments and behavior may affect mobility can only be addressed through further inquiry into the complexities seen in cross-cultural research. In particular, the potential variability represented in the broader geographical extent and temporal depth of the archaeological record demands methodological and theoretical advances in addressing mobility and its systemic roles and tactics.

In light of these considerations, it is easy to imagine the difficulties that await scholars trying to infer mobility strategies from the archaeological record. As a detective needs to retrace the path of interactions that produced observed clues, the archaeologist needs to establish whether particular components of sites are indicative of specific kinds of mobility and to investigate connections between patterned human behavior and independently established stimuli. Similar behaviors encountered even in apparently identical contexts, however, may have been caused by different events. Evaluating plausible explanations requires rigorously informed comparative methodologies that can measure mobility with consistency. It also requires controlled understanding of the principles governing human movement derived from ethnographic data.

It is well recognized that the record of the past is not directly reflective of behaviors that we can see in the present. But information in the record is not only about the things that are preserved for us to study. The movement of stone raw material, for example, may be more reflective of variable currencies used in social alliances or situational collection embedded within other critical activities, rather than being a measure of a procurement system. The need for appreciation of actual interactions that can be determined independently of the archaeological record prompted the inclusion in this volume of studies focusing on aspects of mobility derived from behavioral ecology, ethnoarchaeology, and biological anthropology.

The study of mobility is at the core of many important archaeological inquiries (including migration, territory size and location, settlement patterns, and subsistence strategies). Because questions about mobility are essential to so many archaeological problems, we chose in this volume to be inclusive rather than exclusive. In the words of noted historian Lytton Strachey:

> It is not by the direct method of scrupulous narration that the explorer of the past can hope to depict a singular epoch. If he is wise, he will adopt a subtler strategy. He will attack his subject in unexpected places; he will fall upon the flank and rear; he will shoot a sudden revealing searchlight into

obscure recesses, hitherto undivined. He will row out over the great ocean of material, and lower down into it, here and there, a little bucket, which will bring up to the light of day some characteristic specimen, from those far depths, to be examined with a careful curiosity. (Strachey, *Eminent Victorians* [New York: Modern Library, 1918], vii)

In order to cast light on mobility from multiple angles, this collection of essays contrasts ethnoarchaeological and archaeological approaches.

Testing and refining hypotheses are the means by which scientific knowledge moves toward a greater ability to describe relationships and predict their outcomes. Ethnographic information is derived from a different scale of observation than is archaeology. As long as we recognize that it represents a separate but relevant frame of reference for structuring data and describing organizational relationships that are the object of archaeological inquiry, ethnographic information can augment appreciation of potential variability in behavior. Ethnography offers details about the interactions between activities and provides controls for understanding how events are interrelated. Modern human behavior is a rich source for modeling the past and seeking relationships for the development of hypotheses that can be tested using archaeological data.

Ethnoarchaeological research is an additional link between observable modern behavior and the kinds of remains preserved in the archaeological record. While much ethnoarchaeology has been artifact-focused, the chapters in this volume are informed by human behavioral ecology. Several seek associations between significant ethnographic activities and ways to investigate their dynamic interactions through the static components in archaeological sites. Rather than providing a narrow material focus, this recognizes the potential for archaeology and ethnography to do complementary research by emphasizing the behavioral parameters that influence different aspects of mobility. Ethnoarchaeological observations expand archaeological interpretations by examining economic dynamics, organization of living and workspaces, indigenous geography, interactions with neighboring communities, and the influences of environments on the ways in which populations and task groups move within different scales of territory. The results are detailed data sets with enormous potential to contribute to our understanding of past human organization and modern behavioral variability.

Archaeology, however, provides the largest—if sometimes mysterious—record of human adaptation. The time depth that the archaeological record encompasses cannot be ignored. Indeed, this time-transgressive knowledge is critical to creating a historical and evolutionary framework for improved insight about modern human behavior.

By bridging the gap between archaeological and ethnographic inquiries, this volume provides a venue for new insights into raw material acquisition and use, site structure, settlement organization, group composition, and changes in subsistence. We feel that the results presented here invigorate the quest for improved scientific understanding of mobility in human adaptations, technology, and social organization. This synthetic methodological presentation expands anthropological knowledge about a diverse range of mobility-related activities from the archaeological and ethnographic records.

The twelve essays that follow cover a broad topical and geographical landscape. The ethnoarchaeological chapters explore relationships between human settlement, subsistence, and tool use behaviors and their potential archaeological signatures.

The first two chapters serve partly as a warning to archaeologists. They show the limits of some of the analytical constructs commonly used to evaluate mobility in the archaeological record. Lewis R. Binford questions the validity of the archetypal notion of "band" as employed by many anthropologists and suggests that such a hypothetical level of organization does not exist among most mobile hunter-gatherers. He finds that families are generally the fundamental units of decision-making and the largest culturally defined social unit. Gustavo Politis explores detailed behavioral data, ideology, material culture, and campsite information for the Nukak of Colombia. He exposes multiple layers of analysis that are sometimes ignored by ecological or archaeological approaches to mobility.

The next four chapters provide innovative methods for approaching mobility from an ethnographic point of view. Nathan Craig and Napoleon A. Chagnon use GIS analysis to create a landscape view of settlement dynamics among the Yanomamö of Venezuela. Their study compares data on village and garden locations, demography, and population movements. Robert L. Kelly, Lin Poyer, and Bram Tucker examine the mobility of the Mikea of Madagascar through characteristics of household architecture and associated features that would be archaeologically visible. Michael S. Alvard addresses sustainability and transitions from hunting to agriculture among the Wana hunter-horticulturalists of Sulawesi, using GPS data on house sites, fields, and hunting areas. Finally, Russell D. Greaves examines biseasonal changes of residence and subsistence mobility of the Pumé foragers of Venezuela, linking behavioral activity budgets to archaeological patterns of landscape use.

The archaeological section of this book offers glimpses into the organization of raw material procurement, technological strategies, and the effects of mobility on tools during and after use. These chapters also explore archaeological signatures of mobility through analyses of settlement systems and direct osteological evidence. They provide a wider temporal and geographical depth that complements the ethnoarchaeological chapters.

Marsha D. Ogilvie examines past human activity levels for transitional horticulturalists and committed agriculturalists from the American Southwest. Her analysis of lower limb bone morphology connects gracilization to decreasing mobility, with an earlier decline among females in transitional populations. Claudia Chang's study of pastoral archaeology in southeastern Kazakhstan presents a case evaluation of equestrian mobility and agricultural sedentism. Pei-Lin Yu investigates the shift from atlatl to bow and arrow in North America, Spain, and Japan. She surmises a correlation, with decreased mobility options related to population packing in productive environments. The next two chapters investigate connections between raw material economy at archaeological sites and behavioral inferences. They both advocate contextualization of lithic activities in order to extract mobility-related information from stone tools. Frédéric Sellet illustrates, with an example from a Paleoindian site, how transported tools and local manufacture can be examined independently of raw materials to address long-term planning, risk minimization, and mobility. In a complementary study, Paul T. Thacker argues that comparisons of lithic raw materials often overemphasize the role of exotic sources in reconstructing past mobility strategies. He demonstrates the relevance of ecological modeling of local toolstone use in landscape analyses with a discussion of Portuguese Pleistocene/Holocene sites. Finally, Peter Veth explores the colonization of the Australian Western Desert, showing that transformations in later Pleistocene settlement and mobility are related to the onset of hyperaridity.

All of these examples emphasize the value of varied comparative approaches in improving comprehensive knowledge of mobility for deciphering the archaeological record. If we admit that our understanding of human mobility is imperfect and perhaps even rudimentary, then the potential arenas of interest and relevance are enormous.

Human behavioral variability and cultural differences are the essence of human adaptability. The capacity to inhabit an immense range of physical habitats with unique periodicities and ranges of climatic and biomass variation is at least partially dependent on strategies of population and labor movement within these environments. The complexity of a number of these systems is ingenious. Although some cases are moderately well understood, our comparative sense of system differences is as yet minimally developed. The value of this open-ended set of comparative studies should be obvious to researchers interested in using all tools at our disposal to address issues of modern human mobility and learning to read the vast archaeological record of environmental adaptations. These studies are offered with the hope that they can stimulate the use of many domains of knowledge to address issues of human strategic use of landscapes in the past and observable present.

I

Ethnoarchaeology of Mobility

Bands as Characteristic of "Mobile Hunter-Gatherers" May Exist Only in the History of Anthropology

LEWIS R. BINFORD

This chapter focuses on relationships between the described world of hunter-gatherers and the synthesizing concepts that anthropologists have employed to classify or characterize them. I make use of knowledge gained and methods of analysis developed during my recently completed nine-year comparative study and analysis of hunter-gatherer ethnography (Binford 2001). It is particularly germane to the studies assembled in this book that this study embraces a few aspects of mobility as a suspected major conditioner of organizationally variable social forms found among hunter-gatherers.

BANDS: A LITTLE HISTORY

Central to the materials to be developed is the way in which anthropologists have used the concept of "band." Perhaps the most often stated characteristic of hunter-gatherers is that they live in bands (see Binford 2001: 12–16, 114–115, 244–245; Kent 1996: 12–14, 69–70, 91, 102, 196, 298, 303; Kelly 1995: 10–15, 276–277; Lee and Daly 1999: 1–11). While bands are generally described as small social groups with coherent membership, we need to go back to an earlier time to gain a more accurate appreciation for the concept. Julian Steward (1970: 115) describes how his concept of "band" changed dramatically over the course of his anthropological career. His educational years were influenced by the examples used by his teachers, and his model for the "band" was drawn from the organization of the North American Plains Indians mounted hunters. His subsequent reading was strongly influenced by descriptions of the Boreal Forest Athapaskan speakers of western Canada and Alaska by the Canadian missionary Father Adrien-Gabriel Morice, who described such units as on average composed of about 250 persons, a number consistent with the Plains Indian model for a band. To my knowledge, Steward does not appear to have deviated from the "large population" model for "band" until the late 1960s, when he encountered June Helm's work on Athapaskan "bands" (Helm 1968).

In the late 1930s Julian Steward cited the lack of continuity in group membership as fundamental to the recognition of a "family level" of social organization among hunter-gatherers. He stated that the Western Shoshone, probably Southern Paiute, and perhaps some Northern Paiute fall outside the scope of his previous generalizations (Steward 1936) which were too inclusive. These groups lacked bands and any form of land ownership. The only stable social and political unit was the family. Larger groupings for social and economic purposes were temporary and shifting (Steward 1938:260).

Steward reached the view that he had not observed "bands" in the Great Basin,[1] because at that time bands were conceptualized as composed of a stable set of persons, integrated into landholding groups.[2] What he encountered in the Great Basin were families variously associating with other families in camp groups but exhibiting little continuity in the composition of such associations from camp to camp. Similarly, the idea of land ownership was exclusionary, and the Great Basin people did not seem to have such strategic goals.

Perhaps the most widely read conservative response to Steward was by Elman Service (1962: 97–98), who argued that a band was a cellular or socially bounded unit whose composition was only affected by births and the moving in and out of newly married spouses. Service (1962: 75) generalized that "band exogamy and virilocal residence make a patrilocal band." He went so far as to claim that "proper" hunter-gatherers lived in bands and that the materials cited by Steward as warranting evidence for his family level of integration were all distortions arising from culture contact and acculturation. Resistance to Steward's ideas dominated the 30 years between their original publication and any substantial support for his arguments.

Once the results of field researches conducted after World War II began to accumulate, a very interesting point was commonly made: namely, that hunter-gatherer bands were "flexible." This was one of the conclusions drawn at the "Man the Hunter" conference (Woodburn 1968; Lee and DeVore 1968; and Turnbull 1968). By "flexible" the discussants meant that the composition of "bands" was unstable; individuals and families regularly moved in and out of local camps. As previously pointed out, flexibility was not a new idea, yet Steward's ideas were not generally included in the discussions of flexibility at the conference. His conclusions were nevertheless based on the same phenomena that the postwar field workers had observed as "flux" (Turnbull 1968) or "group flexibility" in composition.

Older ideas regarding bands were under attack empirically in the 1960s. This attack did not focus only on American or African cases; discussions at the "Man the Hunter" conference challenged a long-standing misunderstanding regarding the composition and kinship basis of Australian Aboriginal groups. The "normative" Australian "patrilocal" band or "horde" was challenged, and the issue of "flexibility" in Australian camp or local group composition was demonstrated

(Meggitt 1962; Hiatt 1962, 1965, 1966, 1968). The importance of this discussion was simple: what was being described contrasted dramatically with the stability in composition of "bands" that had commonly been assumed in anthropological discussion prior to the conference.

In this setting B. J. Williams (1968) found it difficult to defend the more traditional-normative idea of "band" that he advocated at the "Man the Hunter" conference. Long-term field studies of Australian hunter-gatherers were yielding information that strongly challenged traditional views of the band. In retrospect, Williams was correct about the Birhor of India, now recognized as a mutualist group living in the broader context of a caste system (see Binford 2004). Nevertheless, at the time the Birhor were being taken as an organizational model for hunter-gatherers in general, much as Service (1962) had done earlier. This attitude was very common at the "Man the Hunter" conference. Fortunately, long-term field studies of Australian hunter-gatherers and astute observations regarding "flux" and "flexibility" were yielding information that strongly challenged the traditional views of the band as the basic form of organization among all hunter-gatherers.

RESEARCHING HUNTER-GATHERER SOCIAL ORGANIZATION

The basic issue addressed in this chapter is the exploration of organizational principles that stand behind variability among hunter-gatherers in social structure, as documented by recent research. In turn, such findings are used to address the old debates surrounding the concept of band briefly outlined above.

One of the more fundamental differences that my recent research documented (Binford 2001: 312–314, 367–368, 377, 383–385, 422–423) was between mobile peoples and near-sedentary groups. Bands might well be an appropriate form of organization among the latter. Before considering such possibilities, however, we need to understand strategic variability in the use of mobility as a means to ensure a secure subsistence base. In addition, we need to consider the role of mobility in securing adequate nonsubsistence-related resources. Finally, we need to explore the impact of differing scales of mobility on other organizational features, which may be synergistically linked to mobility.

The dominant source of foods is another major conditioner of variability among hunter-gatherers (Binford 2001: 209–210, 214–216, 298–299, 385–386, 389–390). We can expect organizational variability among hunter-gatherers to differ minimally with respect to dimensions of mobility and subsistence base, so I focus my initial exploration of ethnographic case material on hunter-gatherers who differ in dramatic ways in regard to mobility. For the purposes of this chapter, mobile peoples are defined as hunter-gatherers with population densities of less than nine persons per 100 km^2 (see Binford 2001: 555, for generalizations regarding the packing threshold). Nonmobile hunter-gatherers are defined as cases where population density exceeds the threshold value of about nine persons per 100 km^2.

The decision to direct attention in this chapter to peoples who differed dramatically in population density was based on the knowledge that population density and mobility are strongly related, as demonstrated below (Figure 1.1). The goal here is to demonstrate relationships among measures of mobility, population density, and subsistence base as a baseline for discussing social practices, which are believed to have implicated multiple forms of group organization. More particularly, my initial idea was that the presence or absence of "flexibility," widely reported among hunter-gatherers, was related to tactical differences in the role of mobility as well as differences in the organizational context for making decisions to move and associate with various suites of people. These properties should contrast strongly with a "cellular" strategy, designed to restrict categories of people from joining the group and thereby sharing access to products of the area controlled by the group. In short, do hunter-gatherers differ significantly in an organizational sense, and if so are such differences related to fundamental variables such as population density, mobility, and subsistence base?

Of the 339 cases of hunter-gatherer societies that I have studied, 161 may be considered mobile, while 178 are nonmobile. Among the 161 mobile cases, mean population density is 3.5 persons per 100 km^2 for peoples dominantly dependent upon terrestrial plants and aquatic resources. Among those gaining the bulk of their food from terrestrial animals, a mean density of 2.0 persons per 100 km^2 is noted. The mean distances moved residentially are greatest among terrestrial animal-dependent peoples (548 km annually), followed by terrestrial plant-dependent peoples (373 km annually). The least residential movement is seen among aquatic resource–dependent peoples, who cover only 289 km annually. While the mean values orient our thinking, the relationship between kilometers moved and population density among mobile peoples is in fact an inverse and linear one: the higher the population density, the lower the annual residential mobility. This is illustrated in Figures 1.1, 1.2, and 1.3, where very similar distributions are shown among all three subsistence-based classes.[3] A significant difference, however, is notable in the number of aquatic resource-dependent cases (Figure 1.3), with less than 150 km moved residentially per year prior to the packing threshold. This feature constitutes a statistically significant difference when compared to the graphs for the terrestrial plant and animal-dependent peoples (Figures 1.1 and 1.2).

Among peoples who are dependent on terrestrial plants and animals, the 150-km threshold is achieved essentially at the packing threshold of approximately

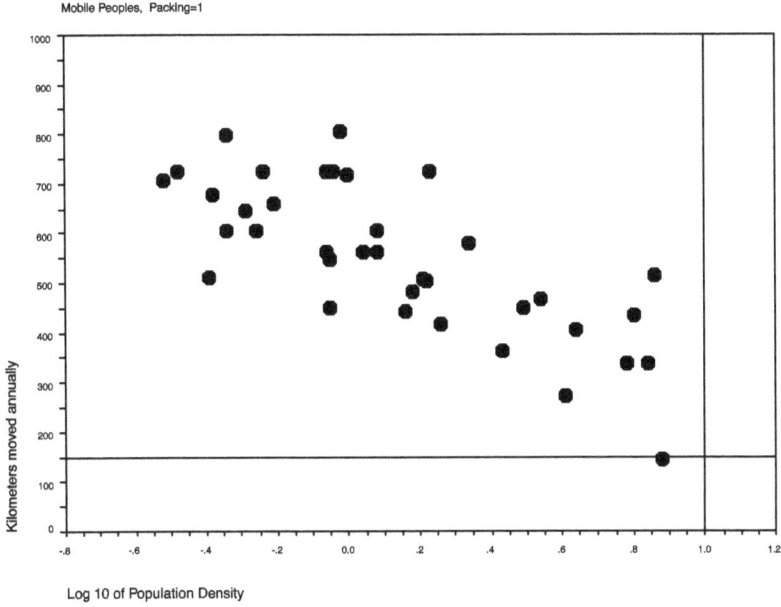

Figure 1.1. Log$_{10}$ density by kilometers moved annually for mobile terrestrial animal-dependent cases.

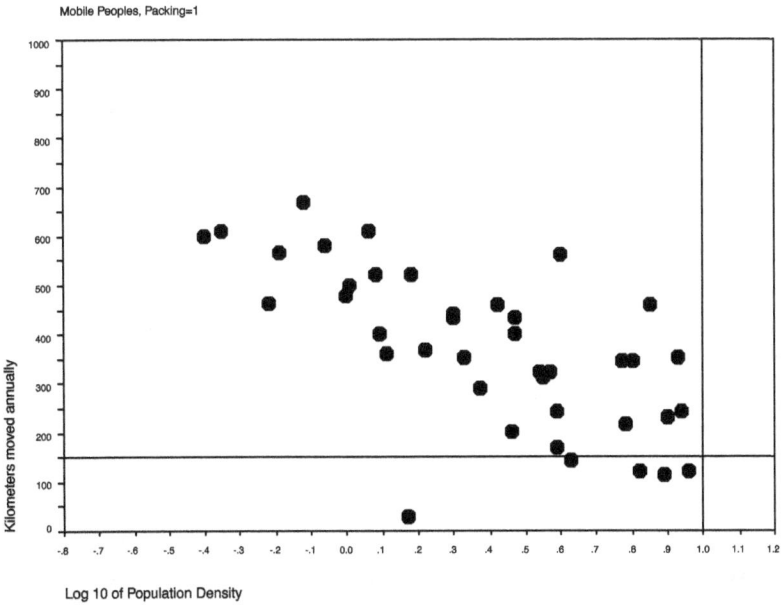

Figure 1.2. Log$_{10}$ density by kilometers moved annually for mobile terrestrial plant-dependent cases.

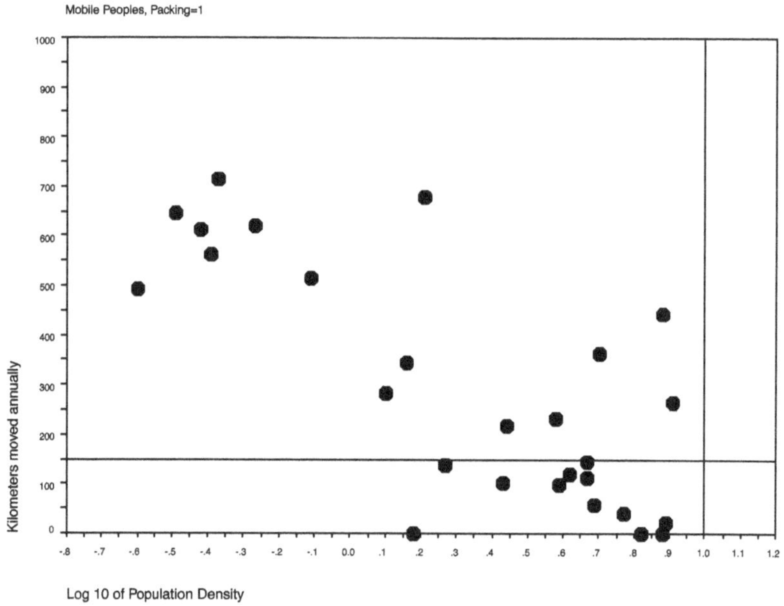

Mobile Peoples, Packing=1

Log 10 of Population Density

Figure 1.3. Log$_{10}$ density by kilometers moved annually for mobile aquatic resource-dependent cases.

9.0 persons per 100 square kilometers.[4] Consistent with this difference is the fact that packed or "nonmobile" aquatic resource-dependent peoples continue to exhibit an annual residential mobility pattern of only up to about 100 km. This is not true of peoples who depend on either terrestrial animals or terrestrial plants. In spite of the seeming similarity in the three graphs, the two graphs for terrestrial resource dependence show two additional meaningful differences.

First, what has not been made clear above is that no ethnologically studied groups remain dominantly dependent upon terrestrial animals when population density exceeds about nine persons per 100 km^2 (Binford 2001: 348, 375–376, 381, 385). This is demonstrated nicely in the comparison of Figure 1.1 with Figures 1.2 and 1.3; however, there are two cases to the right of the packing threshold in Figure 1.1. The Achumawi (recognized as an exception in Binford 2001: Figure 10.06, 382) were found to have been misclassified as to subsistence base. They are aquatic resource-dependent peoples (see Olmsted and Stewart 1978: 225). The Honey Lake Paiute, the second exception, are within the range of measurement error for population density and are therefore not considered a significant exception.

Viewing the situation documented here for animal-dependant peoples somewhat differently, the hunters, who without exception require more space to be successful, face the obsolescence of hunting as a reliable subsistence strategy at the lowest population densities.

The second exception (noted above) has reference to terrestrial plant-dependent peoples (Figure 1.4) and their responses to the packing threshold.[5] It was previously argued that intensification or increasing the food yield per unit area was only possible for the hunter-gatherers shown in Figure 1.2 (dominantly terrestrial plant-dependent peoples) by shifting to increased dependence upon aquatic resources if available and/or intensifying the exploitation of their traditional area. The latter would necessarily increase their diet breadth and/or their specialization in plant foods with more "elastic" productive potentials. This is the only possibility that would at the same time reduce the mobility costs of the more intensive niche being developed.

One could interpret the patterning immediately following the packing threshold among terrestrial plant-dependent cases (Figure 1.4) as reflecting groups employing such options. There is a regular scatter of cases—which extend past the packing threshold, located at slightly less than $\log_{10}$ of population density = 1.09, with regular annual mobility values reaching 300 and almost 400 km (Figure 1.4). This is considerably higher than the 150-km value that marks the upper limit of residential mobility among aquatic resource-dependent peoples (Figure 1.5) achieved after they passed the packing threshold.

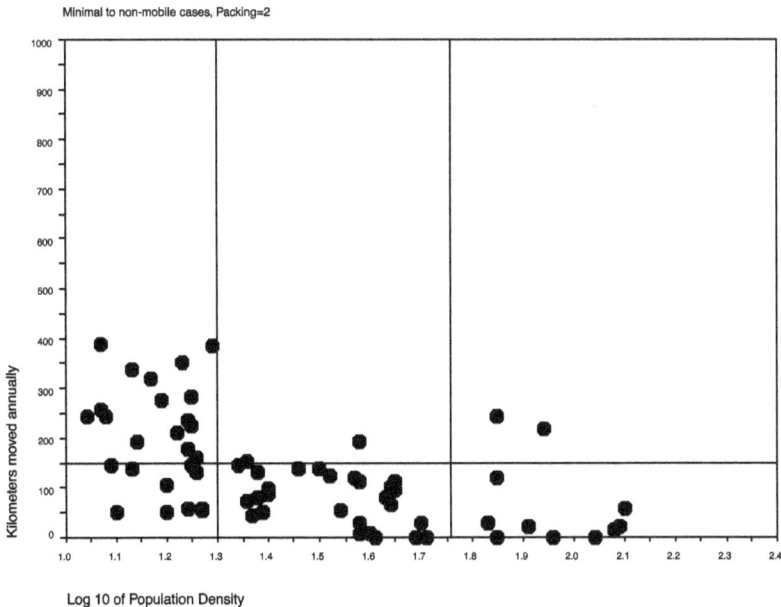

Figure 1.4. $\log_{10}$ density by kilometers moved annually for minimally mobile/nonmobile terrestrial plant-dependent cases.

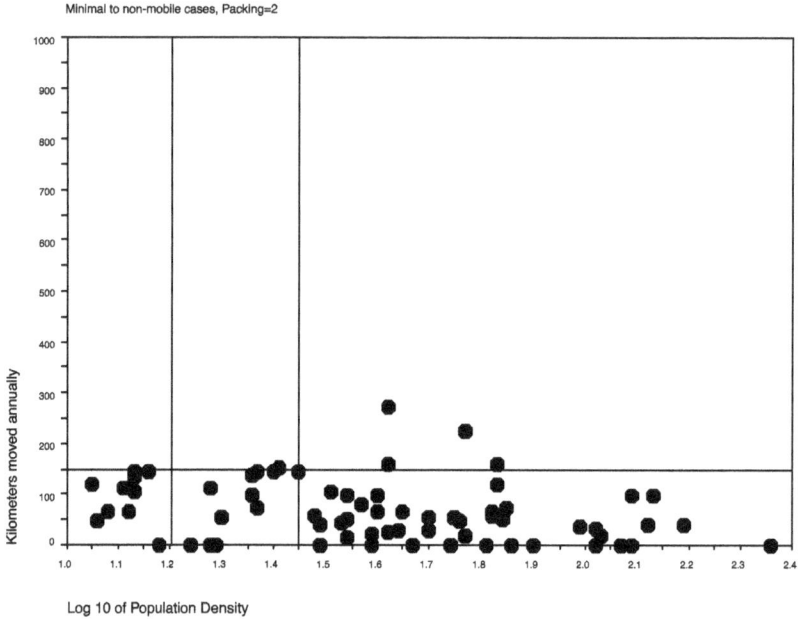

Figure 1.5. Log$_{10}$ density by kilometers moved annually for minimally mobile/nonmobile aquatic resource-dependent cases.

This interesting response among terrestrial plant-dependent peoples begins before the packing threshold and lasts until a precipitous drop in mobility at a log$_{10}$ of population density level of 1.30, or about 19.95 persons per 100 km^2. It could be seen as an intensification response that maintains a substantial dependence upon terrestrial animals in spite of a dominance shift in favor of terrestrial plants. The switch from making a single seasonal circuit through the subsistence range to covering the same area more than once per year, however, is an alternative that could also accompany reductions in terrestrial animal dependence. If correct, this could only work for a short time, because after the packing threshold the mobility drops precipitously to approximately 150 km or less moved annually at a log$_{10}$ of population density value of 1.30 (19.95 persons per 100 km^2). I checked the identity of the cases involved, which revealed nothing that would conflict with the tactical alternatives suggested above.

The terrestrial plant array (Figure 1.2) exhibits a line of five cases below the 150-km mobility reference line prior to reaching the packing threshold between 9 and 10 persons per 100 km^2. A very similar line of cases is seen among aquatic-dependent peoples (Figure 1.3), differing only in the lower mobility values among the latter. It is tempting to view the plant-dependent cases as ones where aquatic resources were a realistic intensification response. The line is depicting higher mobility for the terrestrial plant-dependent cases, which is consistent with this possibility. Nevertheless, when the identities of the plant-dependent cases making up this line were obtained, all were instances that did not significantly exploit aquatic resources. These five cases are the Railroad Valley people of the Great Basin, the !Kung of the Kalahari, the partially agricultural Nambikara of South America, and two examples of mutualism: the Aka pygmies of Africa and the Paliyans of south India (see Binford 2004).

I conclude that my interpretive analogies to the distributions on Figure 1.3 are completely unwarranted. At the same time, one could cite at least four of the five "exceptional" cases listed above as supporting my earlier suggestions that intensification within "their territory" could have occurred. One case had added horticulture; two cases had become mutual specialists, while one additional case had intensified around waterholes.

Turning away from the fascinating relationships of density, mobility, and intensification responses, this chapter also focuses upon the organizational implications of both increased density and decreased mobility as conditioners of varying social strategies used in integrating hunter-gatherer social units.

FOCUSING ON SOCIAL STRUCTURE

Several additional suggestions regarding mobile hunter-gatherers are considered warranted: (a) many examples are commonly organized socially with respect to egocentric networks; (b) basic units (such as families) move and differentially associate with other families in terms of these links to specific persons; and (c) when local camps "move" they may simply disassemble and later reassemble at a new venue, frequently with a different set of constituent families. This is what Steward documented earlier for the Great Basin hunter-gatherers. The identity of the occupants from camp to camp may overlap somewhat, but such camping associations are not generally long-lasting with respect to any given sequence of camps inventoried. Over longer periods, however, some families may tend to camp together frequently, particularly when adult heads of family are between 25 and 40 years of age (Binford 1991: 94–102).[6]

Networks may reach far beyond "local groups," linking persons over potentially very wide geographic regions. The networks in which the movement of goods, persons, and information may be organized are both inherited and "built," in that there are ways of creating "kin" beyond the commonly recognized birth and marriage events. It is even more interesting that networks may be created in terms of phenomena other than traditionally considered kinship. I have in mind the dreaming tracks that traverse the geography of Australia, to which persons are conventionally assimilated; thus personal identities are enhanced, strictly speaking, "outside" of kinship. These dream-time identities are in turn integrated with network lattices generated through conventional ways of ordering kin. A timeless, unchanging geographic framework of meaning and responsibility to which persons are conventionally assimilated that is directly articulated with the less stable changing conventional behaviors of people in social association through marriage and reproduction ensures some conceptual complexities. Of course, it is in terms of these intersecting latticeworks that groups move and differentially associate with other people across the landscape. Such a dual basis forms a complex, extensive, and comprehensive way of integrating peoples from distant places across vast spaces.

In Australia, there is a complete augmenting conventional system structured semi-independently of the egocentric kinship system (see Myers 1986: 71–102 for an excellent description of a network-organized system). If augmentation (or kinship "built" by referential conventions outside of the common events of birth and marriage) is present, it is generally initiated by individuals. It also remains in effect during each "partner's" lifetime (Heinrich 1963). This is the most common form of augmentation, although Australian cases and some others show that it is not universal (see Houseman and White 1998 and Viveiros de Castro 1998). Conventions for augmenting or expanding kin networks include adoption or gift-children (Gubser 1965: 146–147; Guemple 1972), same-name relationships (Marshall 1976: 238–242), shared initiation mates (G. Silberbauer, personal communication, 1993), simple sexual intercourse (Burch 1975: 207), trading partnerships (Burch and Correll 1972: 25), and cognitive elaboration of causes of birth (Berndt and Berndt 1964: 120–123).

Proposition

Augmenting criteria should be common when network expansion increases an individual's options for marriage, movement, exchange of goods and information, and choices of persons for organizing cooperative labor arrangements and co-residence.

Such functional links are interesting, but my main motive for exploring augmentation is its potential as a diagnostic for recognizing network organized systems as opposed to finite cellular "band" societies. Archaeologists should know the difference between network and "band" or cellular systems. For instance, it is not uncommon for archaeologists to suggest that rock art represents huntergatherer "bands" who are marking their territory! This simple equation is difficult to imagine among network-organized hunter-gatherers.[7] A more provocative feature, however, derives from this examination of kinship augmentation. It will be recalled that my initial reasoning in regard to the context where "expanded kin" linkages would be advantageous was that it would be where mobility was high. A strong set of relationships exists between population density and annual

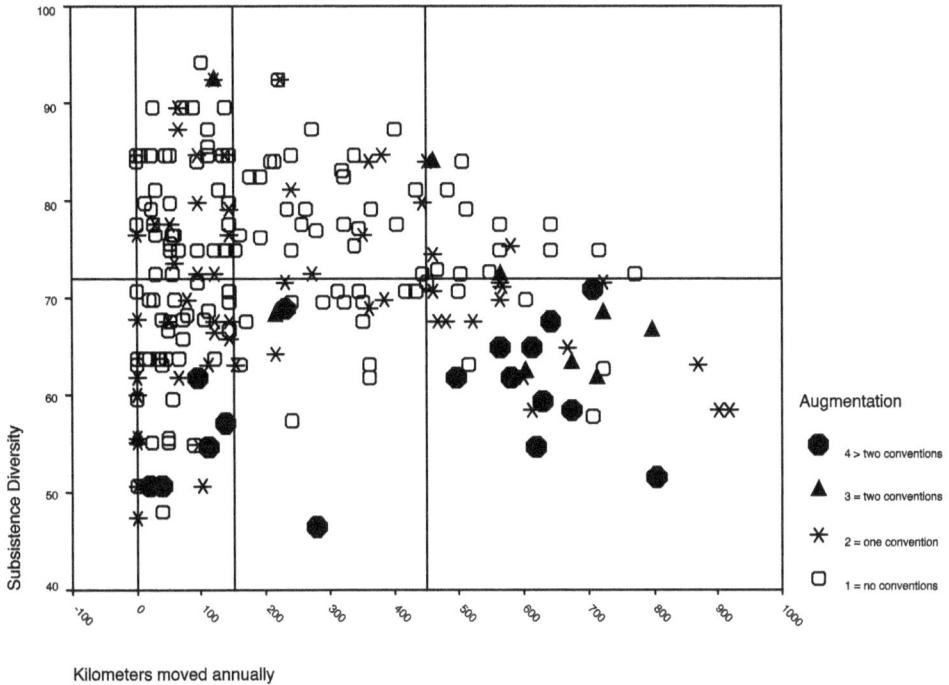

Figure 1.6. Number of conventions for augmenting kinship marked on plot of kilometers moved annually by subsistence diversity.

residential mobility (see Figure 1.2). What we want to know now is this: is there a strong relationship between mobility and augmentation?

In Figure 1.6 mobility is measured on the *X* axis by total kilometers moved residentially during a year (Binford 2001: 117). The *Y* axis is scaled with respect to subsistence diversity (Binford 2001: 404–405). Values for subsistence diversity were obtained by subtracting from one the standard deviation obtained by calculation of the percentage values for dependence upon terrestrial animals, plants, and aquatic resources. This value was then multiplied by 100. Subsistence diversity was chosen because it had previously proved useful in the analysis of hunter-gatherer variability (Binford 2001: 402–403, 426).

The following generalization should be clear from an examination of Figure 1.6:

Generalization #1
Low subsistence diversity (values = 50–60) conditions increases in kin augmentation regardless of the scale of mobility.[8]

It was suggested earlier that the scale of egocentric networks should be most developed when mobility was high! As Figure 1.6 demonstrates, when mobility is high (200–800 kilometers annually) and subsistence diversity is high (exceeds a value of 71.5 or 72), augmentation is generally absent; when it is present, only a single convention is utilized. I conclude:

Generalization #2

Extensive kin networks as indicated by large-scale kin augmentation appear to be unrelated in any linear manner to residential mobility per se. It should be apparent that when subsistence diversity measures are below 71.5, subsistence diversity is more of a conditioner for augmented kin networks than is mobility per se. Under conditions of low subsistence diversity, high augmentation is associated with specialization either on particular trophic levels within a terrestrial biome or on the aquatic biome itself.

The relationships between networks and subsistence diversity shown in Figure 1.6 are probably most understandable by referencing differing environments with distinctive spatial patterns among resources.

Cases where the expansion of both niche and diet breadth is feasible (subsistence diversity is high) tend to exhibit little investment in the augmentation of kinship conventions. Subsistence security, under those conditions, is most likely ensured by alternative tactics and alternative food locations available in different biomes within any given territory. In situations where subsistence diversity is low (limited perhaps by a poverty of species or a poverty of reliable trophic levels and/or biomes), maximizing one's access to expanded territory through maximally extended kinship networks is the most obvious way of buffering against periodic local resource failure, other things being equal. We may see kin augmentation and extensive egocentric networks as tactical buffers against subsistence insecurity, arising from high-risk mobility due to uncertainty when anticipating the accessibility of foods at various times and locations. To put it another way:

Generalization #3

Egocentric networks, frequently expanded by augmenting conventions, provide enhanced tactical options to decision-making units in the face of uncertainty regarding the reliability of information used in making mobility decisions.

This generalization leads us back to the environmental conditions that favor specialization and in turn offer few incentives for increasing diet breadth. Short of direct environmental investigations, I decided to summarize the data on augmentation with respect to gross subsistence categories, because they have been shown to relate in rather direct ways to environmental variables (Binford 2001: 215–216, 222, 276–279, 306–307).

Table 1.1 shows the percentage of cases arranged by mobile versus nonmobile

Table 1.1. Augmentation

	Mobile	Nonmobile
	(Density < 9.098)	(Density > 9.098)
I. Terrestrial animal-dependent cases	($n = 48$)	($n = 6$)
absent	68.9%	83.3%
one convention	10.4%	16.7%
two conventions	20.3%	0.0%
II. Terrestrial plant-dependent cases	($n = 52$)	($n = 82$)
absent	42.5%	65.3%
one convention	50.0%	28.0%
two conventions	3.8%	6.0%
III. Aquatic resource-dependent cases	($n = 31$)	($n = 86$)
absent	19.4%	81.4%
one convention	35.5%	17.4%
two conventions	45.2%	1.2%

sets, with each set differentiated by additional subsets based upon the dominant sources of food. One striking feature of this tabulation is that kin-augmenting conventions occurring in all subsistence subsets are more common among mobile cases. This shows that kinship augmentation and linked network enhancing techniques are associated with mobility.

As mobility phases out as the major means of adjusting to productive variability in the habitat, kinship augmentation is reduced. In sum, with increased sedentism the cases are becoming more intensified and thus more like non-hunter-gatherers as they are typically conceived.

Table 1.1 verifies that augmentation is most important among mobile peoples who are primarily aquatic resource-dependent. Terrestrial plant-dependent peoples are second. This is consistent with what appears to be a bias among leaders of terrestrial animal-dependent peoples to have more distant marriage partners—both spatially and in terms of kinship distance—than do normal group members. This differentiation fits with reductions of egocentric networks among "hunters." Both of these properties indicate the obsolescence of an ancient hunting-dependent way of life. We might therefore expect:

Generalization #4
When the geographic scale of egocentric networks diminishes among hunter-gatherers, there should be a correlated decrease in the contribution to the diet of terrestrially hunted large to moderate body–sized animals.

This very important contrast is not the only one of note. Visible camp leaders among mobile peoples who are dependent on terrestrial animals and aquatic resources contrast rather markedly with the lack of strong leadership characteristic of mobile peoples who are dependent on terrestrial plants. Instead of visible leaders being a rare phenomenon, they occur in between 77 and 83% of the cases among people who are dependent on aquatic and terrestrial animals. This does not mean that they have a bounded or cellular form of society, only that the duration for having a core group of "followers" around a good hunter and a wise decision-maker (see Henriksen 1973: 40–72 for a provocative description) tends to stabilize group membership over a longer period. Given these conditions, "flexibility" is less visible over a relatively short period of field observation. The stability of such hunting groups is considerably greater than might be observed among truly mobile terrestrial plant-dependent peoples. Failure by leaders, however, may cause a very quick reorganization around an alternative leader or even in support for others competing for leadership recognition (see Binford 1991).

These contrasts in leadership roles carry over into other features, such as different scales of exogamy for leaders versus "followers," as previously noted. Among the plant-dependent peoples, exogamic distance scales with mobility such that even second-cousin marriage is commonly forbidden among mobile peoples, but nonmobile groups may frequently allow marriage between first cousins. Such linked organizational shifts that scale with population density and mobility are important. Nevertheless, the point of this limited discussion is to emphasize that bands with culturally defined "members" are a consequence of increased population density and associated intensification in land use within ethnic ranges. It is perhaps a bit ironic that the !Kung, who were advanced as the archetype for hunter-gatherer social organization, were packed or very nearly so at the time of initial study. The !Kung-speaking Ju/'hoansi may represent a form of hunter-gatherer organization; but they hardly provide a model for hunter-gatherers living at low population densities across a variety of environments, as was originally suggested (Lee 1968: 43).

Summary of Findings

Bands may be found among intensified hunter-gatherers who are dominantly dependent upon terrestrial plants and aquatic resources. Bands will not be found among foot-mobile terrestrial animal-dependent peoples. Network systems with small family units as the largest integrated unit that has any substantial duration are characteristic of "specialized" hunter-gatherers with low subsistence diversity. Larger groupings of such units are short-term camping associations and are not structured in any permanent manner. Bands do not exist among low-subsistence-diversity hunter-gatherers. Julian Steward was right that, among some hunter-gatherers, families are the fundamental units of decision-making. They are the largest culturally defined social unit. Camps are associations of such units, and the families make independent decisions regarding mobility. Such hunter-gatherers do not live in bands.

Preliminary investigations indicate that among mobile peoples living under modern environmental conditions extensive networks are a phenomenon of polar and near-polar habitats. Importantly, extensive network systems dominate dry desert and near-desert settings among nonpacked hunter-gatherers (see Binford 2001: 375–385). "Cellular" systems, however, dominate modern types of environments of the temperate zones, subtropical islands, and small land mass regions as well as most coastal zones under packed demographic conditions. Network-based systems are not unique to mobile hunter-gatherer societies. I suggest that with increased population density and packing (see Binford 2001: 347, 380, 385–386, 418–424, 427–428, 434) networks differentiate, perhaps into several kinds.

Geographically extensive egocentric networks are found among highly mobile and nonpacked hunter-gatherers. More regionally intensive networks may be developed among comparable-"status" individuals representing social units such as lineages and/or bands themselves when residential mobility is low and population densities are relatively high. Regular patterns of marriage "alliance" or circulating "sister exchange," as Claude Lévi-Strauss was fond of postulating, might hint at forms of a network system where the basic units being articulated are larger than individuals and families and roughly sedentary.[9]

What is interesting about these suggestions is that there are no known packed cases of hunter-gatherers in the central and western interior regions of Australia, which are characterized by extensive macroregional network systems. There are a substantial number of packed cases along the north and east coast of Australia, however; these cases do have network systems, which are "intensive" or more local in scope, coupled with network organized means of adjudicating disputes and so forth. The patterning reported here helps us to see in a new light some of the documented variability among mobile hunter-gatherers and provides hints to further studies that could yield clues as to how we might discriminate spatially "extensive" from "intensive" network systems.

One final point cannot be stressed too strongly: networks (as discussed here) and bands (as commonly defined by "traditional" ethnologists) are likely to be only two of a larger number of organizational forms that existed in the past and that are recorded in the ethnographic literature but remain organizationally unrecognized. The cultural past of hunter-gatherers was not a "monolithic" organizational world. The task of anthropology is to recognize variability and to explain it. Trying to make the past organizationally homogeneous simply denies the fundamental problem that anthropology seeks to address—variability and the explanation thereof.

I therefore argue that the archetypical band to which many anthropologists refer does not exist among most mobile hunter-gatherers. Julian Steward was correct in his judgment that, for some hunter-gatherers, families are the fundamental units of decision-making and the largest culturally defined social unit with membership integrity. Hunter-gatherer camps are commonly associations of such units, and the families largely make independent decisions regarding mobility.

One wonders why anthropologists have tried for so many years to reduce the captivating diversity in human social organization to static archetypes, which may only exist in the anthropological literature! In so doing, knowledge of the reality of hunter-gatherer variability is left to confound the archaeologist. Regrettably, the archaeologist commonly adopts the ethnographic characterizations as guides to the interpretation of the archaeological record, thereby obscuring from view the potentially fascinating variety of the past.

NOTES

1. I have traced Steward's thought on these issues through to his recognition, after his exposure to June Helm's (1968) work on bands, of why he had the idea that bands were large social units and the Great Basin cases did not seem to have such units (Binford 2001: 12–16).

2. The contrast that Steward noted was with his understanding of the term "band." His view was taken from studies of the Plains Indians, where "bands" were large (175–280 persons). The ethnography of the Athapaskan speakers of the north was largely drawn from Father Morice, who described the Boreal Forest people as living in bands of 250 or more people, consistent with Steward's received knowledge regarding the use of the term "band." (See Steward 1970: 115 for an account of his learning the difference between "band" as received knowledge and "band" as it was being used for hunter-gatherers later in time.)

3. Many of the illustrations include cases that appear to be exceptions to the generalizations in this chapter. These are important and will be dealt with in a forthcoming paper on the subject of the scientific conventions of generalizations being dependent upon specified initial conditions and the phrase "other things being equal."

4. This difference is probably related to the fact that aquatic resource-dependent peoples tend to exhibit linear settlement patterning and not uncommonly an ordered size hierarchy along coasts and rivers. This pattern is totally at odds with the shape of the land-use pattern developed largely from empirical generalizations based on terrestrial

plant-dependent peoples (Binford 2001: 374), which was used in reasoning to the value of 9.098.

5. This was discussed earlier (Binford 2001: 375–385).

6. It is a bit ironic that the long-term emotional controversy between Julian Steward and Omer Stewart, played out dramatically before the U.S. Indian Claims Commission during the 1940s and 1950s (see Ronaasen et al. 1999), would be repeated relative to such far-away places as the Arctic and Australia. The issue was simply that no Western ethnographer has recognized what I think of as framework systems. In that type of network, kin- and quasi-kin-based articulations to places provide the integrative framework for truly "flexible" group associations among persons as well as articulations of heterogeneous groups of persons to places. Western anthropology has been committed to the idea of territorially defined societies. Based on my work in Australia I became convinced that extensive cognitive integration of "places" could provide a different and effective basis for organizing social relationships that were not strictly stabilized in terms of proximity and/or mutual long-term association, as might be imagined with the idea of a band. It is too bad that Steward's ideas were not taken into account in the "flexibility" discussions of the 1960s. It is equally ironic that the authors in the 1999 volume (edited by Clemmer, Myers, and Rudden) were unaware that the controversy between Steward and Stewart was not unique to the Great Basin, with the expectation that one was right and the other wrong.

7. A good example of how complicated this may be when networks are involved is found in the master's thesis produced by Jennifer Galindo and submitted to the Department of Anthropology at the University of Nebraska in December 1997, "Scales of Human Organization and Rock Art Distributions: An Ethnoarchaeological Study among the Kunwinjku People of Arnhem Land, Australia."

8. Subsistence diversity is used as an indicator of niche breadth. It is the standard deviation of the percentage values calculated among the three recognized sources of food: terrestrial animals, terrestrial plants, and aquatic resources. This value is then standardized to a scale of 1–100 and subtracted from 100, resulting in positive values for high subsistence diversity. Empirically, it was found that a value of 71.5 represented a major threshold where distributions of cases tended to differentiate (Binford 2001: 417–428). It actually represents the median value for the distribution (roughly between 41 and 100) of empirically observed cases. This threshold value is shown in a number of the illustrations produced for this chapter.

9. I do not mean to suggest that the particular arguments of Lévi-Strauss are accurate, only that alliance in a very broad sense may be organized among differentiated but neighboring ethnic groups. In fact, near-sedentism and high population densities are probably a prerequisite for alliance systems, something that Lévi-Strauss apparently thought was characteristic of the Central Desert Australian aborigines (see Leach 1970: 40, 113–114, for an excellent evaluation of the works of Lévi-Strauss).

REFERENCES

Berndt, R. M., and C. H. Berndt
1964   *The World of the First Australians*. University of Chicago Press, Chicago.

Binford, L. R.
1991   When the Going Gets Tough, the Tough Get Going: Nunamiut Local Groups, Camping Patterns, and Economic Organization. In *Ethnoarchaeological Approaches to Mobile Campsites: Hunter-Gatherer and Pastorals Case Studies*, edited by W. A. Boismier and C. S. Gamble, pp. 25–137. Ethnoarchaeology Series No. 1. International Monographs in Prehistory, Ann Arbor.

2001   *Constructing Frames of Reference*: University of California Press, Berkeley.

2004   Niche: A Productive Guide for Use in the Analysis of Cultural Complexity. In *Processual Archaeology: Exploring Analytical Strategies, Frames of Reference, and Culture Process*, edited by Amber L. Johnson, pp. 297–314. Praeger Publishers, Westport, Conn.

Burch, E. S., Jr.
1975   *Eskimo Kinsmen: Changing Family Relationships in Northwest Alaska*. West Publishing, St. Paul.

Burch, E. S., Jr., and T. C. Correll
1972   Alliance and Conflict: Inter-Regional Relations in North Alaska. In *Alliance in Eskimo Society*, edited by L. Guemple, pp. 17–39. American Ethnological Society, Seattle.

Clemmer, R. O., L. D. Myers, and M. E. Rudden (editors)
1999   *Julian Steward and the Great Basin: The Making of an Anthropologist*. University of Utah Press, Salt Lake City.

Gubser, N. J.
1965   *The Nunamiut Eskimos: Hunters of Caribou*. Yale University Press, New Haven.

Guemple, L. (editor)
1972   *Proceeding of the American Ethnological Society 1971, Supplement*. University of Washington Press, Seattle.

Heinrich, A. C.
1963   Eskimo Type Kinship and Eskimo Kinship: An Evaluation and a Provisional Model for Presenting Data Pertaining to Inupiaq Kinship Systems. Ph.D. dissertation, University of Washington. University Microfilms, Ann Arbor.

Helm, J.
1968   The Nature of Dogrib Socio-territorial Groups. In *Man the Hunter*, edited by R. B. Lee and I. DeVore, pp. 118–125. Aldine Press, Chicago.

Henriksen, G.
1973   *Hunters in the Barrens: The Naskapi on the Edge of the White Man's World*. Newfoundland Social and Economic Studies No. 12. Institute of Social and Economic Research, Memorial University of Newfoundland, St. John's, Canada.

Hiatt, L. R.

1962   Local Organization among the Australian Aborigines. *Oceania* 32: 267–286.

1965   *Kinship and Conflict: A Study of an Aboriginal Community in Northern Arnhem Land*. Australian National University, Canberra.

1966   The Lost Horde. *Oceania* 37: 81–92.

1968   Ownership and Use of Land among the Australian Aborigines. In *Man the Hunter*, edited by R. B. Lee and I. DeVore, pp. 99–102. Aldine Press, Chicago.

Houseman, M., and D. R. White

1998   Taking Sides: Marriage Networks and Dravidian Kinship in Lowland South America. In *Transformations of Kinship*, edited by M. Godlier, T. R. Trautmann, and F. E. Tjon Sie Fat, pp. 214–243. Smithsonian Institution Press, Washington, D.C.

Kelly, R. L.

1995   *The Foraging Spectrum: Diversity in Hunter-Gatherer Life Ways*. Smithsonian Institution Press, Washington, D.C.

Kent, S. (editor)

1996   *Cultural Diversity among Twentieth-Century Foragers*. Cambridge University Press, Cambridge.

Leach, E.

1970   *Claude Lévi-Strauss*. Viking Press, New York.

Lee, R. B.

1968   What Hunters Do for a Living, or How to Make Out on Scarce Resources. In *Man the Hunter*, edited by R. B. Lee and I. DeVore, pp. 30–48. Aldine Publishing Company, Chicago.

Lee, R. B., and R. Daly

1999   Foragers and Others. In *The Cambridge Encyclopedia of Hunters and Gatherers*, edited by R. B. Lee and R. Daly, pp. 1–19. Cambridge University Press, Cambridge.

Lee, R. B., and I. DeVore

1968   Problems in the Study of Hunters and Gatherers. In *Man the Hunter*, edited by R. B. Lee and I. DeVore, pp. 3–12. Aldine Publishing Company, Chicago.

Marshall, L.

1976   *The !Kung of Nyae Nyae*. Harvard University Press, Cambridge, Mass.

Meggitt, M. J.

1962   *A Study of the Walbiri Aborigines of Central Australia*. University of Chicago Press, Chicago.

Myers, F.

1986   *Pintupi Country, Pintupi Self: Sentiment, Place and Politics among Western Desert Aborigines*. Smithsonian Institution Press and Australian Institute of Aboriginal Studies, Washington, D.C.

Olmsted, D. L., and O. C. Stewart
1978   Achumawi. In *California*, edited by R. F. Heizer, pp. 225–235. *Handbook of North American Indians*, vol. 8. W. C. Sturtevant, general editor. Smithsonian Institution Press, Washington, D.C.

Ronaasen, S., R. O. Clemmer, and M. E. Rudden
1999   Rethinking Cultural Ecology, Multilinear Evolution, and Expert Witnesses: Julian Steward and the Indian Claims Commission Proceedings. In *Julian Steward and the Great Basin: The Making of an Anthropologist*, edited by R. O. Clemmer, L. D. Myers, and M. E. Rudden, pp. 170–202. University of Utah Press, Salt Lake City.

Service, E. R.
1962   *Primitive Social Organization*. Random House, New York.

Steward, J. H.
1936   The Economic and Social Basis of Primitive Bands. In *Essays in Anthropology Presented to Alfred Louis Kroeber*, edited by R. H. Lowie, pp. 331–350. University of California Press, Berkeley.

1938   *Basin-Plateau Aboriginal Sociopolitical Groups*. Bureau of American Ethnology Bulletin No. 120. Smithsonian Institution Press, Washington, D.C.

1969   Postscript to Bands: On Taxonomy, Processes, and Causes. In *Contributions to Anthropology: Band Societies*, edited by D. Damas, pp. 288–295. National Museums of Canada Bulletin No. 228. Anthropological Series No. 84. National Museums of Canada, Ottawa.

1970   The Foundations of Basin-Plateau Shoshonean Society. In *Languages and Cultures of Western North America: Essays in Honor of Sven S. Liljeblad*, edited by E. H. Swanson, Jr., pp. 113–151. Idaho State University, Pocatello.

Turnbull, C. M.
1968   The Importance of Flux in Two Hunting Societies. In *Man the Hunter*, edited by R. B. Lee and I. DeVore, pp. 132–137. Aldine Publishing Company, Chicago.

Viveiros de Castro, E.
1998   Dravidian and Related Kinship Systems. In *Transformations of Kinship*, edited by M. Godelier, T. R. Trautmann, and F. E. Tjon Sie Fat, pp. 332–385. Smithsonian Institution Press, Washington, D.C.

Williams, B. J.
1968   Some Comments on Band Organization and Data on the Birhor of Hazaribagh District, India. In *Man the Hunter*, edited by R. Lee and I. DeVore, pp. 126–131. Aldine Publishing Company, Chicago.

Woodburn, J.
1968   Stability and Flexibility in Hadza Residential Groupings. In *Man the Hunter*, edited by R. B. Lee and I. DeVore, pp. 103–110. Aldine Publishing Company, Chicago.

2

The Different Dimensions of Mobility among the Nukak Foragers of the Colombian Amazon

GUSTAVO G. POLITIS

Hunter-gatherer mobility has been a central theme in contemporary ethnoarchaeological research. In the past 20 years this aspect of culture has been studied from a global perspective in terms of the relationship between hunter-gatherer behavior and ecological factors (Binford 1980; Bettinger 1991; Kelly 1995). Nonetheless, the existing information on the mobility of hunter-gatherers from the tropical rainforest is limited, with the bulk of the data coming from Africa (Fisher 1987; Hewlett 1995). The treatment of the subject is even more incomplete for South America. This chapter contributes original information on Nukak mobility and attempts to put to the test the models commonly used for interpreting the archaeological record of hunter-gatherers in general and of tropical rainforest groups in particular.

Mobility models of prehistoric foragers have been developed from ethnographic cases (Binford 1980; Hayden 1981; Kelly 1983). Most of these models, however, are not able to capture the complexity of the relationship between multiple types of mobility and environmental, social, and cognitive factors. This failure is fundamentally the result of the enormous difficulty in recording data—through archaeological eyes—on different types of mobility in contemporary forager societies. By the time ethnoarchaeologists arrive, foragers often have already changed their mobility patterns and are moving toward sedentism. Sometimes co-varied factors such as technology or subsistence also have been heavily impacted by Western contact (see Yellen 1977; Gould 1968, 1969 for exceptions). This is especially true of South America, where good-quality ethnographic and ethnoarchaeological data on foragers were collected only after the groups studied had already become partly or totally sedentary (the Hupdu Makú and Bará Makú, Guayaki, Pumé, and others). The Sirionó from Bolivia maintained a "traditional" way of life during Allan R. Holmberg's (1950) fieldwork, but information on their mobility is scarce and nonquantitative.

The aim of this chapter is to present fine-grained qualitative and quantitative information on the multiple dimensions of territory and the different kinds of Nukak mobility and, finally, to discuss their archaeological implications. Three types of mobility are considered here: residential and logistical mobility (as defined by Binford 1980) and the daily foraging trips that are classified as the movements of individuals or small task parties to and from the residential camp, within a single day.

The different types of mobility are closely related to the multidimensioned nature of territory. In recent decades territorial analyses among hunter-gatherers have taken two contrasting directions. First, they have been closely allied with the analysis of Cartesian space, in relation to the physical reality of a territory, and the analysis of how these characteristics—especially the structure of food resources and raw materials—in turn shape human adaptation. Within these lines of research, territoriality has been recognized as a property of all human populations (Dyson-Hudson and Smith 1978) and is defined as the behavioral system that controls and maintains the more or less specific use of a particular area (Lanata 1993: 10). Conversely, there has been a reaction against this kind of spatial analysis, which has been accused of being environmentally deterministic. It has been pointed out that the first approach "reduce[s] human action to a series of numerical variables suitable for understanding the relationship between the friction of distance and economic and social behavior" (Gosden 1999: 154). In the second line of research, emphasis has been placed both on the way in which nonindustrial societies perceive their land and more broadly on the ontology of landscape (Gosden 1999; Ingold 2000; Ucko and Layton 1999).

The Multiple Dimensions of Territory

During our fieldwork the Nukak ethnic group effectively occupied 10,000 km² in the Amazonian tropical rainforest, in the Department of Guaviare in Colombia (Figure 2.1). During fieldwork (185 days between 1990 and 1996) this group, linguistically related to the Makú (Reina 1989; Mondragón n.d.; Cabrera et al. 1999), maintained a foraging economy and an egalitarian sociopolitical organization integrated by highly mobile small bands (for more information about the Nukak, see Politis 1996a, 1996b, 1999; Cabrera et al. 1999).

The Nukak's concept of territory is quite complex. It articulates spatial and ideational elements that go beyond the conditions of defensibility and resource use—one of the factors commonly employed to explain the territoriality of hunter-gatherer groups. The Nukak perceive, use, and conceptualize space through five juxtaposed dimensions. The first dimension of the Nukak landscape is the band's territory. It is defined as the regular and favorite area (but not exclusively so) exploited by a band. Most residential relocations, camps, or horticultural plots are found within this territory. The procurement of resources from the residential camps is carried out within this space, as are the great major-

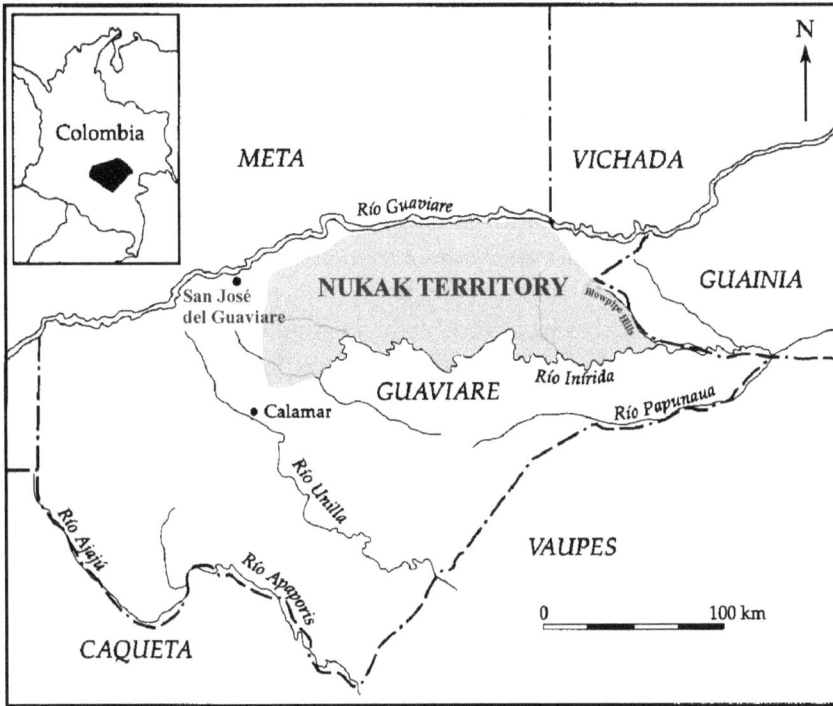

Figure 2.1. Location of the Nukak territory in the Department of Guaviare (Colombia).

ity of activities (both in- and off-camp) that make up the daily life of the Nukak. Bands have certain usage rights over this territory. Yet its borders are vague. This landscape is shaped by the members of the band and their ancestors. After generations of use and management of the tropical rainforest, they have succeeded in modifying it and in making it more productive (Politis 1999). They also have impregnated it with traces of themselves and inundated it with their symbolism. Consequently, each person has a profound and detailed knowledge of the band's territory, around which resource exploitation, residential relocations, logistics, and ritual activity are planned.

The size of a band's territory is difficult to define due to the looseness of its boundaries. It is possible to speculate, however, that the territory of each band oscillates between 200 and 500 km². This figure is tentative and of course varies from band to band. The northwestern bands are closer to the lower limit of this range because of the lure of the colonization of Sabana de la Fuga, which may have caused the overlapping of traditional territories. The southeastern bands have larger territories (some even greater than 500 km²), which may be the result of the recent depopulation that followed contact with Westerners. Based on the recent history of the Nukak and reconstructed demographic changes over the last two decades, we can speculate that band territories before Western contact were in the range of 400–500 km^2.

The second dimension of Nukak territory is the territory of the regional groups (Figure 2.2). These groups (Wayari, Meu, Tákayu, and Muhabeh) are also associated with a particular space (or *munu*), the boundaries of which are even less clear. Each of these *munu* extends from the watershed to either the Guaviare River or the Inírida River, which delimit Nukak land to the north and south. Members of the bands can travel without restriction within the space shared by the regional group. They can visit other bands (under certain conditions determined by kinship after having carried out the appropriate rituals) and exploit the resources available in the area during these activities. Regional groups do not act as endogamous units, but a high proportion of unions are formed by members of two different bands within the same *munu*. Thus every couple has a detailed knowledge of the territories of at least two bands within the *munu*. The surface area of this territory is extremely difficult to calculate, given the looseness of the borders and the flexibility of the concept itself. It can be estimated, however, to be between 1,000 and 2,000 km^2.

Beyond the band or regional affiliation group territory, the Nukak travel to distinct regions occupied by bands with whom (in general) they have little contact, although they of course know of their existence. This represents the third dimension of Nukak territory: a distant space that is known but only rarely visited. The reasons for these infrequent journeys are diverse, ranging from the gathering of canes for blowpipes to the search for potential spouses or curiosity about colonist farms and villages. The limits of this territory are even harder to estimate. It is larger than the regional group territory and may include several thousand square kilometers, even incorporating the entire territory currently occupied by the Nukak ethnic group.

The fourth dimension of territory includes the places whose existences are known to the Nukak, who have never or seldom actually visited them. This outside landscape is known through the flow of information between bands or through oral tradition but usually not through direct experience. Within this broad and diffuse territory, the Nukak recognize the existence of territories occupied long ago by the ancestors. The lands at the headwaters of the Unilla and Itilla Rivers can be included in this dimension. Their existence is known through oral tradition. The band that appeared in Calamar in 1988 was journeying toward this location in search of the Kawka. The eastern land that the Nukak originally came from (according to their oral tradition) is also found within this territorial dimension. The ethnic group considers this area Nukak territory, although at present they do not know whether there are other Nukak occupying it and information relative to it is scarce and relatively old.

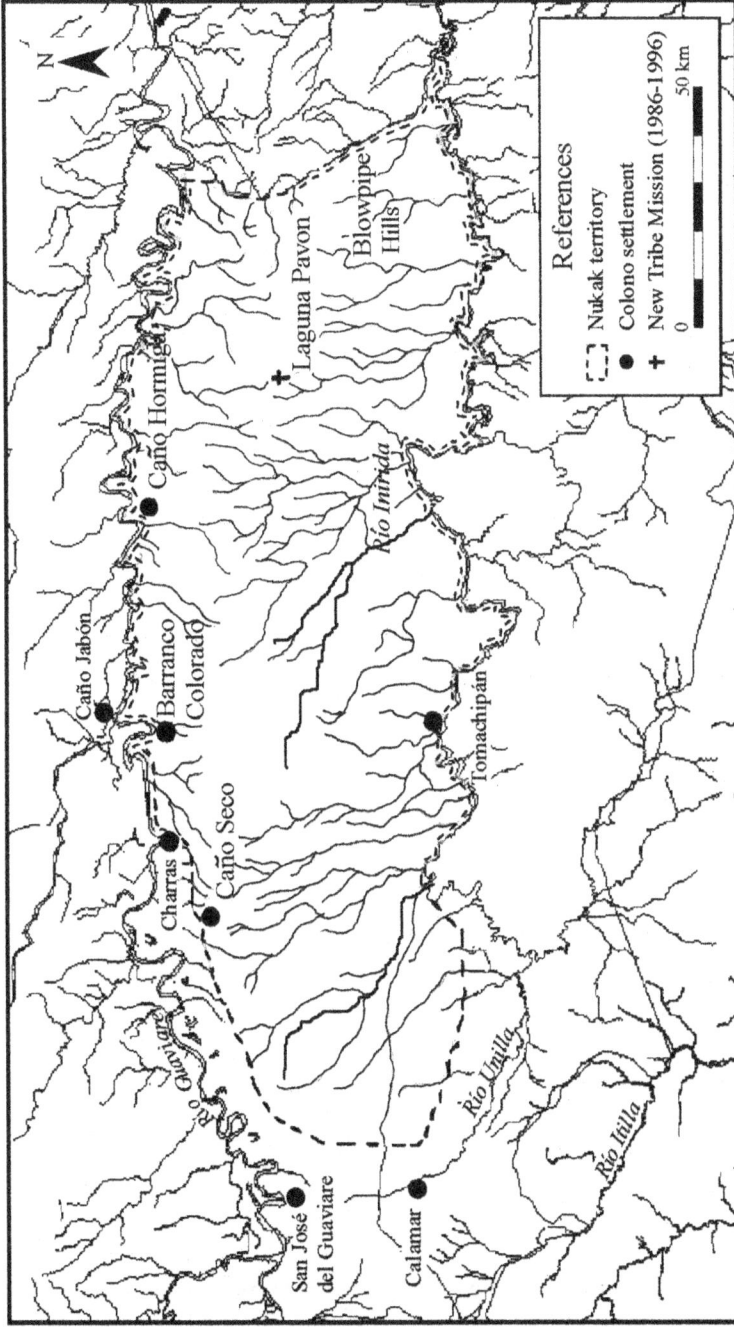

Figure 2.2. Approximate location of the four main regional groups (*munu*).

The fifth territory dimension is the mythic and ideological landscape and exists within the framework of Nukak cosmology. This territory is seen as real and tangible, despite its supernatural condition. It is formed by three flat, overlapping strata. These stratified worlds have particular physical and environmental characteristics and are interconnected by markers in the landscape (such as hollows and paths) as well as by spirit-ancestors who travel up and down the layers (occasionally on winged animals), producing climatic phenomena. This fifth dimension interacts with the other four and gives the natural, human, and supernatural spaces a flat and stratified character.

The Nukak view these dimensions as continuous; the separation between the real and physical territory and the ideological and mythic is nonexistent. The New Tribe missionary Kenneth Conduff told me an extremely revealing story in this respect (personal communication, 1995). He had heard it from the Nukak on an occasion when the missionaries were boring holes for water. The Nukak warned them about piercing through to the lower world. When they found some fine sand after several meters, the Nukak said: "See, you have already reached the beach from the river, which runs in the lower world." Later, when water began to emerge from the pipe, they said: "You see, now you have hit the river of the lower world!" This example illustrates how this fifth territorial dimension is conceived. Despite daily life taking place on earth and within the first four dimensions, the existence of other inhabited planes influences the way in which space is conceived and traversed.

The Nukak have an understanding of space beyond their territory. The conceptualization of this "foreign space" has been difficult; it is basically considered the "territories of others" and therefore does not interact in the same way with the other dimensions. The Nukak place cities such as Santafé de Bogotá and Villavicencio within this "territory" and can usually indicate their orientation, although not the distance (regarded as remote and associated with airplanes). They can also locate Mitú (several hundred kilometers to the east-south-east), because the Nukak who were relocated there in 1988 (Azcárate n.d.) passed this information among the Nukak population. In contrast to the other Makú (Reichel-Dolmatoff 1968; Jackson 1983: 148–163), the Nukak currently have no contact with other indigenous groups. Therefore, non-Nukak territory is now basically perceived by them as the land of the *kaweni* (whites).

Residential Mobility

Residential mobility is the displacement of all (or most) members of the band from one camp to another. It implies the definitive abandonment of one camp and the construction of a new one (Figure 2.3). Residential mobility is comple-

Figure 2.3. Construction of a new rainy-season residential camp.

mented by logistical mobility, which involves only segments of the band (Binford 1980), usually small groups of men who travel in search of resources. One of the main characteristics of Nukak bands is their high residential mobility. Residential camps are almost exclusively built in the first territorial dimension, the band's territory. During the period of observation the occupation of camps varied between a single night and 14 days (unless otherwise specified, "camps" hereafter refers to residential camps). Some colonists mentioned occupations of almost a month in camps close to the frontiers of colonization. Gabriel Cabrera et al. (1999: 148) recorded maximum occupations of 35 days. Table 5.1 in Politis (1999) gives an idea of the frequency and dynamic of the residential movements and of their relationship to the camps.

Data about residential mobility have been presented and discussed elsewhere (Politis 1996a, 1999b), so here I review only the main points. Data obtained during fieldwork revealed seasonal patterns of residential mobility. In the winter or rainy season (April to first half of November), the mean distance between camps is 3.85 km ($n = 12$) and the mean camp length of occupation is 4.8 days ($n = 13$). During the summer or dry season (second half of November to March), distances between residential camps produce a pattern with a mean distance of 8.94 km ($n = 13$) and a mean occupation of 3 days ($n = 20$). The overall pattern is that the Nukak remain longer in each residential camp and move shorter distances between camps during the rainy season, while in the dry season they occupy the camp for less time but move greater distances between camps. Combining the

Figure 2.4. Displacement of a band during a residential move.

data from both seasons, it can be estimated that each band makes approximately 100 residential moves per year; this value could be a little high, because I spent more time with the more mobile bands. A more realistic figure is around 70 to 80 relocations per year, covering a total distance between 400 and 500 km. These numbers match those provided by Carlos E. Franky et al. (1995), who arrived at a total of 68.64 moves per year. That represents an occupation average of 5.31 days per camp and an average distance between camps of 6.9 km (this last value is more tentative, because it was calculated based on walking time and not with a pedometer). The total distance traveled is 364 km per band per year.

Residential moves are usually made by the entire co-resident group, following existing well-defined trails into the band's territory in the rainforest, under the canopy (Figure 2.4). They move in two basic ways: either the whole group travels together or they keep some distance between people as they advance through the forest. When they split into several subgroups, there are two main variants. In the first, all members of a family walk together (generally with the adult male at the head), hunting and collecting during the trip. In the second variant, the men travel some distance in advance, procuring complementary food; the women and children walk behind. Larger bands usually split into several subgroups to travel, a practice that is also more frequent during the dry season.

The decision to leave camp is usually taken the day before abandonment or early in the morning before leaving and is based on numerous factors. Some are expressed orally, such as "we go where there is a lot of honey" or "we leave because it smells bad." My grasp of the complete set of factors involved in making the decision to abandon camp, however, is far from comprehensive, because my knowledge of the language is limited and not all the reasons were verbalized. Moreover, some reasons were so obvious to the Nukak that they felt no need to mention them. Among the recorded causes for residential movements are food procurement, the death of a person, the sanitary conditions of the camp, and displacements to perform a gathering ritual with other bands (*baak-wáadn*). Also, when a person becomes sick or is involved in an accident the entire band may move to Laguna Pavón 2 (the now abandoned New Tribe Mission) in search of medical assistance. In recent times the desire to camp close to individual colonists or villages has been an emergent reason for such movements.[1]

LOGISTICAL MOBILITY

In contrast to their high residential mobility, the Nukak have very limited logistical mobility. Logistical mobility is when most members of a band remain in camp (or make short residential moves within the band's territory), while a small group (usually adult males) travels considerable distances beyond the regional group territory. This group covers large stretches each day (more than 10 km) and sets up camps that are occupied for only one night or two.

The journeys to the Blowpipe Hills that men periodically make in order to collect high-quality canes (*ú-baká*) for the manufacture of blowpipes are one of the clearest examples of this type of mobility. Tens of kilometers are covered on these trips, especially by the Meu-munu and Tákayu-munu bands. When they approach the highlands they must "blaze a trail," because there are no more paths. These trips are usually made by most of the adult men in a band, who are at times accompanied by youths and adults of neighboring bands. Once in the hills, they cut the canes (not true canes, as they are actually stems of the palm *Iriartella setigera*). Each person takes about 10 to 15 canes (although one informant mentioned only 5), which are then divided between relatives or stored unprocessed to be transformed into blowpipes later.

I did not participate in these trips;[2] but the distance can be estimated, based on the location of the band territory. The more distant bands had to travel as much as 150 km, while the eastern bands needed to walk only a few tens of kilometers. At times the entire group would approach the Blowpipe Hills, making the men's trip a great deal shorter. It would appear that the logistical groups move along well-established paths. One of these paths apparently follows the watershed, passing close to Laguna Pavón 2. Until a few years ago groups of men also journeyed to the hills south of the Inírida River, between Kinikiarí and Cerro Cocuy, in search of *ú-baká* (Kenneth Conduff, personal communication, 1995). The trip to collect canes combines the social and cognitive realm. Men bring back canes for relatives from each trip and in this way reinforce and maintain the system of reciprocity; the trip is also a "male journey," where young men complete requirements to become full Nukak adults.

Only once did I participate in a logistical excursion during my fieldwork. On that occasion I stayed one night away from camp, 12.03 km to the southeast. The group, which consisted of all the young and adult men from a co-resident group (four in total), had originally planned to visit a chontaduro (*Bactris gascipaes*) orchard and collect fruit. But at 2:30 p.m., after traveling 11.24 km, they hunted a caiman (*Caiman sclerops*) close to the orchards (only 0.77 km ahead). They decided to take the entire animal to the orchard to be butchered, grilled, and eaten. We arrived at the chontaduro orchard at 3:05 p.m., where they unhurriedly processed and cooked the caiman over the course of the next 2 hours and 20 minutes. It was consumed along with some recently collected chontaduro fruits at 5:30 p.m. Afterward we gathered and laid plantain leaves, on top of which we spent the night. Early the next day the remainder of the caiman was eaten. The group left the spot at 7:20 a.m., arriving back at the residential camp at 11:25 a.m.

In recent years, because of the epidemics of the late 1980s (Politis 1996b; Cabrera et al. 1999) and the reduction in the number of potential partners, some men have traveled more frequently beyond the borders of the regional territories to find wives. This has become an important reason for movements within this territorial dimension as well. Also linked to the pressure caused by the colonization are the curiosity to see some of the colonist villages, such as Tomachipán, Guanapalo, Caño Seco, and La Charrasquera, and the need for medical attention and search for industrialized objects. All of these factors have led certain individuals to travel great distances. Sometimes these journeys go beyond Nukak territory, extending as far as San José del Guaviare and other population centers on the road to Calamar. Finally, the New Tribe Mission (Laguna Pavón 2) has been an important focal point. Bands, families, and small groups pass through the territories of other regional groups in order to seek medical attention or to exchange products there.

DAILY FORAGING TRIPS

The Nukak make daily excursions near the camps in order to obtain food, raw materials, and information on a variety of subjects (ranging from the location and condition of resources to the state of specific persons or bands). The number of people involved in these trips registered during our fieldwork ranged from 1

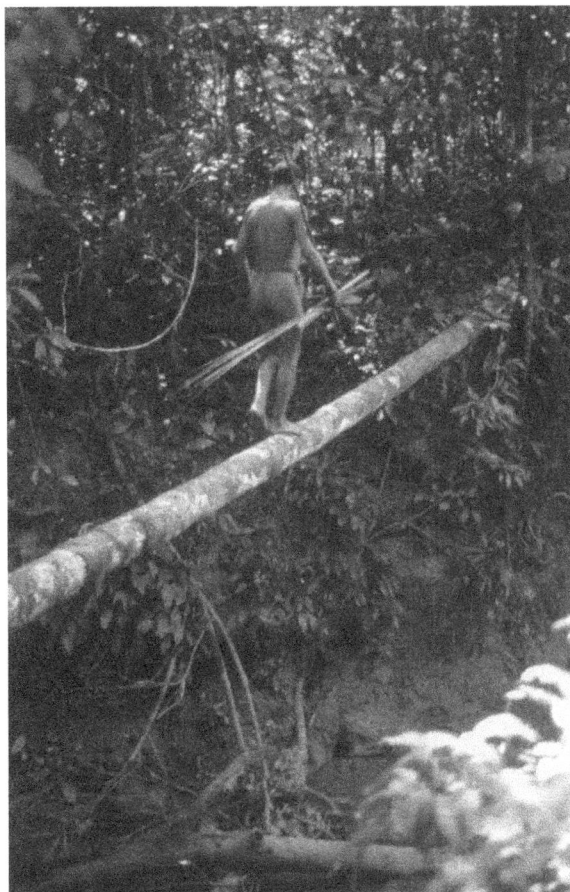

Figure 2.5. Nukak man during a daily foraging trip.

(Figure 2.5) to 11 (Tables 2.1 and 2.2). The groups frequently split up during the foraging trips for some time, however, or encounter other member(s) of the band foraging nearby and return to the camp together.

Daily foraging parties can be sorted by the composition of the party and its objective. I formed the impression (although answers were not clear-cut) that a person or a group leaves camp on a foraging trip not with a single objective (such as to hunt, collect honey, or gather fruits) but with several options related to the area of the trek, the season, the people's needs, the social setting, or specific circumstances. The only exception is the communal hunting of white-lipped peccary. In the two recorded cases the peccary herd had already been found; the group organized the party with a well-known and limited although not exclusive purpose (Figure 2.6).

It has been difficult to estimate the minimum distance traveled on daily foraging trips. Although there is a time-distance continuum, in the present study

Table 2.1. Daily Foraging Trips during the Dry Season

Date	Members	Departure time	Arrival time	Duration in hours	Distance covered (km)
22-1-94	2 Am, 2 Aw, 1 Fa	7:35 a.m.	4:54 p.m.	9:19	11.09
25-1-94	3 Am	8:37 a.m.	1:35 p.m.	4:58	14.56
27-1-94	1 Am, 2 Aw	9:30 a.m.	4:36 p.m.	7:04	7.50
28-1-94	1 Am	1:00 p.m.	4:29 p.m.	3:29	5.31
30-1-94	2 Am	8:05 a.m.	2:19 p.m.	6:14	8.90
31-1-94	1 Am, 1 Fa, 1 b	8:20 a.m.	2:02 p.m.	5:42	5.92
2-2-94	1 Am	12:24 p.m.	2:04 p.m.	1:40	3.06
3-2-94	1 Am	8:39 a.m.	2:30 p.m.	5:51	9.53
4-2-94	1 Am, 1 Aw	8:11 a.m.	10:30 a.m.	2:19	2.18
4-2-94	1 Am, 1 Ma	12:06 p.m.	4:30 p.m.	4:24	10.70
29-1-96	3 Am	7:54 a.m.	3:40 p.m.	7:46	8.32
30-1-96	2 Am	8:40 a.m.	6:40 p.m.	10:00	13.10
31-1-96	2 Am	9:16 a.m.	6:50 p.m.	9:34	9.20

Note: Am = adult man; Ma = male adolescent; Aw = adult woman; Fa = female adolescent; b = boy.

Table 2.2. Daily Foraging Trips during the Rainy Season

Date	Members	Departure time	Arrival time	Duration in hours	Distance covered (km)
31-8-92	2 Am	2:00 p.m.	4:27 p.m.	2:27	4.24
1-9-92	4 Am, 2 Ma, 3 Aw	7:10 a.m.	1:30 p.m.	6:20	7.74
2-9-92	1 Am	8:15 a.m.	12:39 p.m.	4:24	5.12
3-9-92	1 Am	7:49 a.m.	3:45 p.m.	7:56	16.20
4-9-92	1 Am	8:00 a.m.	1:39 p.m.	5:39	8.79
5-9-92	6 Am, 2 Ma, 3 Aw	8:10 a.m.	4:53 p.m.	8:43	8.96
6-9-92	1 Am	1:20 p.m.	5:00 p.m.	1:40	4.30
7-9-92	3 Aw, 2 Am, 1 Ma, 1 Fa, 1 b	9:00 a.m.	1:55 p.m.	4:55	5.70
9-9-92	1 Am	7:35 a.m.	11:02 a.m.	3:27	4.23
11-9-92	1 Am	6:50 a.m.	4:47 p.m.	9:57	15.64
12-9-92	1 Am, 1 Ma	7:15 a.m.	4:00 p.m.	8:45	15.30
15-9-92	1 Am, 1 Ma	9:43 a.m.	6:00 p.m.	8:17	8.40
16-9-92	3 Aw, 2 Fa, 1 Am, 1 b, 1 g	8:10 a.m.	1:05 p.m.	4:55	3.50
17-9-92	1 Am, 1 Ma	9:18 a.m.	4:15 p.m.	6:57	8.20

Note: Am = adult man; Ma = male adolescent; Aw = adult woman; Fa = female adolescent; b = boy; g = girl.

Figure 2.6. Hunting party seeking peccaries.

I have arbitrarily set a radius of 1 km to separate two kinds of off-camp daily activities. I consider the "camp's immediate surroundings" to be the area within a 1-km radius from the residential site. Within this area, camp inhabitants frequently gather fruit, honey, and other beehive products; go fishing; collect leaves for the roof or for making bags; obtain bark and vines for cord; and occasionally (if there is an orchard or *chagra* nearby) also harvest cultivated products. These activities are performed daily and with no formal planning or organization. Basically, the person or persons leave camp for a short while carrying one or two required tools (usually only a machete or an axe) and return soon after with something. Given the dynamic nature and frequency of these excursions, they were difficult to record in detail; but for several weeks we did document a sample of the food and raw materials intake per person per day (see Politis 1996b and Politis et al. 1997).

Beyond the exploitation of the residential camp surroundings, daily foraging trips maintained a similar pattern in both seasons: $x = 8.30$ km (round-trip) in winter ($n = 14$; min. $= 3.50$ km; max. $= 16.20$ km) and $x = 8.41$ km (round-trip) in summer ($n = 13$; min. $= 2.19$ km; max. $= 14.56$ km). The annual average is then 8.36 km ($n = 27$). An important observation is that the distance covered in the foraging trips does not increase in relation to the occupation time of the camp. Neither is there an observable drop in the foraging returns as the occupation of the camp proceeds (see Politis et al. 1997).

Figure 2.7. Nukak man performing a nonacoustic ritual after the hunting of a monkey.

Tables 2.1 and 2.2 provide a rough estimate of these daily foraging trips based on two variables (distance and time) as well as the composition of the party involved. A great range of variation can be distinguished. An imprecise "ritual time" must be included in the calculations, however, in order to obtain a realistic view of the many activities carried out on each trip. By this I mean the time dedicated to performing the ritual of negotiating with the spirit-ancestors for the use of certain resources. This ritual consists of a series of formulas known by all the adults and usually develops around nonacoustic communication (Figure 2.7). It is therefore technically impossible to measure its duration, start, and finish. This "ritual time" is embedded in many of the activities recorded.

Another activity that is hard to record but seems to be common during the daily foraging trips is the sexual relations between couples. Nukak camps lack privacy; although sexual activity takes place as silently as possible, it is frowned upon. Couples routinely take advantage of the daily foraging trips to have relations, although no trip is undertaken specifically for this purpose. Rather, sexual relations take place in between other activities. Obviously, when couples planned to have sex during a particular excursion, they avoided our company.

Observations about Daily Foraging Trips

The vast majority of daily foraging trips are multipurpose. For this reason, the equipment carried is usually appropriate for facing the various potential options. Even when the proper tool is not available, the party members may try to obtain a particular resource by using an expedient tool or a tool designed for another purpose if the occasion presents itself. Examples include hunting a caiman with a sharpened stick, throwing a machete in an attempt to kill a bird, and breaking a honeycomb with a stick and collecting the honey in an expedient leaf bag.

The composition of the foraging trip is fluid and could change during the expedition due to fission of the group or annexation. In addition, people who start the trip together might return together but spend most of the journey apart.

On some trips time is dedicated to the education of children in the many activities involved in hunting, fishing, and collecting.

Activities that have no immediate measurable return are carried out during excursions. These activities nonetheless form part of a sophisticated mechanism of resource management. Examples include the cutting of palms to grow palm grubs, visiting areas with concentrations of certain trees in order to check whether their fruit is ripe, visiting abandoned camps to monitor the regeneration of palms, and checking the location and condition of honeycombs in order to decide whether to relocate the camp.

Another embedded activity is the search for artifacts left in protected places for future use (passive gear). These caches may be in abandoned camps, beside a path, or hidden almost anywhere.

FINAL CONSIDERATIONS

The various types of mobility, the multipurpose nature of daily foraging trips, and their interaction with the different dimensions of territory highlight the complexity of issues such as mobility and territoriality in hunter-gatherer studies. They also demonstrate the difficulty of interpreting the various kinds of mobility through the archaeological record and show the shortcomings of apparently easy and universal correlations between environments and types of mobility. Some such patterns have been identified, but the present study suggests that alternatives must be examined.

In terms of territory, the predominant idea among the Nukak is not defense or ownership but use, construction, and perception. For the Nukak, territory is much more than the physical reality of the portion of land that they occupy or the place where resources to live by can be found. The territory, in any one of its five dimensions, is perceived as tangible, real, and interconnected. It is also the result of the activity of powerful spirits, founding heroes (such as Mauroijumját) and ancestors said to have lived only a few generations ago, who transformed the texture of the earth's surface. They brought trees, planted chontaduro palms, created animals, and made rain and storms. The action of the spirit-ancestors has not ceased, because they are still continuously modifying the landscape.

The multidimensionality of the territory of the Nukak highlights the practical problems that archaeologists face when reconstructing the territory of past hunter-gatherers. Most approaches to the subject involve an explicit or implicit reduction: territory is equated with band territory, and the principal method of analysis is studying the structure of resources. In other words, knowing how and when resources are available and whether or not bands exploit them allows conjectures to be made about numerous cultural aspects, from mobility to postmarital residence (e.g., Kelly 1983; Ember 1975). The Nukak case, however, demonstrates the shortcomings of this type of assumption. The Nukak consider their territory to consist of the four physical, juxtaposed dimensions of territory, all equally important. From the very immediate surroundings of the camp to distant places that were occupied by ancestors generations ago but may be visited by the Nukak at any time, everything is considered to be Nukak territory in some way; one is closer and better known, the other more distant and less frequented. None is exclusive, yet none is alien. How can these interrelated dimensions be differentiated in the archaeological record? If this cannot be done, is it then correct to assume that only the surrounding area represents "the" territory? What happens to the other aspects and their significance in economic, social, and ideational terms?

Some studies have attempted to capture the multiple meanings of the idea of territory, and concepts such as "annual range" (Binford 1980) have been proposed to account for the less used or visited places. Most such attempts, however, fail to realize that the various levels (including the cognitive) operate simultaneously and—together with other factors—shape the territorial behavior and movements of hunter-gatherers. The example of the hole that reached the world below is like others that reveal a similar conception of cosmological territory. The case of the Kabori-Nadöb, in a story related by Mark Münzel (1969–1972: 178), reflects a similar perception. When Münzel explained to them that he lived farther away than the Manau, a Kabori man whose songs Münzel had recorded asked him to play the recording extremely loud when he returned home. From there the Kabori's dead son would hear the song, as "the country of the dead" was also found in that direction beyond the Manau.

Another concept challenged by this study is the idea of defense, which is closely associated with the concept of exclusivity. "Territorial defense" seems to be a notion more appropriate for understanding the behaviors of certain animal communities (Eisenberg 1981) or the manner in which nation-states secure their boundaries than for comprehending attitudes toward territory among foragers. It has been shown that territorial conflicts basically do not exist among the Nukak, and each band freely uses an area of the rainforest with extremely diffuse limits. Presently some areas are rarely visited, and many areas are unoccupied.[3] Accordingly, neither lands nor resources appear to be an issue; therefore, there is no need to "defend" a particular space. One would expect this type of attitude to be present among other hunter-gatherers with similar demographic profiles and resource availability. In any case, the need for a defensible territory (for both past and present foragers) must be considered in each particular instance and not given a universal meaning.

It could be argued that the low population density of the Nukak, whether recent or long-term, prevents the appearance of territorial behavior based on "defense," but this does not appear to be the case. In this example, as in many others, the right to utilize a space is not based on the capacity to defend it but rather on the capacity to construct it; to know and use it; and, finally, to perceive it.

Constructing a territory means that the present landscape is the result of both action by immediate ancestors of the group in question and long-term anthropic activity from the time when humans first settled the place. This not only has modified the resource structure (that is, the physical reality of the land) but also has changed the texture of the landscape. It has been inundated with meanings, charged with symbolism, and sacralized.

Knowing a territory means to inhabit a space and use its resources, which requires a detailed knowledge of the structures and how to exploit them. This is achieved only through daily occupation over a long period, most probably several generations. Hence a band's territory is not merely the area that the band members exploit but rather the area that they know how to exploit. Although this resembles a play on words, it has profound implications. It would be relatively easy to arrive at a place and use the resources that are available, but it is far easier to use this same space over the long term for sustenance, without negatively affecting its productivity. This is only achieved through a detailed knowledge of the resources available in time and space, which in turn is only obtainable through the prolonged daily occupation of a particular area.

The final factor mentioned above is the perception of a territory. Even though resources may not be exploited or a particular area may rarely be visited, many groups (whether hunter-gatherers or villagers, nonhierarchical or ranked societies) perceive the land as possessing sacred connotations or symbolic significance. In other words, the "mental territorial map" of these societies includes areas that are not exploited for resources and not occupied or visited regularly but that contain strong mythical and cosmological significance and are therefore perceived as part of their territory. A clear-cut example of this tendency in the Amazon is the Cerro de los Hombres Chiquitos (Hill of the Little Men), which various riverine villager groups in the Apapopris area perceived as their own, each one bestowing a different meaning on the land and "using" it distinctively (Van der Hammen 1992).

Therefore I propose that territorial defense is by no means a universal condition through which hunter-gatherers maintain their territory; nor is it the most relevant characteristic of their territorial behavior. Without a doubt, the multiple dimensions of territory are difficult to grasp through the archaeological record, but this does not justify their reduction to a single dimension—band territory. We need to move beyond this reduction if we are to gain a full understanding of the complexity of the territorial behavior of forager societies.

The scarcity or even absence of food or goods after a daily foraging trip may mislead an analysis based on the trade-off curve between energy expended and energy obtained. As noted, not all results of foraging trips are of an economic character or measurable in units of energy. Cutting palms to promote the growth of palm grubs and checking the regeneration of palms are examples of nonimmediate economic returns. Collecting a blowpipe for a relative and obtaining information on a neighboring band by visiting its abandoned camp are activities that operate in the social plane. Finally, the ritual for communicating with the ancestor-spirits is also a routine that cannot be represented in an energy equation and must be understood within the cognitive sphere or cosmological framework.

These observations and interpretations highlight the difficulties involved in an archaeological examination of issues such as territory, territorial behavior, and the causal factors of mobility patterns (both logistical and residential). It is not my intention, however, to add a new cautionary tale to the long list of cultural factors supposedly undetectable in the archaeological record. On the contrary, the case presented here suggests that the reductionism of the ecological model leaves aside important features of human behavior. These models cannot capture the wide range of causal factors that affect territory and mobility in past and present foragers and must therefore be modified or replaced by more inclusive and less biased ways of investigating hunter-gatherers.

ACKNOWLEDGMENTS

Initial fieldwork was supported by two grants from the Wenner-Gren Foundation for Anthropological Research and in 1995 and 1996 by the Instituto Amazónico de Investigaciones Científicas (SINCHI, Colombia). Thanks to Benjamin Alberti for helping me with the translation and for his comments.

NOTES

1. One of the Meu-munu bands that had displayed a traditional pattern of mobility during fieldwork in 1991 and 1992 had settled on the edge of a colonist plantation by 1994. The people had opened their *chagra* and had spent months in the same spot.

2. The only non-Nukak person I know who has been on one of these trips is Kenneth Conduff.

3. It is not clear if these unoccupied areas are the product of recent demographic decline or are simply the consequence of the traditional Nukak mode of land occupation.

REFERENCES

Azcárate, L.

n.d.   *Informe de comisión a Laguna Pavón (Guaviare) y Mitú (Vaupés)*. División de Asuntos Indígenas, Ministerio de Gobierno, Colombia, Bogotá.

Bettinger, R.

1991   *Hunter-Gatherers: Archaeological and Evolutionary Theory*. Plenum, New York.

Binford, L.

1980   Willow Smoke and Dogs' Tails: Hunter-Gatherer Settlement Systems and Archaeological Site Formation. *American Antiquity* 45: 4–20.

Cabrera, G., C. Franky, and D. Mahecha

1999   *Los Nukak: Nómadas de la Amazonía colombiana*. Editorial Universidad Nacional, Santafé de Bogotá.

Dyson-Hudson, R., and E. A. Smith

1978   Human Territoriality: An Ecological Reassessment. *American Anthropologist* 80: 21–41.

Eisenberg, J.

1981   *The Mammalian Radiations: An Analysis of Trends in Evolution, Adaptation and Behavior*. University of Chicago Press, Chicago.

Ember, C.

1975   Residential Variation among Hunter-Gatherers. *Behavior Science Research* 3: 199–227.

Fisher, J. W.

1987   Shadows in the Forest: Ethnoarchaeology among the Efe Pygmies. Ph.D. dissertation. University of California, Berkeley.

Franky, C., G. Cabrera, and D. Mahecha

1995   *Demografía y movilidad socio-espacial de los Nukak*. Fundación GAIA, Santafé de Bogotá.

Gosden, C.

1999   *Anthropology and Archaeology: A Changing Relationship*. Routledge, London.

Gould, R.

1968   Living Archaeology: The Ngatatjara of Western Australia. *Southwestern Journal of Anthropology* 24(2): 101–122.

1969   *Yiwara: Foragers of the Australian Desert*. Scribner's, New York.

Hayden, B.

1981   Subsistence and Ecological Adaptation of Modern Hunter/Gatherers. In *Omnivorous Primates: Gathering and Hunting in Human Evolution*, edited by R. S. O. Harding and G. Teleki, pp. 344–421. Columbia University Press, New York.

Hewlett, B.

1995   Cultural Diversity among African Pygmies. In *Cultural Diversity among Twentieth-Century Foragers: An African Perspective*, edited by S. Kent, pp. 215–244. Cambridge University Press, Cambridge.

Holmberg, A. R.

1950   *Nomads of the Long Bow: The Siriono of Eastern Bolivia*. Institute of Social Anthropology, Publication No. 10. Smithsonian Institution Press, Washington, D.C.

Ingold. T.

2000   *The Perception of the Environment: Essays in Livelihood, Dwelling and Skill*. Routledge, London.

Jackson, J.

1983. *The Fish People*. Cambridge University Press, Cambridge/New York.

Kelly, R.

1983   Hunter-Gatherer Mobility Strategies. *Journal of Anthropological Research* 39(3): 277–306.

1995   *The Foraging Spectrum*. Smithsonian Institution Press, Washington, D.C.

Lanata, J. L.

1993   Evolución, espacio y adaptación en grupos cazadores-recolectores. *Revista do Museu de Arqueologia e Etnologia* (São Paulo) 3: 3–15.

Mondragón, H.

n.d.   Estudio para el establecimiento de un programa de defensa de la comunidad indígena Nukak. Informe final presentado al programa de Rehabilitación Nacional (PNR) de la Presidencia de la República de Colombia. Unpublished report. Santafé de Bogotá, Colombia.

Münzel, M.

1969–1972   Notas preliminares sobre os Kaborí (Makú) entre o Rio Negro e o Japurá. *Revista de Antropología* 17–20 (part 1): 137–182.

Politis, G.

1996a   Moving to Produce: Nukak Mobility and Settlement Patterns in Amazonia. *World Archaeology* 27(3): 492–511.

1996b   *Nukak*. Instituto Amazónico de Investigaciones Científicas (SINCHI), Santafé de Bogotá.

1999   Plant Exploitation among the Nukak Hunter-Gatherers of Amazonia: Between

Ecology and Ideology. In *The Prehistory of Food: Appetites for Change*, edited by C. Gosden and J. Hather, pp. 99–125. Routledge, London.

Politis, G., G. Martínez, and J. Rodríguez

1997   Caza, recolección y pesca como estrategia de explotación de recursos en forestas tropicales lluviosas: Los Nukak de la Amazonía colombiana. *Revista Española de Antropología Americana* 27: 167–197.

Reichel-Dolmatoff, G.

1968   *Desana: Simbolismo de los indios Tukano del Vaupés*. Departamento de Antropología de la Universidad de Los Andes, Bogotá.

Reina, L.

1989   Los Nukak: Cacería, recolección y nomadismo en la Amazonía. In *Diversidad es riqueza*, pp. 62–64. ICAN, Bogotá.

Ucko, P., and R. Layton

1999   *The Archaeology and Anthropology of Landscape: Shaping Your Landscape*. One World Archaeology 30. Routledge, London.

Van der Hammen, M. C.

1992   *El manejo del mundo: Naturaleza y sociedad entre los Yukuna de la Amazonía colombiana*. Tropnebos, Bogotá.

Yellen, J. E.

1977   *Archaeological Approaches to the Present*. Academic Press, New York.

3

Locational Analysis of Yanomamö Gardens and Villages Observed in Satellite Imagery

NATHAN CRAIG AND NAPOLEON A. CHAGNON

This chapter examines the placement of Yanomamö horticultural plots as a function of transportation costs between central places (villages) and production sites (gardens) through a modified version of J. H. Von Thünen's marginal value theorem. His theorem, outlined in his *Isolated State* (1966), predicts the intensity of agricultural investment as a function of transportation costs. This theorem has been examined cross-culturally (Chisholm 1968) and has been employed in the interpretation of archaeological settlement data and as a means of site comparison at the regional level. While specific applications of the Von Thünen model have been criticized (Hall 1966: xxxi; Sallade and Braun 1982: 20), criticisms focus on how the model has been applied rather than on the model's fundamental premises as formulated by Von Thünen. Here we use these premises to develop a model that explores the structure of Yanomamö horticultural investment in terms of travel and transportation costs. These costs have been incorporated into diet choice models through the development of central place foraging (CPF), but room for improvement exists with respect to establishing how transportation costs structure the spatial pattern of horticultural investment among tribal peoples. Because a good deal of daily activity takes place in gardens, understanding the relationship between travel and transportation costs and horticultural investment could help improve models of prehistoric activity patterns.

Chagnon (1974) has stressed that (1) settlement is fluid over time, (2) previous settlements are important economic resources, and (3) understanding why villages are located where they are requires a knowledge of political histories. Community locations change through time. Therefore it is not logistically possible for the fieldworker to cover the entire region of study in a single field season. Despite three decades of dedicated effort, not all communities have been located through fieldwork alone. Fortunately, use of remotely sensed imagery from space-borne platforms for photomapping (Birdseye 1940; Petrie 1977; Robinson et al. 1977) remote sensing is a way to record enormous regions in extremely short periods (Jensen 1996; McGwire et al. 1996).

The Yanomamö represent a very different level of technological development and social organization than the agricultural systems that Von Thünen attempted to describe in the *Isolated State*. If a model based these principles can be used to predict garden location with respect to Yanomamö villages, however, it should be possible to make predictions about the layout of tropical forest horticultural plots that are known through archaeological data alone.

In this chapter we show how remote sensing and geographic information systems (GIS) can be used in concert with comparative ethnographic data regarding aspects of mobility, including walking and transportation costs, to develop models that can aid archaeological interpretation. Multispectral space-borne images from the Landsat Thematic Mapper (TM) 5 platform are used as a photobase (Robinson et al. 1977) to document Yanomamö landscape features at a regional scale through photomapping; information on transportation of horticultural products and walking costs is derived from empirical observations reported in ethnographic literature.

The Yanomamö are neotropical tribal-level horticulturalists who practice pioneering shifting cultivation (Chagnon 1966, 1968c). Research described here uses space-borne imagery to inventory landscape features in two areas of Yanomamöland where population appears to be growing rapidly but where some important differences exist. These differences appear to have implications regarding varying potentials for the development of sociopolitical complexity. These two special areas are found in the headwaters of both the Orinoco River and the Mavaca River (one of the Orinoco's major tributaries) in Amazonas State of southern Venezuela.

RELEVANT ETHNOGRAPHIC BACKGROUND OF THE YANOMAMÖ

Yanomamö Physical Location and Population Blocs

The Yanomamö population is or recently was expanding (Chagnon 1966, 1968c, 1974; Chagnon 1983: 54; Hames 1983: 405), with people probably migrating out of the Parima highlands and down into the surrounding foothills and river basins. The Parima massif has been described as a territorial "core" (Smole 1976), and the Parima highlands are a likely source of origin for several genealogically related groups that Chagnon (1966, 1974: 78) termed population blocs. Yanomamö settlements composed of two population blocs located in two different regions, the Upper Mavaca and the Orinoco Headwaters, form the basis of settlement data discussed in this chapter (Figure 3.1). Chagnon (1966, 1974, 1992, 1995) has discussed several distinct population blocs, but his primary focus has been on studies of multiple villages in two of these: the Shamatari and

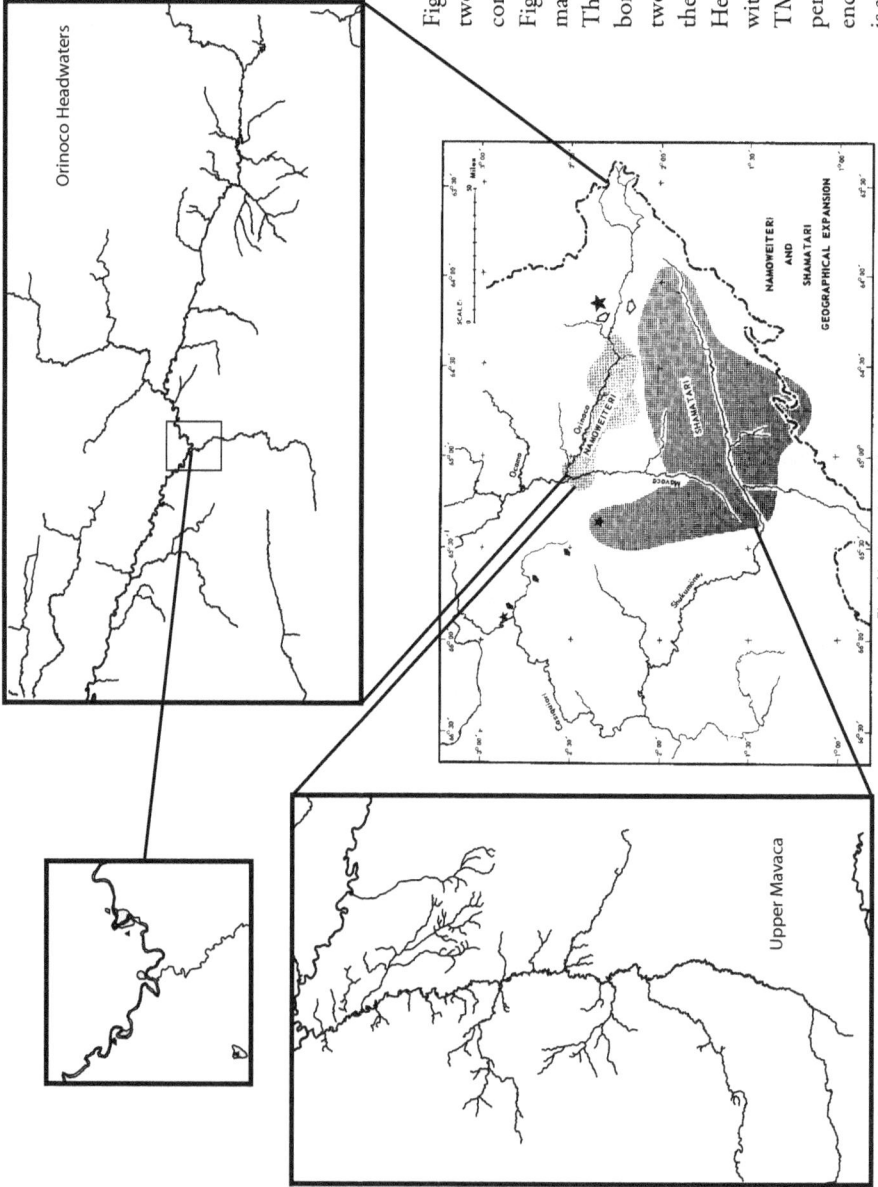

Orinoco Headwaters

Upper Mavaca

NAMOWEITERI
AND
SHAMATARI
GEOGRAPHICAL EXPANSION

Figure 3.1. Map showing the location of the two study areas. The map in the lower right corner is taken from Chagnon (1974: 128, Figure 4.1) and shows the extent of the Shamatari and Namoweiteri population blocs. The dotted line in this map is the political border between Venezuela and Brazil. The two larger maps show the spatial extent of the study areas (Upper Mavaca and Orinoco Headwaters) discussed in this chapter along with hydrology digitized from the Landsat 5 TM imagery. The small inset map in the upper left corner shows the Rahuawa-u confluence in greater detail. This same confluence is also shown in Figures 3.2 and 3.3.

the Namoweiteri. They both originated in the same area north of the Orinoco near the Brazilian border but had become distinct populations by about 1875, when they moved in a generally south and southwesterly direction crossing the Orinoco and into the valleys of the Shanishani River and Mavaca River.

Chagnon (1974: 71) identified areas that he described as "population hearths," where "mother villages" have given rise to what today are widely scattered communities. One area where continued research has revealed this pattern of demographic growth and outmigration is the Shanishani basin. This small tributary of the Orinoco headwaters is an area that Chagnon (1992: 88, 1995: 89) describes as a "demographic pump." In this general area villages tend to grow bigger than normal, largely due to the potential threat of raids from neighboring villages. With internal growth of a community, conflict increases, often leading to fission and eventual outmigration of that community (Chagnon 1975). A significant fraction of the people from Chagnon's census of over four thousand individuals apparently came from the Shanishani basin (Chagnon 1992).

Based on extensive genealogical and demographic research, Chagnon (1974) found that the Shamatari population bloc, located in the Upper Mavaca, was growing and expanding at a faster rate than the Namoweiteri population bloc, located in the Shanishani basin of the Orinoco Headwaters. On the whole, villages from the Shamatari population bloc tended to have more occupants per village, but villages were more widely spaced than villages from the Namoweiteri population bloc.

William Smole (1976: 48) estimated a population density of 0.5 person per square mile for Yanomamöland. Chagnon (1974: 127) suggested a population density of 0.42 person per square mile and a mean intervillage distance of 50 miles (81 km) for the Shamatari population bloc, which is located in the Upper Mavaca. Recognizing that population densities vary in space, Chagnon estimated 0.90 person per square mile and a mean intervillage distance of 30 miles (49 km) for the Namoweiteri population bloc located in this general demographic pump region of the Shanishani basin of the Orinoco Headwaters (Chagnon 1974).

Yanomamö Residential Mobility

Yanomamö live in communities of roughly 40–300 individuals and reside in a domestic structure called a *shabono* (Chagnon 1966, 1968c). Mean *shabono* diameter is about 30 m in the highlands (Smole 1976: 61), with villages in the lowlands tending to be larger (Chagnon 1995). One exceptionally large lowland village had a *shabono* diameter of about 100 m and a population of a little over 300 individuals (Chagnon 1974: 257).

Once a *shabono* location has been chosen, the Yanomamö prefer to remain in the same general area as long as they can before moving. Chagnon (1974) has documented two types of residential relocations that he calls macro-moves and micro-moves. Each mode of residential move is made for unique reasons.

Macro-moves are caused by complex social and historical factors, with the primary social force that stimulates a macro-move being warfare (Chagnon 1974: 129; Hames 1983). The goal of a macro-move is to move far away from an enemy or group of enemies or simply a desire to become hidden in the forest (Chagnon 1974; Hames 1983: 397). Of the 22 macro-moves reported by Raymond Hames (1983: 409, Table 13.2), 11 were caused by raids, the threat of raids, or the abduction of women. The moves in cases where macro-moves were caused by conflict or the threat of conflict were 24 km on average. Five of these same macro-moves reported by Hames (1983: 409) were caused by village fissions, but the distances moved were only 7.8 km on average. Macro-moves are extremely costly, because new village spaces and new gardens must be cleared and residents must travel long distances to the site of the abandoned village to harvest crops and cuttings to be transported back to the new village location. Conflicts with neighbors seem to be the greatest cause for residential mobility in the case of macro-moves.

This type of residential mobility can also be brought about by the fission of a large village, with one of the two split villages moving to a new territory. Recently fissioned villages frequently live "side by side" (*he borarawä*), however, particularly when other neighbors are hostile. Residents of a community prefer to remain in one general area as long as they can and only make long distance macro-moves when the threat of warfare becomes too great and when raids from enemy villages become frequent. The social barriers to migration and residential mobility caused by the presence of hostile neighbors in surrounding territories create a condition that Chagnon (1996: 75–77) calls "social circumscription" and represent an important additional dimension to Robert L. Carneiro's circumscription theory. Macro-moves can also be caused by the desire to live near some other community that has access to trade items. The important point in regard to the causes of macro-moves is that "the relevant 'ecological' variables here are human neighbors, not technology, economic practices, or inherent features of the physical environment as such" (Chagnon 1983: 72).

Micro-movements, by contrast, are most frequently associated with the clearing of new gardens and generally consist of residential moves from 100 m to 1 km (Chagnon 1974: 133, Figure 2, 1983: 70, 1992: 71; Hames 1983: 395). Yanomamö *shabono*s are not always moved or relocated when new gardens are cleared, but often the domestic structure becomes dilapidated or infested with insects. Chagnon (1983: 69) also observes that "the movement of a garden a few hundred yards might also be the occasion to move the *shabono* as well—to keep it located conveniently near the food crops." Hames (1983: 416, Table 13.3) provides the causes given for 34 micro-moves among the Haiyamo population bloc: the most frequent reason (35%) given was garden land being too distant. Thus we see a level of residential mobility that is stimulated in part by a desire to be located near horticultural investments.

Hunting and Foraging

The Yanomamö are now widely seen as mobile horticulturalists, though this has not always been the case (Chagnon 1995: 65; Lizot 1980). Some early observers of the Yanomamö saw them as nomadic hunter-gathers (e.g., Zerries 1955; Wilbert 1966). They are listed in George P. Murdock's survey of the current status of hunter-gatherers (1968: 19) as the Sanema and Shirawa under the category of "Interior Marginal Tribes of Southern Venezuela." Chagnon (1966, 1968a, 1968b, 1968c) described the Yanomamö as highly mobile pioneering shifting cultivators.

Later research on Yanomamö subsistence has demonstrated that they obtain food from a mixture of horticulture and foraging as well as silviculture (Chagnon 1983; Lizot 1980; Smole 1976; Hames 1983). Old horticultural investments play an important role in long-term subsistence practices. For example, garden plots are usually planted with tree crops that continue producing long after the site has been abandoned. Food produced in old gardens also attracts animal species that the Yanomamö prey upon. Treks, extended camping trips for food, or what might be considered a form of logistical mobility are a common strategy used by the Yanomamö to supplement a purely horticultural economic mode. This kind of logistical mobility involves extended hunting trips and foraging on seasonally available plant foods. Helena Valero's account of her twenty-five years living among the Namoeteri Yanomamö is replete with references to logistical forays away from a *shabono*, during which individuals were subsisting on "wild" plant foods from the forest (Biocca 1996: 88, 89). While many of these plant foods may in fact be wild, at least some of them may be remnant or feral products from abandoned gardens that are no longer cultivated (Chagnon 1996; Hames, personal communication, 1998). For the Yanomamö of the Upper Padamo, Hames (personal communication, 1998) has observed that a considerable amount of hunting occurs in and around garden plots that are no longer under direct cultivation. Similar patterns have been noted for other Amazonian groups as well (Balée 1998).

Thus while the Yanomamö are highly mobile, it seems plausible to characterize at least some of this activity as logistical extensive exploitation of previous, intensive horticultural investments from an earlier date. Abandoned plots continue to provide returns long after the period of intensive labor investment has ended.

Horticulture

Yanomamö gardens are best described as polyvariety cropping systems (Hames 1983). Some few crops may be grown together, but for the most part individual crop types are planted together as sections within a garden. Clearings are made by groups of individuals, but plots within the clearing are worked by individuals from specific households from a given community (Hames 1983). Plantain is the principal crop, though the Yanomamö have been recorded as growing a very large number of crops, including mapuey, manioc, maize, cotton, tobacco, papaya, arrow cane, gourds, hot peppers, and peach palm (*rasha*) among many others.

New gardens are cleared when persistent thorny and weedy growth begins to dominate garden plots after about three years, a pattern very similar to the one that Carneiro (1960, 1961) observed for Kuikuru horticulturalists of the Brazilian Xingu. Only in a few cases reported by Hames (1983) were new gardens cleared due to declining yields caused by soil depletion.

The general pattern of labor among Amazonian tribal horticulturalists is that males clear gardens while females do most of the regular garden work. Hames (1978: 470), however, states that "among the Yanomamö and Machiguenga, men do nearly as much weeding, planting, and harvesting as do women. Heavy male participation in gardening among the Yanomamö is due to warfare. The connection may seem odd, but when one realizes that most kills during raids take place near the village or in gardens and the Yanomamö habitually kidnap women, heavy male participation in gardening becomes sensible." Thus, for the Yanomamö, gardens are a focal point of activity for both males and females.

Hames (1978) estimates the distance from the village (7 of 23 gardens) or travel time (11 gardens) as part of his time allocation research on Ye'kwana and Yanomamö from the village of Toki. In all of these cases gardens were about 500 m or about a 15-minute walking distance from the village of Toki. While this meticulous empirical work by Hames represents a major contribution to studies of horticultural labor, his survey only estimates distances for some gardens at one community. Our approach in this chapter does not achieve the level of behavioral detail that ethnographic fieldwork can achieve, but it does allow collection of regional-scale data.

MOBILITY, TRANSPORTATION, AND HORTICULTURE

Mobility is a property of individuals (Kelly 1992: 44), who vary in the amount that they travel from a home base (Hewlett et al. 1982: 422; Hawkes et al. 1995;

Kelly 1992: 44) and how often residences are moved. At least among hunter-gatherers, people tend to travel more as they get older (Hewlett et al. 1982: 424; Hawkes et al. 1995). Although studies of Aka mobility indicate individual variation in patterns of travel within age and sex categories, individual variability did not appear to alter the shape of the distribution (Hewlett et al. 1982: 422). Thus it would seem that patterns defined to describe the group as a whole account well for the actions of individuals. While there is individual variability in movement and travel as well as labor, measures of central tendency appear to be valid. This warrants the application of more general statements about travel and mobility that can be applied to archaeological interpretations when there is recognition that the broad patterns useful for cross-cultural comparison and model building may mask some individual variability.

Mobility varies in many ways, but the fundamental mechanics and energetic requirements of human walking should remain largely the same for all individuals. As long as foot travel is the primary source of transportation, one should expect some similarities in system structure arising out of the costs of transportation in the context of a limited energy budget. Some age categories might be expected to travel and transport more than others, but the costs of travel and transport for all members of a given age and sex group should be about the same.

In human behavioral ecology (HBE) minimization of movement costs plays a principal role in studies of production. The diet-breadth model is one of the most commonly used HBE studies of production and probably has the widest application to archaeological data. This model has its theoretical foundation in the diet-choice models developed by Robert H. MacArthur and Eric R. Pianka (1966) and later elaborated by E. L. Charnov (1976). In the diet-breadth model subsistence alternatives are ranked by the energetic return relative to the time investment once a resource is located. Increasingly lower-ranked resources are added to the diet so that pre-encounter search time (or one-way travel cost) and post-encounter handling times optimize energetic returns for overall foraging time. Subsequent HBE studies of human foraging demonstrate that post-encounter travel and transportation costs play an important role in foraging decisions (Orians and Pearson 1979; Foley 1985; Layton and Foley 1992; Hawkes and O'Connell 1992; Winterhalder 1983, 1993; Winterhalder and Smith 2000). Unlike many animal species that forage primarily "on the fly," human beings generally transport foods back to a residential base for consumption. Inclusion of transportation costs in human foraging studies is important for this reason.

Central place foraging (CPF) models incorporate round-trip travel times as additional costs in foraging equations. When round-trip travel costs are included, minimum acceptable prey size generally increases in relation to travel time (Bettinger et al. 1997: 888; Schoener 1971). Resources have a maximum acceptable transport distance, and the weight of a resource can be used to predict the distance it will be transported. Humans have developed post-foraging behaviors like field processing that help reduce post-encounter resource transport to central place costs (Kaplan and Hill 1992; Metcalfe and Barlow 1992; Barlow and Metcalfe 1996; Barlow et al. 1993; Bird and Bird 1997; Bettinger et al. 1997: 887). CPF logic would suggest that to reduce transportation costs home-base residences should be located near principal resource patches (Bettinger et al. 1997; Winterhalder and Smith 2000).

For the Yanomamö, time allocation research by Hames (1992: 230) shows that most gardening is done in the morning and evening, because individuals are seeking to avoid heat stress, which reduces the efficiency of labor. Thus in this case both transportation costs (to and from gardens) and the costs of garden labor (another aspect of movement) are being minimized.

Transportation Costs in Horticulture—Marginality and Exploitation Territories

Travel cost minimization also plays a prominent role in anthropological studies of agriculture and horticulture. Much of this work is grounded in Von Thünen's (1966) economic theories of agriculture developed during the mid-1800s. Von Thünen's marginal value theorem lies at the foundation of diet-choice or optimal foraging models (Charnov 1976) and by extension central place foraging models as well. Von Thünen's model of land use is an attempt to explain and predict the location and intensity of agricultural activity through the isolation of single conditioning factors (Sallade and Braun 1982: 20). As defined in *The Isolated State* (Von Thünen 1966) the model has two basic parts.

Part 1 asserts that a crop may be cultivated under different methods, some of which may involve higher costs but also higher returns. The intensity of production will depend on the returns from cultivation. Return rates will depend on transport costs that are partially a function of the distance between a field and the town. Given these two relationships, intensive methods of production for some crop will occur in plots closer to the town.

Part 2 adds that the kind of crop grown in a plot can vary. In a system involving multiple crop types, decisions must be made as to what kinds of crops to plant in a given field. Crop types will differ in their overall costs relative to plot distance; these costs include labor costs per unit of area as well as transportation costs as a function of mass yield per unit of area. Crops will displace one another in a field based on these relationships.

A general statement of Von Thünen's principles can be expressed as the *time-distance factor* used in territorial analysis developed by the paleoeconomic school of thought (for example, Jarman et al. 1982). The time-distance factor is described as "the complex of forces which combine to determine exploitation thresholds" (Jarman et al. 1982: 31). Territorial analysis is a theoretical expression of the norms of exploitation territory, and the time-distance factor is constructed to define a zone of maximum economic effort and return (Jarman et al. 1982: 38). Optimal foraging theory (OFT) and CPF models incorporate travel costs into production equations, but these models are largely aspatial and generally express travel costs units of time or calories rather than distance. Territorial analysis, however, uses Von Thünen's principle of marginality to express labor and travel costs in terms of distance traveled; thus expectations have a more clearly stated spatial component, estimating and describing areas of habitual exploitation.

M. R. Jarman et al. (1982) attempt to operationalize territorial analysis to define exploitation territories of early agricultural settlements in Europe. They observe that precision in the calculation of an exploitation territory is difficult to achieve (Jarman et al. 1982: 33) for three reasons. First, it is difficult to establish the time-distance threshold correctly; they assume a walking rate of 5 km/hour. Studies of walking and horticultural activity described below can be used to develop predictions about the extent of exploitation and investment territories. Second, Jarman et al. (1982) note that chronological estimates are coarse in archaeology, so it is difficult to determine whether sites are contemporaneously occupied (Martin and Plog 1973). Third, because of this, it is difficult to tell whether sites have overlapping exploitation territories (Dewar 1991: 606). Use of remotely sensed imagery to document Yanomamö horticultural settlements avoids the contemporaneity problem, making it possible to determine empirically if sites have overlapping exploitation territories.

One limitation of a remote sensing approach, nonetheless, is the detector's spatial resolution. Transit sites (Jarman et al. 1982: 36) or logistical camps (Binford 1980: 13) modify exploitation territories, and the degree to which these sites impact exploitation territories should depend upon the use frequency. Use of remote sensing data to inventory transit sites or logistic camps in the tropical forest is difficult, because these are generally small and ephemeral sites where human activity does not produce a significant enough impact on forest canopies to leave a lasting and detectable trace. This problem is similar, however, to the problem faced by survey archaeologists when they are attempting to locate small camps in a large region. It is highly likely that these small sites will leave little or no trace in the remnant settlement pattern; finer transect intervals are required to find them.

Walking, Transportation, and Agricultural Home Ranges

Studies of walking and energy expenditure provide information that is helpful in developing estimates of the time-distance threshold used in territorial analysis. G. M. O. Maloiy et al. (1986: 668, Figure 1) compared energetic costs of walking among females from two African groups who carry loads on their heads slightly differently. The Luo carry loads directly on top of their heads, while the Kikuyu carry loads with a strap or tumpline across their foreheads. The Kikuyu mode of transport is similar to that used by Yanomamö females (Chagnon 1996: 124, Figure 4.2). In both the Luo and Kikuyu cases, the optimal rate of travel was a little over 3 km/hr^{-1}. This speed had the lowest oxygen consumption (ml O_2 per g h) and the lowest stride frequency (steps/min^{-1}) for speed (km/h^{-1}) (Maloiy et al. 1986: 669). As loads increased, efficiency decreased, but not until the women were carrying 20% of their body weight. At higher loads, decreases in efficiency were proportional to weight increase. These findings are consistent with Darna L. Dufour's (1984: 45) earlier observation that there did not seem to be significant decreases in energy expenditure with moderate load increases among the Yapú.

These bioanthropological studies suggest that walking rate should be estimated at about 3 km/h^{-1}. The 5 km/hr^{-1} walking rate used by Jarman et al. (1982) is an overestimation of walking rate for horticultural transport. This value may be appropriate for some hunter-gatherer walking rates for travel to and from foraging locations but probably not for agriculture or horticulture. A travel rate of 3 km/hr^{-1} seems to be a much more appropriate estimate for horticultural transport in this case. The optimal walking rate reported by Maloiy et al. (1986) is actually a little higher than 3 km/hr^{-1}. Return travel frequently involves the transport of horticultural products or firewood; children are frequently taken on trips, and infants are carried by women to and from gardens. All of these factors also lower walking rates (Dufour 1984).

Michael Chisholm (1968) found that agricultural activity is concentrated within a 1–2 km radius of settlement. He states that activities beyond this zone are strongly influenced by travel times and suggests that agricultural activity often terminates at about 5 km from a settlement. While Glenn D. Stone's (1991: 347) study of Kofyar horticulture demonstrates that social factors of production influence the shape of horticultural territories, some general principles for a home range do exist. His study of Kofyar labor shows that 73% of all trips to fields were within 1 km (Stone 1991). Trips increase linearly up to 700 m from a residence; beyond that, distance increases in total trips drop off sharply. Dufour's (1984: 41–42) study of energy expenditure among the Yapú indicates that foot travel to and from horticultural plots averaged about 1.5 hours, which was about 40% of the total time that women spent acquiring food. Walking time to gardens ranged from only 15 minutes to 2.5 hours by trail for one garden. Among the Yapú nearly all travel to and from fields is done by trail, and travel time is about 1.43 hours a day (Dufour 1984: 44).

Table 3.1. Walking Times and Distances Traveled for Five Horticultural Groups

Culture group	Average two-way walking time (h)	Distance traveled (km)	Reference
Yapú	1.43	4.29	(Dufour 1984)
Kaul (coastal)	1.56	4.68	(Norgan et al. 1974: 334; reported in Dufour 1984: 44)
Lufa (highland)	2.35	7.05	(Norgan et al. 1974: 334; reported in Dufour 1984: 44)
Pari (highland)	1.65	4.95	(Hipsley and Kirk 1965; reported in Dufour 1984: 44)

From these ethnographic examples it seems that walking time to gardens in habitual-use zones should be about an hour one-way. It is not always clear whether walking times reported in the ethnographic literature are for two-way or one-way travel. In most cases, however, studies appear to be reporting the total walking time, which suggests that two-way travel times are being reported (see Table 3.1). Dufour (1984), in her study of the Yapú, reports one-way travel times from 15 minutes to over two hours in one case.

Model Predictions

In light of the comparative ethnographic information discussed above, economic models based on the principle of marginality permit the formulation of some predictions about Yanomamö site location and land use in terms of travel and transportation costs.

Relative location of garden types: in order to minimize travel and transportation costs actively cultivated gardens should be closer to residential bases than gardens that are not cultivated.

Habitual-use zones: using an estimated walking rate of 3 km/hr with a habitual-use zone one-way travel time of 1 hour, most actively cultivated gardens should be within 3 km from a residential base and should drop off quickly beyond this distance.

Settlement clusters: human neighbors are the salient environmental variables when it comes to Yanomamö macro-move decisions: people seek to live near friendly neighbors and far from enemy villages. On a regional scale these social forces should produce a settlement pattern consisting of clusters of mutually friendly communities, but distances greater than habitual-use zones should separate these clusters.

Intracluster distances: settlements should be spaced within clusters such that communities are on the edges of exploitation territories. While Yanomamö seek to live close to communities with which they are on good terms, the tribal level of political organization involves little or no lasting multicommunity organization. Because villages are politically autonomous, there should be some spacing between settlements. The separating distance should most likely be about the same as the habitual-use zone.

Settlement Studies and Remote Sensing

Testing these predictions requires an inventory of settlements at a regional scale of analysis, much like settlement data collected by a survey archaeologist. For this project Yanomamö cultural features located in two regions (the upper Mavaca and the Orinoco Headwaters) are inventoried through analysis and inspection of Landsat 5 Thematic Mapper images captured in 1991. Basic metadata for these images are reported in Table 3.2.

Yanomamö settlements (villages and gardens) leave distinct signatures on the landscape (Smole 1976; Chagnon 1995). Based on simulations of feature measurement at different spatial resolutions, K. McGwire et al. (1996: 104, Figure 6.2) conclude that the Landsat 5 Thematic Mapper detector has a spatial resolution capable of discriminating Yanomamö villages and gardens.

Remote sensing is a powerful tool for examining regional patterns, but it has limitations for understanding many nonspatial parameters. Extracting local details like political relations and histories between communities is not possible, which therefore limits the scope of analysis for purely remote sensing studies. Nonetheless, remotely sensed data do provide a unique regional perspective that is not available to the field researcher, providing opportunities to refine other forms of analysis. Furthermore, because of information extraction through techniques of photomapping, remotely sensed data have value even if it is not possible to examine the entire study area from the ground or if the images were captured prior to the start of fieldwork for a given project. Historic imagery has significant value for studies of settlement change.[1]

Image-Processing Procedures

The two Landsat 5 TM images were imported into the Earth Resources Data Analysis System (ERDAS) Imagine GIS software package and georeferenced to their respective corner coordinates that had been provided in metadata reported by the U.S. Geological Survey (USGS). Once a real world coordinate system had been applied to the data, a number of image transformations and classification algorithms were applied to the images to improve the definition of features like villages and gardens (Craig 1996). After extensive experimentation with image-processing techniques, two analytical transformations (Principal Components

Table 3.2. Landsat Thematic Mapper Imagery Metadata

Landsat 5 Thematic Mapper	WGS: Path 1 Rows 58 and 59
Image Acquisition: 10/7/1991 13:59:45	Image Dimensions: 7420 × 11648
Pixel Spacing: 28.5 m × 28.5 m	Sun Elevation and Azimuth: 57.5, 102.93
TopLeftLatLong: 3.78, −66.43	TopRightLatLong: 0.50, −64.58
BottomLeftLatLong: 0.84, −66.86	BottomRightLatLong: 0.58, −65.00

Analysis and the Tassel Cap) were relied upon for most image interpretation. These transformations were used to aid image interpretation by allowing the inspection of various factors present in the multispectral data.

The first principal component (PC1) expresses an enormous amount of variability in passive multispectral data (Jensen 1996: 179; Ribed and Lopez 1995; Sabins 1987). Non-nadir illumination angles of the sun shining on mountains and valleys create topographic lighting effects described as multiplicative gains, affecting all bands of the image data. The difference between self shading caused by the complex limb architecture of climax canopies compared to the flatter, more reflective canopies of swidden plots is a key relationship used for feature discrimination of human-caused disturbance that is well expressed in PC1.

The Tassel Cap (TC) transformation is derived by rotating principal components around a set of coefficients (Christ and Cicone 1984; Jensen 1996: 185;

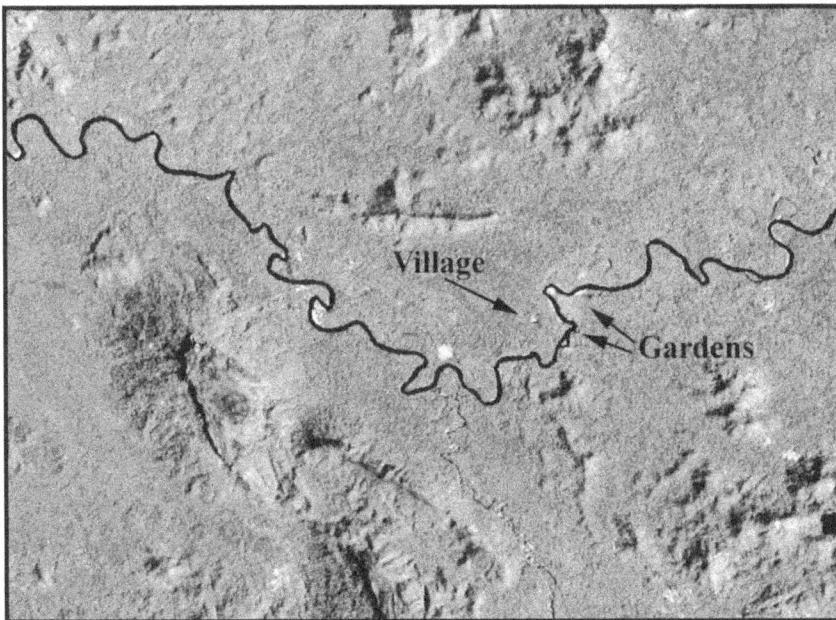

Figure 3.2. Tassel Cap transformed brightness factor image, indicating the location of select cultural features visible in the subscene.

Kauth and Thomas 1976). Tassel Cap produces three new factors that are useful for the interpretation of vegetation communities and the discrimination of vegetation from soil and water: Brightness, Greenness, and Moisture. Brightness and Greenness are both quite effective at depicting variation in forest cover and defining the location of cultural features based on canopy texture and soil exposure.

During image interpretation and feature identification, each transformed image was inspected simultaneously using the "flicker" technique (rapidly alternating the display of the image transformations for each of the two study areas) to find features that were consistent in both transformations. The existence of functionally uniform relationships with regard to canopy disturbances and energy reflectance permits the development of analogies necessary for image interpretation (Steiniger 1996; Adams et al. 1995; Mertes et al. 1995). For example, exposed dirt and thatched roofs are highly reflective in PC1 because they are flatter, more reflective surfaces than climax forest canopy. Village clearings appear red in a TC-transformed image because they have low values in the plane of greenness and are relatively flat and therefore bright (Figure 3.2). Gardens are distinct from climax forests due to their flatter, more reflective canopies, making them brighter and greener than the surrounding forest. When gardens fall out of use, there is less investment in weeding; and forest regeneration begins to occur. As regeneration occurs, larger trees take hold and forest canopies become more complex. When this happens, less radiant energy is reflected back to the detector; pixels in these regions begin to take on the darker shades of climax canopy. Active gardens are the patches of brightest vegetation pixels, while old gardens are defined by patches of pixels that are transitional between values of climax canopy and recent clearings.

Implementation of a Horticulture/Mobility-Transport Model in a GIS Context

Once cultural features were identified, they were digitized on screen as vector objects and attributed accordingly. Digitized features were then counted, and the area of each digitized polygon was computed. In addition, the nearest distance from a village to the following features was computed: (1) another village, (2) a garden, and (3) the nearest old garden. Village features can be represented as points by taking the centroid, so it was also possible to compute the nearest neighbor statistic (Clark and Evans 1954), using the extent of the image as the sample universe. If nearest neighbor determines that the data are clustered, then the number of clusters based on spatial position alone is estimated. Archaeologists frequently want to know whether aggregation or settlement clustering is occurring during some occupational period; but an archaeologist studying prehistoric settlement data is generally confronted with the task of interpreting the presence and configuration of settlement clusters based on locational data alone. A considerable amount of social and behavioral information is known about the settlements in this study, making it possible to evaluate how well clustering algorithms estimate the presence and configuration of real settlement clusters. In this chapter coordinate data (x, y) serve as the cluster criteria. Five hierarchical clustering algorithms based on differences in linking method are used (Aldenderfer and Blashfield 1984: 35–45). Linkage methods are average, centroid, ward's, single, and complete.

In addition to these counts, measures, and basic spatial statistics it was also possible to develop an isotropic model for exploring the distribution of Yanomamö horticultural investment using GIS. From the centroid of village polygons, multiple concentric buffer rings were constructed as an interval of 500 m out to a distance of 10 km. This approximates the maximum distance that most Yanomamö will generally travel to hunt (Hames 1983: 416). Therefore 10 km serves as a good maximum distance of daily travel without the construction of a logistic camp and an overnight stay. Once constructed, the buffer rings were intersected with polygons that had been interpreted as gardens and old gardens. Intersecting the concentric buffer rings with garden and old garden polygons subdivides these two kinds of gardens into consecutive 500-m distance classes. After this subdivision into distance classes up to 10 km by the concentric 500-m buffer rings, garden area could be recomputed and summed within each of the 20 distance classes defined by the buffer rings. Thus it is possible to determine how much garden area or old garden area falls within a given distance zone from a village (Figure 3.3).

Before garden areas can be compared between distance classes, garden area for each class must be divided by the area of the buffer ring for that distance class. As the classes increase in distance from a village, the area of the annular zone increases at a much higher rate. When the rates of increases are compared between distance and area, we find that the slope of the line is 3,123. This means that as the zones increase in distance from a village they become large at an increasingly greater rate. As distance increases, the space created by the buffer is much larger at greater distances, thus increasing the possibility of encountering a garden. To make garden areas comparable across zones, garden area was divided by the area of the buffer ring.

An additional consideration must be taken into account in the development of the model. Most garden or old garden plots are intersected with more than one village's set of buffer rings. Thus the total area of the intersected garden polygons can be greater than the total area of garden polygons prior to intersection. The aim, however, is to examine how distance explains the degree of horticul-

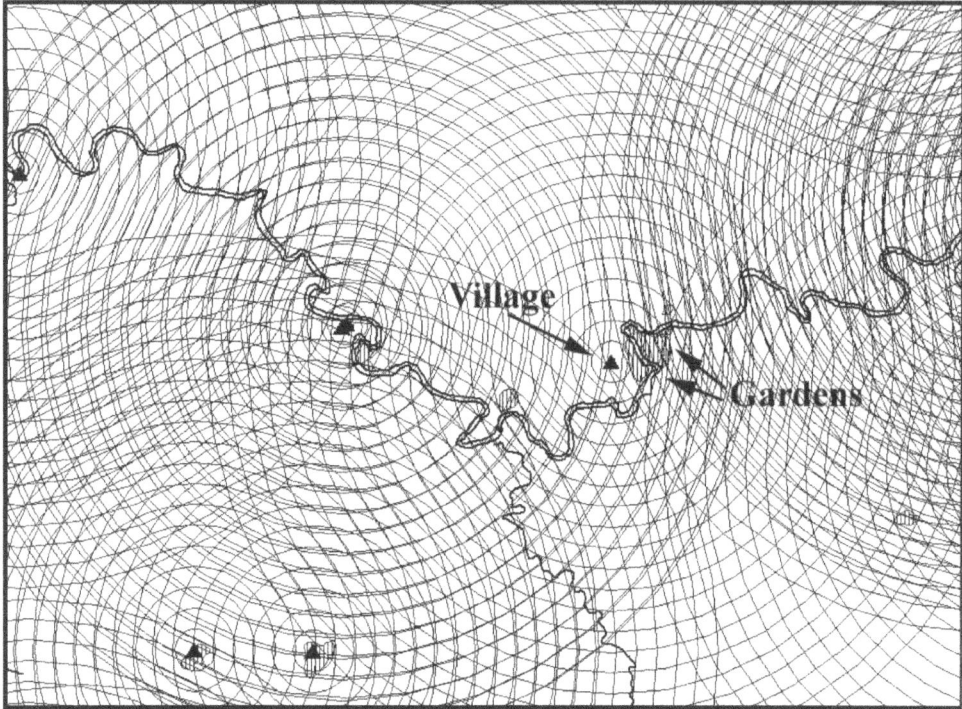

Figure 3.3. Digitized cultural features identified in the transformed Landsat 5 TM data and the 500-m buffer rings used to explore the distribution of garden area. Villages are represented as triangles. Gardens are represented as hatched lines. In addition to the village and gardens annotated in the figure, two other villages that have closely associated gardens can be seen in the lower left portion.

tural investments as represented by cultivated space from a village. Therefore the GIS was allowed to re-intersect a given polygon if it was within a 10-km buffer of more than one village. If villages are spatially associated, for example, it is not possible to determine which plots a given village's residents cultivate. Stone's (1991) study of Kofyar horticulture showed that individuals will visit neighbors and work in their gardens. Hence all cultivated space is examined in relation to each village.

RESULTS

The Orinoco Headwaters image covers a much larger area than the Upper Mavaca image. The total area of the Upper Mavaca Basin image covers an extent of 8,769 km² (3,386 square miles), while the Orinoco Headwaters image extends over 14,362 km² (5,545 square miles), making the Orinoco Headwaters image

Table 3.3. Summary Results of Village and Garden Distributions from Both Study Areas

Variable	Orinoco Headwaters	Upper Mavaca
Settlements	56	36
Gardens	161	97
Minimum distance		
between villages	mean: 2,529 m; s.d.: 7,789 m	mean: 1,436 m; s.d.: 3,827 m
	median: 378 m	median: 240 m
between village and garden	mean: 240 m; s.d.: 168 m	mean: 62.7 m; s.d.: 168 m
	median: 4 m	median: 0 m
between village and old garden	mean: 11,205 m; s.d.: 9,866 m	mean: 5,589 m; s.d.: 3,247 m
	median: 10,116	median: 4,932 m
Total garden area < 500 m		
from villages	49%	33%
Total old garden area < 500 m		
from villages	14%	2%

161% larger than the Upper Mavaca image. Only a small fraction of either study area has been modified in such a way as to be visible in TM imagery. Including all categories of clearing, 0.88% of the study area in the Upper Mavaca image consists of visible patches of some form of disturbance, while 1.05% of the Orinoco Headwaters image consists of visible patches of disturbance. Summary values for key land cover classes in the Upper Mavaca and Orinoco Headwaters images are reported in Table 3.3. Specific patterns are described below.

Village Spacing

Calculation of the minimum distances between patches interpreted as villages in the Upper Mavaca image produced a mean of 1,436 m, with a standard deviation of 3,827 m. The median minimum distance between villages in the Upper Mavaca is only 240 m. The mean minimum distance between villages in the Orinoco Headwaters image was 2,529 m, with a standard deviation of 7,789 m. The median distance between villages in the Orinoco Headwaters image is 378 m. In both cases, only a few villages are located exceptionally far from any other village. The distance that separates these villages from any other is so great, however, that they strongly leverage the mean and generate large standard deviations in the set. The median is a better measure of central tendency in both cases. Although in each case a few villages are located a great distance from any other village, most villages are located relatively close together. This would seem significant, considering that only about 1% of either image shows evidence of humanly caused forest clearing.

Results of the nearest neighbor statistic run on the location of villages and potential villages from both the Upper Mavaca and Orinoco Headwaters images produced a p value of 0.01, indicating that the set of points exhibits significant clustering. Settlement distributions are almost always clustered; but evidence for clustering is interesting in this case, because such a small fraction of either image shows evidence for canopy modifications.

Each cluster linkage method produced natural breaks at consistent numbers of groups in each image. Each linkage method produced the same number of clusters within each respective study area, but in every case clusters in a study area consist of slightly different sites, due to differences in linkage method. Inspection of the clusters showed that formal clusters constructed from spatial coordinates on an isotropic plane do not take isotropic characteristics of the terrain into account. The nonisotropic dimension of terrain places important constraints on travel, because mountains and rivers can increase travel costs significantly. Thus techniques for grouping sites that do not take landscape travel constraints into account are unlikely to produce results that relate to real social groupings. Fortunately, a variety of new sources of data for constructing digital elevation models from satellite imagery should provide promising new directions for analysis of site groupings in a nonisotropic context. Time did not permit the incorporation of these data in this analysis, however. An informal approach to defining clusters based on judgmental grouping of sites was employed as another means to evaluate within and between group site spacing. While this technique suffers from observer bias problems and is not based on objective quantitative techniques, human observers are extremely effective at detecting patterns.

Informal grouping of villages produced results that appear to be far more consistent with expectations and minimum distance results reported above. Sites in the Upper Mavaca are generally spaced within a group from about 0.7 km to about 2 km. Spacing between village clusters ranged from about 4 km to about 9 km; in one instance a distance of about 24 km separates the two closest groups. A similar pattern was observed in the Orinoco Headwaters: sites within groups generally appear to be less than about 1.5 km apart, and the distance between site clusters is generally about 7 km.

Visual inspection of judgmental settlement clusters suggests that village groups are rather linear in shape, undoubtedly because settlements are mapped onto river systems. Slightly more circular settlement clusters can be observed at river confluences, where the largest settlement clusters are found. The presence of missions in both images probably also affects intervillage spacing, primarily by decreasing the spacing between communities near missions. Villages spatially associated with a mission have a decreased threat from raids as well as an increased reliance on a cash economy and bartering of native items (for example, for metal tools). These factors together seem to result in a change in residential pattern. Rather than living in a large *shabono* that provides improved protection from outsiders but fewer opportunities to hide private property, Yanomamö living near missions opt to live in enclosed single-household structures scattered along river edges. These single-family structures are less defensible but provide superior protection from insects and also offer greater privacy. The transition to living in dispersed single-household structures near missions increases the counts of village class polygons as well as decreasing the intervillage distances.

Minimum Distance to Gardens and Old Gardens from a Village

Mean minimum distance from a village to a garden is 240 m in the Orinoco Headwaters image and 62.7 m in the Upper Mavaca image; in both cases the standard deviation is 168 m. Median distance to a garden is 4 m in the Orinoco Headwaters, while it is 0 m (adjacent) in the Upper Mavaca. In most cases on both images garden clearings appeared to be adjacent to villages (a distance of 0 m). It is difficult to believe that some cultivation is not always taking place immediately adjacent to village clearings, even if the scale of cultivation is not detectable in Landsat imagery. Mean minimum distance from a village to an old garden is 11,205 m (with a standard deviation of 9,866 m) in the Orinoco Headwaters image, while it is 5,589 m (with a standard deviation of 3,247 m) in the Upper Mavaca image. Median distance is 10,116 m in the Orinoco Headwaters and 4,932 m in the Upper Mavaca. The high standard deviation combined with a large difference in the mean and the median suggests again that a few outlying cases are leveraging the mean, making the median a better representation of central tendency.

Garden and Old Garden Area by Isotropic Buffer Rings from a Village

Results of the isotropic horticultural investment model show that in both the Orinoco Headwaters and the Upper Mavaca images garden area decreases rapidly with distance (Figure 3.4). The rapid decrease in garden area stops at about 2 km in the Orinoco Headwaters image, which would work out to a walking distance of about 0.6 hour; it drops off at about 3.5 km or a walking time of a little over 1 hour in the Upper Mavaca image. In both cases over 30% of total garden area is within the 500-m buffer zone, and over 80% of total garden area is within the 3-km habitual-use zone of villages. In the Orinoco Headwaters nearly 50% of total garden area is within 500 m of a village—almost 20% more than in the Upper Mavaca. In both cases garden area decreases quickly as a function of distance from a village; but it appears to decrease more quickly in the Orinoco Headwaters than it does in the Upper Mavaca.

A very different pattern is seen in the distribution of old garden area by distance (Figure 3.5). In the Orinoco Headwaters image nearly 14% of total old

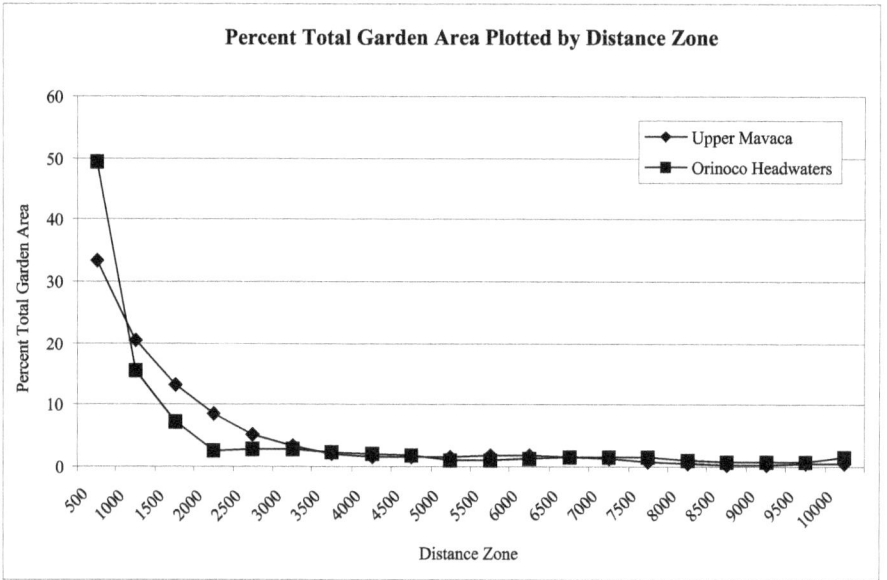

Figure 3.4. Mean garden area divided by buffer area across all distance buffer classes.

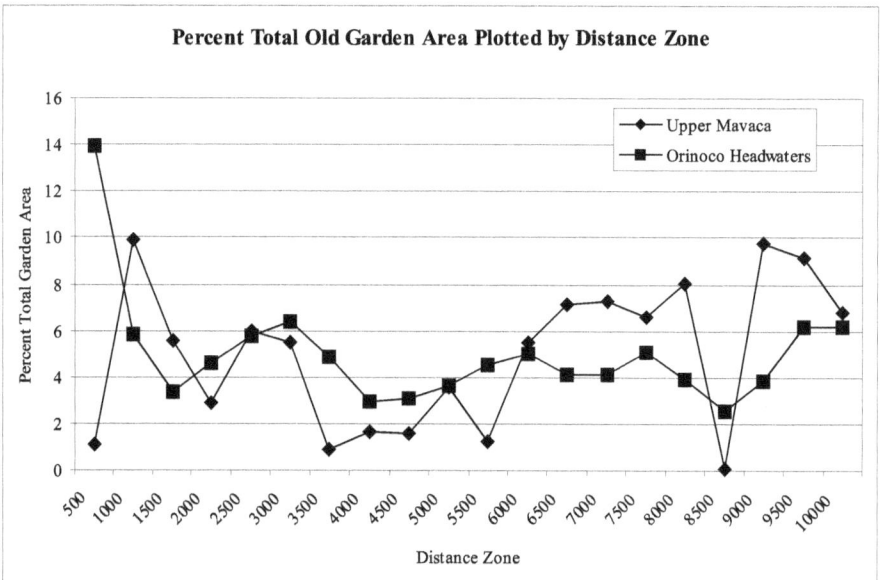

Figure 3.5. Mean old garden area divided by buffer area across all distance buffer classes.

garden area lies within the 500-m buffer zone; it drops off very quickly, with less than 6% of total old garden area located within the 1,000-m buffer ring. Less than 2% of total old garden area is within 500 m of a village, but this rises to slightly less than 10% of total old garden area in the 1,000-m buffer ring. Interestingly, old garden area rises and falls rather unpredictably and does not seem to produce an overall trend with respect to distance, except that nearly as much old garden area is beyond the predicted habitual-use zone as is within this zone. When comparing the distribution of gardens to old gardens, a pattern is clear. A much greater fraction of total active garden area is located near villages, while the fraction of gardens that appear older is much lower near villages. Newer garden distribution seems to be strongly conditioned by distance from a village, but the distribution of older gardens does not.

ANALYSIS

The results presented above are derived from a remote sensing inventory of Yanomamö landscape features and are documented through the use of space-borne remote sensing in a GIS context. These data can be used to evaluate the predictions defined earlier, based on Von Thünen's theory of marginality considered in light of ethnographic research on walking and transportation.

Relative location of garden types: the prediction was that horticultural investments expressed as gardens should be distributed in order to minimize transportation costs. Thus actively cultivated gardens should be closer to settlements than gardens that are not actively cultivated. Results show that the area of new gardens is strongly influenced by distance, while the area of older gardens does not seem to be influenced by distance. Actively cultivated gardens have higher labor costs than older gardens, but the returns are higher. These costs include clearing weeds and other activities that take place in gardens; but investment of labor requires travel to and from gardens. Hence all garden labor inputs have associated transportation costs. The combination of labor requirements and transportation costs works to structure the distribution of recently cleared gardens. Older gardens, in contrast, do not have high labor requirements; therefore output is considerably lower. Although marginality theory would lead us to expect that older gardens should not be located near settlements, this does not exactly seem to be the case. The area of old gardens does not increase drastically with distance in both cases. It does increase in the Upper Mavaca, but it decreases in the Orinoco Headwaters.

Habitual use zones: it was predicted that a walking time of 1 hour at 3 km/hr from a village should define a zone that accounts for the majority of total garden area and that few cultivated gardens should exist beyond that distance.

Garden location and garden areas are highly structured by distance from a village. Results obtained here regarding the distribution of horticultural gardens are quite consistent with other results reported for the Ye'kwana (Hames 1978), Yapú, Kaul, Lufa (Dufour 1984), and Kofyar (Stone 1991). The walking rate of 3 km/hr at one-hour one-way travel time predicted quite accurately when garden area drops off. This suggests that for horticultural populations a walking rate of 5 km/hr is probably high. In light of the bioanthropological research (Maloiy et al. 1986) on walking, 5 km/hr is probably high for hunter-gatherer and pastoral populations as well.

A greater fraction of total garden area is within 500 m of villages in the Upper Mavaca image. Both images show considerable variability in the minimum distance to a garden, however, suggesting that differences in the distributions may not be significant.

For comparative purposes, normalized garden area in each image was cube root transformed to produce normal distributions that permit comparison by the paired t-test. This comparison showed that differences between the two distributions are not significant and could easily have been produced by chance ($t = 0.56$, df = 19, p = 0.58). While chance cannot be ruled out as the cause of differences in the two distributions, in both cases 3 km defined a distance where there is a distinct dropoff in garden area. This defined a habitual-use zone that appears to be similar to what has been reported for other tropical horticultural peoples.

Settlement clusters: human neighbors are the key environmental variables with regard to Yanomamö macro-movement decisions, which should produce clusters of mutually friendly villages. Distances greater than habitual-use zones should separate clusters of villages. The two images show little difference in the number of settlement per area; in both cases, the nearest neighbor statistic indicates that settlements are highly clustered. Formal clustering techniques using only coordinate data are insufficient to define socially relevant groups of villages, because these methods do not take into account physical barriers to transportation and movement. This is significant: archaeologists seeking to define structure and patterning in regional survey data might use similar clustering methods to help define groups of sites during a period in order to examine some aspect of settlement pattern. Defining groups of sites based on these techniques is unlikely to produce analytically useful results. If formal clustering techniques can incorporate the nonisotropic nature of terrain, then it is far more likely that socially meaningful clusters could be defined. Until data terrain data such as digital elevation models can be incorporated into clustering techniques, the analyst will most likely have to rely upon judgmental methods to define groups. These methods permit greater sensitivity to the local topographic context. Informal grouping based on visual inspection of the distribution of village indicates that sites within clusters are generally spaced about 0.7–1.5 km apart. In both cases, about 60% of garden area is less than 1 km from a village. This places villages within a cluster inside each other's habitual-use zone; but they are spaced far enough apart that the area of overlap is at a distance where horticultural investment is starting to drop off considerably.

Intracluster distances: while there are likely to be clusters of mutually friendly settlements, the Yanomamö have no lasting multicommunity organization. Thus there should be some spacing between settlements within a cluster. It is most likely that this separation will be about the distance of horticultural-use zones. Mean and median minimum distances between villages indicate that nearest neighbors are spaced well within easy travel distances from each other. Informal grouping based on visual inspection of the distribution of village sites, however, shows that clusters tend to be about 7–9 km apart. This distance places villages on the edge of settlement clusters at separate but nearly overlapping habitual-use zones, as determined by the distribution of garden area in relation to distance from a village. It may be that village clusters are determined more by hunting zones (distances that hunters are willing to travel) than by the demands of horticulture.

While based on informal grouping techniques, these data suggest some potential for refining interpretation of archaeological settlement maps using a modified form of territorial analysis. The contemporaneity problem is significant in archaeology and can confound spatial analysis of settlement distribution maps. These data, however, indicate that habitual-use zones between settlement clusters should be expected for tribal-level political groups with no permanent multicommunity organization and a subsistence economy based on horticulture.

Discussion

The distribution of Yanomamö settlement data generated through inspection of space-borne remotely sensed imagery produced patterns generally consistent with Von Thünen's model of land use and marginality theory. The distribution of cultivated garden area is highly patterned with regard to distance from a village. A critical distance of 3 km for dropoff of garden area is consistent with ethnographically derived predictions about the habitual-use zones of other horticultural people.

These results are promising for archaeological studies of settlement patterns of horticultural people. Once people become dependent on the production of food, several preharvest labor requirements arise (including clearing, planting, and weeding). People must travel to gardens or resource patches to carry out weeding and garden maintenance on a fairly regular basis. These activities are conditioned by transportation and travel costs, which should be relatively constant across populations. Defining the resulting spatial distribution that these constraints place on garden location through ethnographic examples should help to improve our understanding of archaeological settlement patterns. This kind of research should make it possible to produce reasonably accurate estimates of habitual-use zones for horticultural societies that are known through archaeological data alone.

Ethnographic field research provides a level of detail about behavior and individual variation that is critical for expanding interpretation of archaeological deposits. Modern ethnographic studies of exploration ranges, hunting territories, and patterns of residential mobility help to develop an inferential framework that expands and strengthens the kinds of archaeological interpretation that are possible. While this ethnographic work is crucial, studies are often limited to residents of a single community or a small group of communities. Thus it becomes difficult to envision how the behaviors recorded by ethnographers and ethnoarchaeologists play out at the regional scale. While remote sensing studies admittedly do not capture the level of individual detail that field ethnographers can provide, data collected from space-borne vehicles make possible the systematic application of locally derived observations at a broad regional scale and help to express them in material and spatial ways that can aid archaeological research.

Note

1. Most recent remote sensing studies combine remotely sensed data with "ground truthed" data from the field (Adams et al. 1995; Mertes et al. 1995). Field research strengthens lines of argument improving image interpretation and is always preferable to purely remote studies when it is possible. A significant number of remote sensing studies, however, have been done with no possibility of visiting the study site. For example, the 10-year inventory of deforested areas in the Amazon Basin by David Skole and Compton Tucker (1993: Figure 1; http://www.bsrsi.msu.edu/rfrc/rfrc.html) relied heavily on interpretation of band 5 TM data. Untransformed single-band imagery was interpreted to find patches of deforested land. Areas interpreted as disturbed were digitized for analysis as vector shapes in a GIS. The methods of data collection through image interpretation and digitizing used for generating settlement inventories in this research are very similar to those employed by Skole and Tucker (1993).

REFERENCES

Adams, J. B., D. E. Sabol, V. Kapos, R. A. Filko, D. A. Roberts, M. O. Smith, and A. R. Gillespie

1995   Classification of Multispectral Images Based on Fractions of Endmembers: Application to Land Cover Change in the Brazilian Amazon. *Remote Sensing of Environment* 52: 137–154.

Aldenderfer, M. S., and R. K. Blashfield

1984   *Cluster Analysis*. Sage University Paper, Beverly Hills.

Balée, W. (editor)

1998   *Advances in Historical Ecology*. Columbia University Press, New York.

Barlow, R. K., P. R. Henriksen, and D. Metcalfe

1993   Estimating Load Size in the Great Basin: Data from Conical Burden Baskets. *Utah Archaeology* 6: 27–36.

Barlow, R. K., and D. Metcalfe

1996   Plant Utility Indices: Two Great Basin Examples. *Journal of Archaeological Science* 23: 351–371.

Baxter, M. J.

1994   *Exploratory Multivariate Analysis in Archaeology*. Edinburgh University Press, Edinburgh.

Bettinger, R., R. Malhi, and H. McCarthy

1997   Central Place Models of Acorn and Mussel Processing. *Journal of Archaeological Science* 24: 887–899.

Binford, L.

1980   Willow Smoke and Dogs' Tails: Hunter-Gatherer Settlement Systems and Archaeological Site Formation. *American Antiquity* 45(1): 4–20.

Biocca, E. (editor)

1996   *Yanoama: The Story of Helena Valero, a Girl Kidnapped by Amazonian Indians*. Kodansha, New York.

Bird, D. W., and R. Bird

1997   Contemporary Shellfish Gathering Strategies among the Meriam of the Torres Strait Islands, Australia: Testing Predictions of a Central Place Foraging Model. *Journal of Archaeological Science* 24: 39–63.

Birdseye, C. H.

1940   Stereoscopic Photographic Mapping. *Annals of the Association of American Geographers* 30(1): 1–24.

Carneiro, R. L.

1960   Slash and Burn Agriculture: A Closer Look at its Implication for Settlement Patterns. In *Men and Cultures: Selected Papers of the Fifth International Congress of Anthropological and Ethnological Sciences*, edited by A. Wallace, pp. 229–234. University of Pennsylvania Press, Philadelphia.

1961   Slash and Burn Cultivation among the Kuikuru and Its Implications for Cultural

Development in the Amazon Basin. In *The Evolution of Horticultural Systems in Native South America, Causes and Consequences: A Symposium*, edited by J. Wilbert, vol. 2, pp. 47–67. Antropológica Supplement Publication, Caracas.

Chagnon, N. A.

1966   Yanomamö Warfare, Social Organization and Marriage Alliances. Ph.D. dissertation. Department of Anthropology, University of Michigan, Ann Arbor.

1968a   The Culture-Ecology of Shifting (Pioneering) Cultivation among the Yanomamö Indians. *International Congress of Anthropological and Ethnological Sciences* 3: 249–255.

1968b   Yanomamö Social Organization and Warfare. In *War: The Anthropology of Armed Conflict and Aggression*, edited by M. Fried, M. Harris, and R. Murphy, pp. 109–159. Natural History Press, Garden City.

1968c   *Yanomamö: The Fierce People.* 1st ed. Case Studies in Cultural Anthropology. Harcourt Brace College Publishers, San Diego.

1974   *Studying the Yanomamö.* 1st ed. Studies in Anthropological Method. Holt, Rinehart and Winston, New York.

1975   Genealogy, Solidarity and Relatedness: Limits to Local Group Size and Patterns of Fissioning in an Expanding Population. *Yearbook of Physical Anthropology* 19: 95–110.

1983   *Yanomamö.* 3rd ed. Case Studies in Cultural Anthropology. Holt, Rhinehart, and Winston, Austin.

1988   Life Histories, Blood Revenge, and Warfare in a Tribal Population. *Science* 239: 985.

1992   *Yanomamö.* 4th ed. Case Studies in Cultural Anthropology. Harcourt Brace College Publishers, San Diego.

1995   GPS and Documenting Political and Military Aspects of a Contemporary Agricultural Revolution. Paper presented at the Trimble Surveying and Mapping Users Conference and Exposition, Sunnyvale, California.

1996   *Yanomamö.* 5th ed. Case Studies in Cultural Anthropology. Harcourt Brace, San Diego.

Chagnon, N. A., M. V. Flinn, and T. F. Melancon

1979   Sex-Ratio Variation among the Yanomamö Indians. In *Evolutionary Biology and Human Social Behavior*, edited by N. Chagnon and W. Irons, pp. 290–320. Duxbury Press, North Scituate, Mass.

Charnov, E. L.

1976   Optimal Foraging, Marginal Value Theorem. *Theoretical Population Biology* 9(2): 129–136.

Chisholm, M.

1968   *Rural Settlement and Land Use: An Essay in Location.* Hutchinson University Library, London.

Christ, E. P., and R. C. Cicone
1984   Application of the Tasseled Cap Concept to Simulated Thematic Mapper Data. *Photogrammetric Engineering and Remote Sensing* 50(3): 343–352.

Clark, P. J., and F. C. Evans
1954   Distance to Nearest Neighbor as a Measure of Spatial Relationships in Populations. *Ecology* 35(4): 445–453.

Craig, N. M.
1996   Discussion of Image Processing Methods Applied to Multispectral Landsat 5 Thematic Mapper (TM) Data for Identification of Yanomamö Settlements, pp. 1–71. Manuscript in possession of N. Craig.

Dewar, R.
1991   Incorporating Variation in Occupation Span into Settlement-Pattern Analysis. *American Antiquity* 56(4): 604–620.

Dufour, D.
1984   The Time and Energy Expenditure of Indigenous Woman Horticulturalists in the Northwest Amazon. *American Journal of Physical Anthropology* 65: 37–46.

Flannery, K. V. (editor)
1986   *Guila Naquitz: Archaic Foraging and Early Agriculture in Oaxaca, Mexico.* Academic Press, New York.

Foley, R.
1985   Optimality Theory in Anthropology. *Man* 20(2): 222–242.

Garget, R., and B. Hayden
1991   Site Structure, Kinship, and Sharing in Aboriginal Australia: Implications for Archaeology. In *The Interpretation of Archaeological Spatial Patterning*, edited by E. M. Kroll and D. T. Price, pp. 11–32. Plenum, New York.

Hall, P.
1966   Introduction. In *Von Thünen's Isolated State*, pp. ix–liii. Translated by C. Wartenberg. Pergamon Press, Oxford.

Hames, R.
1978   Behavioral Account of the Division of Labor among the Ye'kwana Indians of Southern Venezuela. Ph.D. dissertation. University of California at Santa Barbara.

1983   The Settlement Pattern of a Yanomamö Population Bloc. In *Adaptive Responses of Native Amazonians*, edited by R. Hames and W. Vickers, pp. 393–427. Academic Press, New York.

1992   Time Allocation. In *Ecology, Evolution, and Behavior*, edited by B. Winterhalder and E. A. Smith, pp. 203–236. Aldine de Gruyter, Chicago.

Hawkes, K., F. O'Connell, and N. G. Blurton-Jones
1995   Hadza Children's Foraging: Juvenile Dependency, Social Arrangements, and Mobility among Hunter-Gatherers. *Current Anthropology* 36(4): 688–700.

Hawkes, K., and J. O'Connell

1992   On Optimal Foraging Models and Subsistence Transitions. *Current Anthropology* 33(1): 63–66.

Hewlett, B., J. M. H. van de Koppel, and L. L. Cavali-Sforza

1982   Exploration Ranges of Aka Pygmies of the Central African Republic. *Man* 17(3): 418–430.

Hipsley, E. H., and N. Kirk

1965   Studies of Dietary Intake and Expenditure of Energy by New Guineans. Technical paper no. 147. South Pacific Commission, Nouméa, New Caledonia.

Jarman, M. R., G. N. Bailey, and H. N. Jarman (editors)

1982   *Early European Agriculture: Its Foundations and Development.* Cambridge University Press, Cambridge.

Jensen, J. R.

1996   *Introductory Digital Image Processing: A Remote Sensing Perspective.* Prentice Hall Series in Geographic Information Science. Prentice Hall, Upper Saddle River, N.J.

Kaplan, H., and K. Hill

1992   The Evolutionary Ecology of Food Acquisition. In *Evolutionary Ecology and Human Behavior,* edited by E. A. Smith and B. Winterhalder, pp. 167–201. Aldine de Gruyter, New York.

Kauth, R. J., and G. S. Thomas.

1976   The Tasseled Cap: A Graphic Description of the Spectral-Temporal Development of Agricultural Crops as Seen by Landsat. Paper presented at the Proceedings of the Symposium on Machine Processing of Remotely Sensed Data, West Lafayette, Indiana.

Kelly, R. L.

1992   Mobility/Sedentism: Concepts, Archaeological Measures, and Effects. *Annual Review of Anthropology* 21: 43–66.

Layton, R., and R. Foley

1992   On Subsistence Transitions: Response to Hawkes and O'Connell. *Current Anthropology* 33(2): 218–219.

Lizot, J.

1980   La agricultura Yanomami. *Antropológica* 53: 3–94.

MacArthur, R. H., and E. R. Pianka

1966   On Optimal Use of a Patchy Environment. *American Naturalist* 100(916): 603–609.

Maloiy, G. M. O., N. C. Heglund, L. M. Prager, G. A. Cavagna, and C. R. Taylor

1986   Energetic Cost of Carrying Loads: Have African Woman Discovered and Economic Way? *Nature* 319 (February): 668–669.

Martin, P. S., and F. Plog

1973   *The Archaeology of Arizona: A Study of the Southwest Region.* Doubleday Press, Garden City, N.Y.

McGwire, K., N. A. Chagnon, and C. B. Carias
1996   Empirical and Methodological Problems in Developing a GIS Database for Yanomamö Tribesmen Located in Remote Areas. In *Anthropology, Space, and Geographic Information Systems*, edited by M. Aldenderfer and H. Maschner, pp. 97–106. Oxford University Press, New York.

Mertes, L. A. K., D. Daniel, L., J. M. Melack, B. Nelson, L. A. Martinelli, and B. R. Forsberg
1995   Spatial Patterns of Hydrology, Geomorphology, and Vegetation on the Floodplain of the Amazon River in Brazil from a Remote Sensing Perspective. *Geomorphology* 13: 215–232.

Metcalfe, D., and R. K. Barlow
1992   A Model for Exploring the Optimal Tradeoff between Field Processing and Transport. *American Anthropologist* 94: 340–356.

Murdock, G. P.
1968   The Current Status of the World's Hunting and Gathering Peoples. In *Man the Hunter*, edited by R. Lee and I. Devore, pp. 13–22. Aldine Publishing Company, Chicago.

Norgan, N. G., A. Ferro Luzzi, and J. V. G. A. Durnin
1974   The Energy and Nutrient Intake and Energy Expenditure of 204 New Guinea Adults. *Philosophical Transactions of the Royal Society of London* 268(893): 309–348.

Orians, G. H., and N. E. Pearson
1979   On the Theory of Central Place Foraging. In *Analysis of Ecological Systems*, edited by D. J. Horn, G. R. Stairs, and R. D. Mitchell, pp. 155–177. Ohio State University, Columbus.

Petrie, G.
1977   Orthophotomaps. *Transactions of the Institute of British Geographers* 2(1): 49–70.

Ribed, S. P., and M. A. Lopez
1995   Monitoring Burnt Areas by Principal Components Analysis of Multi-Temporal TM Data. *International Journal of Remote Sensing* 16(9): 1577–1587.

Robinson, A. H., J. L. Morrison, and P. C. Meuehrcke
1977   Cartography 1950–2000. *Transactions of the Institute of British Geographers* 2(1): 3–18.

Sabins, F. F.
1987   *Remote Sensing Principles and Interpretation*. 2nd ed. W. H. Freeman and Company, New York.

Sallade, J., and D. Braun
1982   Spatial Organization of Peasant Agricultural Subsistence Territories: Distance Factors and Crop Location. In *Ethnography by Archaeologists*, edited by E. Tooker, pp. 19–42. American Ethnological Society, Washington, D.C.

Schoener, T. W.

1971   Theory of Feeding Strategies. *Annual Review of Ecology and Systematics* 2: 369–404.

Skole, D., and C. Tucker

1993   Tropical Deforestation and Habitat Fragmentation in the Amazon: Satellite Data from 1978 to 1988. *Science* 260(5116): 1905–1910.

Smole, W.

1976   *Yanomama Indians: A Cultural Geography*. University of Texas Press, Austin.

Steiniger, M. K.

1996   Tropical Secondary Forest Regrowth in the Amazon: Age, Area and Change Estimation with Thematic Mapper Data. *International Journal of Remote Sensing* 17(1): 9–27.

Stone, G. D.

1991   Agricultural Territories in a Dispersed Settlement System. *Current Anthropology* 32(3): 343–352.

Von Thünen, J. H.

1966   *The Isolated State*. Translated by C. Wartenberg. Blackie and Son, Ltd, Bishopbriggs, Glasgow.

Wilbert, J.

1966   *Indios de la Región Orinoco-Ventuari*. Fundación La Salle de Ciencias Naturales, Caracas.

Winterhalder, B.

1983   Opportunity-Cost Foraging Models for Stationary and Mobile Predators. *American Naturalist* 122(1): 73–84.

1993   Work, Resources, and Population in Foraging Societies. *Man* 28(2): 321–340.

Winterhalder, B., and E. A. Smith

2000   Analyzing Adaptive Strategies: Human Behavioral Ecology at Twenty-five. *Evolutionary Anthropology* 9(2): 51–72.

Zerries, O.

1955   Some Aspects of Waica Culture. *Proceedings of the Inernational Congress of Americanists* 31: 73–88.

4

Mobility and Houses in Southwestern Madagascar

Ethnoarchaeology among the Mikea and Their Neighbors

ROBERT L. KELLY, LIN POYER, AND BRAM TUCKER

It has been clear for some time now that the transition from nomadism to sedentism is not necessarily quick, irreversible, or pervasive. Archaeological research in the western United States and a growing number of ethnographic cases suggest that the transition to sedentism results in a landscape that can become a mosaic of adaptations. Sometimes sedentary villagers lived cheek-by-jowl with nomadic groups; sometimes people moved back and forth seasonally or, on a longer time scale, between nomadic and sedentary settlement options; sometimes the same people kept a permanent residence and one or more temporary residences (for example, Ames 1991; Eder 1984; Kelly 1992; Kent 1992; Madsen and Simms 1998; Schlanger 1991). Thus, for a given region, archaeologists could be seeing remains produced by a range of settlement options implemented concurrently and/or sequentially.

Susan Kent (1991, 1992, 1993; Kent and Vierich 1989) argues that certain archaeologically recoverable variables can sort out occupations of different lengths of stay. For Kent, the key variable is not actual length but the *anticipated* length of stay. Her point is well taken: people construct houses and organize activities spatially depending on how long they anticipate remaining in a settlement. As Kent points out, however, anticipated and actual mobility should be correlated. Indeed, her data suggest a strong correlation between the two.[1] We focus on actual mobility but recognize that anticipated mobility plays a role in site structure at the time scale of ethnography.

In this chapter we discuss ethnoarchaeological research in southwest Madagascar among a population that practices a mix of horticulture, pastoralism, foraging, wage labor, and craft production (Dina and Hoerner 1976; Fanony 1986; Kelly et al. 1999; Molet 1958, 1966; Poyer and Kelly 2000; Stiles 1991; Tucker 2001; Yount et al. 2001). Households are mobile to maintain a mixed economy. Most maintain several houses at a time, one in each of the microenvironments they exploit. As the household allocates its supply of labor to different activities, it moves its members among different houses, from villages to forest hamlets to foraging camps.

First we introduce southwest Madagascar and the types of settlements encountered there and present data on differences in the settlements in terms of house size, feature diversity, trash disposal, and intensity of house construction. We conclude with a discussion of the influences of social organization and food sharing on those variables that record change in length of occupation.

THE STUDY AREA

Madagascar's dry southwest is environmentally heterogeneous (Figure 4.1). A 35-km east-west transect traverses mangrove swamps, coastal mudflats, dunes, dry lake beds, thorn forest, dense deciduous forest, anthropogenic clearings, savanna, and savanna woodland (Seddon et al. 2000). Rainfall is seasonal in occurrence (85 to 95% falls in the months of December through March) and unpredictable in amount, with annual precipitation varying from 200 mm to 1,700 mm (Tucker 2001). The people who live in this region exploit the environmental heterogeneity to counter the climatic unpredictability. They practice a diversified economy, with different activities taking place in each microenvironment. On the coast they collect marine products and tend small gardens; in the forest they forage for wild tubers, honey, and small game, grow maize in slash-and-burn fields, and raise livestock; in the savanna they cultivate manioc and sweet potatoes, herd cattle, and participate in village markets.

Households move their members among different houses in different microenvironments flexibly, depending on labor requirements, subsistence payoffs, and social obligations. Settlement types vary according to their lengths of stay. We classify these (from longest to shortest stay) as villages, forest hamlets, seasonal hamlets, and foraging camps.

For the most part, the people described in this chapter may be called Mikea, although identity formation in this part of Madagascar is complex (see Yount et al. 2001; Poyer and Kelly 2000; Tucker 2003). Three identity terms are commonly used: Mikea, Masikoro, and Vezo. Informants claimed that identity is not tribal or ethnic but is based on performance of economic activities: Mikea are forest-dwelling foragers, Masikoro are savanna-dwelling agropastoralists, and Vezo are coastal fisherpeople. People do claim different identities in different social circumstances, however, justified with various sets of rules that sometimes seem to contradict one another. Mikea, Masikoro, and Vezo speak the same dialect of Malagasy. Most households practice some combination of fishing, foraging, farming, and retailing; they move freely among coast, forest, and savanna and may own houses in each zone; and they are historically, genealogically, socially, and commercially interrelated (Astuti 1995a, 1995b; Poyer and Kelly 2000).

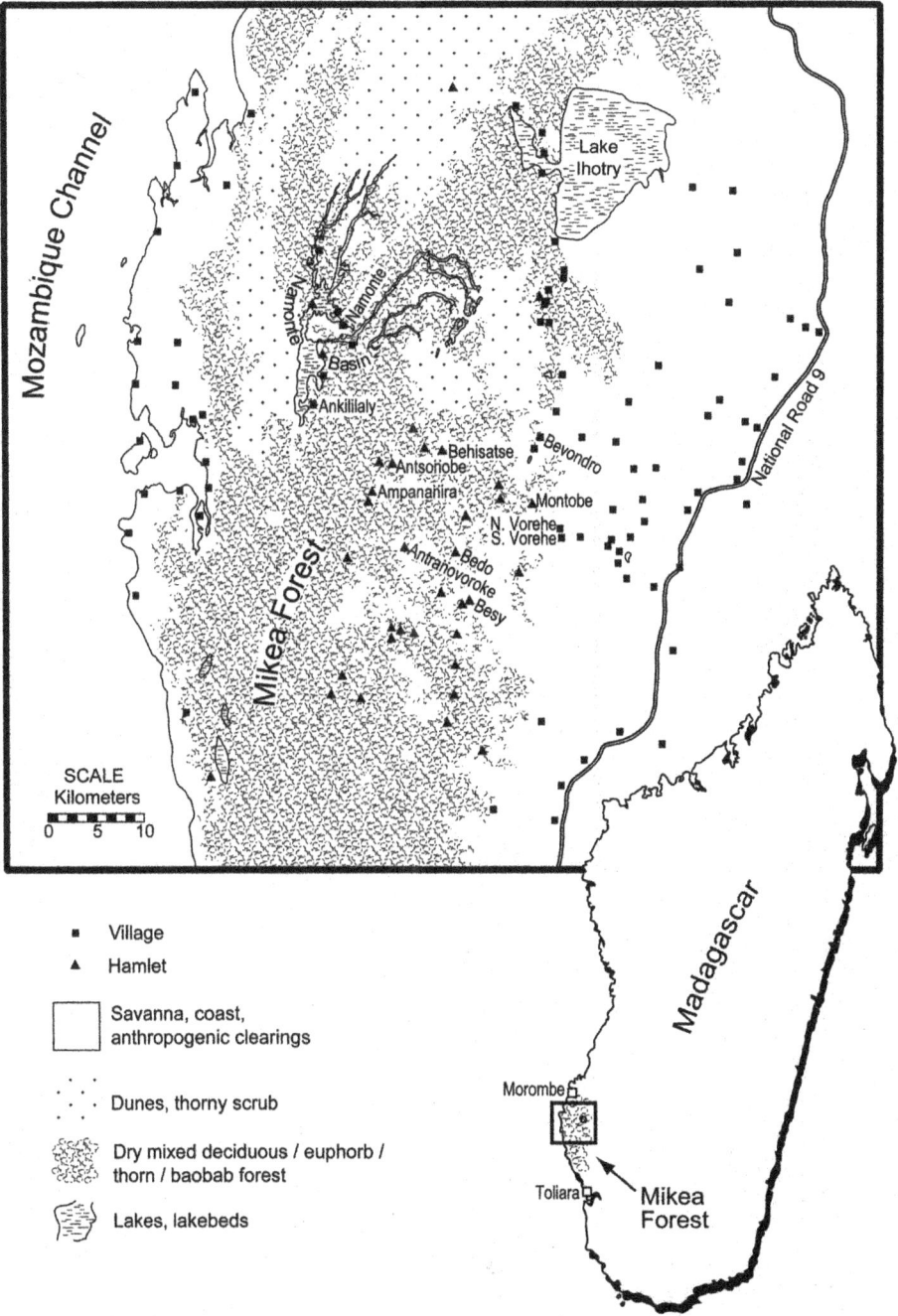

Figure 4.1. Map of the northern part of the Mikea Forest, showing all villages and hamlets. Forest extent by James Yount based on 1994 Landsat TM and 1999 SPOT imagery.

Oral histories indicate that Mikea were once Masikoro or Vezo but fled into the forest during the past four or more centuries to avoid tribute demands and threats of violence and slavery from the Masikoro kings of the Andrevola dynastic lineage and later to escape French colonial forced relocation schemes, head taxes, and mandatory labor projects (Yount et al. 2001; Poyer and Kelly 2000; Tucker 2003). As a result, some village inhabitants are more or less Vezo (particularly on the coast) or Masikoro (on the savanna). When these villagers exploit the forest for part of their subsistence, however, they are considered Mikea. Meanwhile the descendants of those who fled into the forest in the past continue to forage, farm, and herd there, although most also have houses in one or more villages; they too are considered Mikea. Because the identity scheme is potentially confusing for our study of mobility and house construction—they claim to be three separate populations of economic specialists but are actually one population of highly mobile generalists—we avoid labeling settlements as Mikea, Vezo, or Masikoro and use the term "Mikea" in its most general sense. The ethnographic present in this chapter is June to August of 1993, 1994, and 1995, when two of us (Kelly and Poyer) conducted our fieldwork and collected most of the data reported herein. Many of the sites in our sample were no longer inhabited or had changed from one settlement type to another, by the time Tucker did his fieldwork in 1997–1999.

Villages

Permanent, sedentary villages with several hundred to several thousand inhabitants border the Mikea Forest on the western coast and the eastern savanna and within the lakebeds of the Namonte Basin (Figure 4.1).

To the west of the forest is the coast, consisting of dunes, mudflats, mangrove swamps, and the shallow bays of Fañemotse and Tsingilofilo. Inhabitants of coastal villages exploit the sea, catching fish and octopi and gathering shellfish, crabs, and sea cucumbers, which are often sold to mobile retailers for eventual marketing in the savanna. They also grow manioc and maize in small gardens and herd cattle, goats, and swine. Coastal villages include Afeza, Andavadoake, Ampasilava, Befandefa, Ankindranoke, and Ampanonga.

The forest is bordered on the east by savanna and savanna-woodland. Savanna villagers cultivate manioc, sweet potatoes, and maize in rain-fed fields and irrigated rice in the floodplains of the Iovy, Androka, and Befandrea Rivers. They also herd cattle, goats, and swine. Savanna villages include Andohasakoa, Andranodehoke, Marolinta, Bevondro, Basibasy, and Vorehe. National Road 9 (mostly unpaved) connects these villages and their markets to each other and to the cities of Morombe and Toliara.

Within the northwestern portion of the forest is an area labeled on government maps as the Namonte Basin (FTM 1968a, 1968b), a region of many lakebeds, channels, and dunes. The lakebeds are a series of flat, grassy pans and channels that flood during some years; a few contain standing water year-round. The villages of Ankililaly, Antanimena, Antrañotelo, Ankilimiavotse, Ampijilova, and Namonte consist of 20 to 100 reed houses. Inhabitants herd cattle and goats, grow manioc and sweet potatoes in the lakebed mud, and forage for freshwater fish, birds, and honey.

Traditionally two kinds of houses predominate in villages—wattle-and-daub and reed thatch. In recent years an increasing number of houses have been made from planks on the coast and adobe brick in the savanna. Wattle-and-daub houses are made from posts set upright about 70–75 cm (about the distance from an adult's hand to the armpit) into the ground, with smaller, more supple poles woven between them. On the coast daub is often made from a white clay mixed with crushed, burnt seashells, occasionally chinked with small limestone rocks; these houses are called *trañ osokay*. In the savanna daub is made from red sandy clay mixed with water (straw and dung are not used); these houses are called *trañ ofotake*. A hole 2 m in diameter and up to 2 m deep is sufficient to provide the mud plaster for a small house. With proper maintenance, wattle-and-daub houses may last for 20 to 40 years (Figure 4.2a). The floors of the houses are normally compacted earth, although in recent years some have been made of poured concrete. Floors are covered in mats woven from the fronds of a palm (*Hyphaena shatan*). Houses normally have only one door and one or two shuttered windows, fashioned from planks. The door usually has a manufactured lock purchased in a market. Furniture may include beds, tables, chairs, and storage boxes. The houses that we entered were always filled, often packed, with belongings, material goods, and food stores. Very few belongings remain outside of village houses.

The roof is critical in determining the lifespan of the wattle-and-daub house. It normally extends for a meter or so around the entire edge of the house, forming a veranda that protects the mud walls from the weather and promotes their longevity. Secondary functions of the veranda are providing shade for outdoor work and social activities and serving as a platform for sun-drying manioc. The roofs are generally thatched with reeds or savanna grasses; a few villagers can now afford corrugated tin. A strip of baobab bark, plastic, or metal is sometimes laid on the roof keel to make it waterproof. A thick, well-made roof can last up to 20 years.

These wattle-and-daub houses can require more than two months to gather the requisite materials and to construct them. The gathering of the materials required to build these more permanent houses may be the biggest cost in terms

Figure 4.2. Architecture in villages and hamlets. The first four photos show different house constructions: (a) mud house or *trañ ofotake* in Vorehe; (b) reed thatch house or *trañ ovondro* in Ankililaly; (c) grass thatch house or *trañ oakata* in Besy; (d) bark thatch house or *trañ oholits'hazo* in Bedo. Notice the *kitrely* platforms in front of the houses in (c) and (d). Photo (e) is Montobe, a settlement that would be classified as a village based on intended settlement length alone but is more like a forest hamlet in most respects. Photo (f) shows a freshly swept "trash ring" in Behisatse.

of time or money (some individuals specialize in collecting these materials for sale).

The other common house type in villages, and the only one in Namonte Basin villages, is reed thatch houses or *trañ ovondro* (Figure 4.2b). These are similar to wattle-and-daub houses, with poles set in the ground (no wall trench is excavated for either wattle-and-daub or pole-and-thatch houses) but with bundles of reeds tied to the latticework. *Trañ ovondro* tend to be smaller than wattle-and-daub houses and consequently use thinner posts. They also tend not to have verandas; if a veranda is present, it is only on one side (usually the side with the door). These houses take two to four weeks to build. After three years the reed thatch (if not the poles) is replaced. Like wattle-and-daub houses, reed houses are sometimes furnished with beds, tables, chairs, and storage boxes and are often packed with personal belongings. Floors are covered with large palm mats.

Villages have few outside shade structures; people lounge inside their houses, under their narrow verandas, in the shade of tamarind trees, or along the east side of their houses late in the afternoon.

In savanna villages, where the clay eventually forms a hard surface in most areas, extramural areas are swept clean of trash. It is then deposited either in the pits that were excavated to provide clay for the walls or at the edge of the village, which may be 40 m or more away.

These more permanent locations also include many different exterior features, such as outside cook houses, privacy screens for combined latrines/washrooms, and storage facilities. Villagers construct fenced enclosures around gardens, houses, and sets of houses and outbuildings whose owners/inhabitants are patrilineally linked, to restrict access of people and livestock. Individuals who practice spirit possession (*tromba*) have enclosures made of upright logs set close together behind or around their houses. Such enclosures have a shrine in which an individual's possessing spirit is said to live when not inhabiting his or her host.

FOREST SETTLEMENTS

Between the coast and the savanna is the Mikea Forest or Añalamikea, a swath of dry forest 20 to 40 km wide. The Mikea Forest is a mosaic of dense, dry, deciduous forest choked with vines, with patches of thorn forest, groves of baobab trees, slash-and-burn maize fields (*hatsake*), and old clearings. These clearings have been colonized by grasses in some places, creating savanna. In other places the dry forest is slowly regenerating.

Slash-and-burn maize fields are cut in July and August, burned in October, and planted in late November or December. The maize is harvested in March and April; it is eaten, dried and stored, kept as seed, and sold to traders.

The forest is a source for wild foods. People forage for wild tubers, especially *ovy* (*Dioscorea acuminata*), *babo* (*Dioscorea bemandry*), and *tavolo* (*Tacca pinnatifida*); wild cucurbits; honey; and small game animals, particularly three species of tenrecs (*Tenrec ecaudatus, Echinops telfairi, Setifer setosus*), feral cats (*Felix sylvestris*), tortoises (*Pyxis arachnoides*), and occasionally lemurs (*Microcebus murinus, Chierogaleus medius, Lepilemur ruficaudatus*). No large game animals are found in the Mikea Forest, except for the exceedingly rare wild boar (*Potamocorus larvatus*).

The forest is also used as pasture for cattle and goats and more rarely for swine. While living in the forest, Mikea may also do wage labor, such as cutting fields for village men or guarding their cattle.

Forest Hamlets

We identified at least three settlement types within the forest: forest hamlets, seasonal hamlets, and foraging camps. Forest hamlets usually consist of three to twenty houses located near *hatsake* fields. Some households live most of the year in these settlements, while others move in seasonally to tend their *hatsake* or to pasture livestock in the surrounding forest. Most forest hamlets are occupied by kin who also have houses in a village. Foraging, especially for wild tubers, is a daily activity in these settlements.

Informants claimed that forest hamlets were occupied for three to five years, although we have witnessed considerable variability. A hamlet founded in the Antaolandambo region in 1996 had been abandoned by 1998. The hamlet of Behisatse (where much of our research took place) has been more or less continuously occupied for over a decade.[2]

Forest settlement houses are usually fairly small, with about 5 m² of floorspace and eaves 1 to 2 m high—often not tall enough to stand up in. They are made with a set of upright posts (though fewer in number than in village houses) set into the ground about 45 cm deep. Pliable sticks are woven through them to form the lattice that holds the wall thatch. The houses are thatched with reeds (*trañ ovondro*), bark (*trañ oholits'hazo*), or grass (*trañ oakata*) (Figure 4.2c, d, and e). More rarely, walls are made from upright logs or planks.

Different types of thatch have varying costs and benefits too. Reeds are the most solid and do not attract bugs; but they do not grow within the forest and must be transported from elsewhere or purchased from someone who specializes in harvesting and delivering them. Tree bark is easy to acquire but requires a lot of maintenance. It tends to crumble and fall out of the lattice and is usually replaced once or twice each year. Grass houses require less maintenance but tend to attract roaches. As houses age, people patch holes in walls and roof with different materials, including thatch, old woven mats, and plastic bags. Roofs are often but not always thatched with the same material as the walls. Broad slabs of baobab bark (30–40 cm wide, 2–3 cm thick) are a preferred roofing material, because they are quite waterproof. Grass roofs may have a strip of baobab bark along the keel to keep out rainwater. Baobab bark is rarely used for walls, because goats eat it.

It normally takes a man about one to two weeks to gather the material and construct the house. Houses require annual maintenance. The house is often destroyed after a few years, and a new one is built in its place, with some of the poles, lattice, and baobab bark slabs being reused. When hamlets are moved to a new location (which may only be a kilometer or two away), existing structures are sometimes dismantled and the materials reused in the new location.

Another common structure in hamlets is the *kitrely*, a platform 2 m or more above the ground. It is composed of four to six large posts (sometimes more if they are especially big) set into the ground about 75 cm deep. *Kitrely* platforms serve two primary functions: storage on top, especially for maize and tools; and shade below, for work and social activities. They are often erected just outside the front doors of homes (Figure 4.2c, d, e).

In nearly all cases, houses have an exterior hearth located about 2 m in front of the doorway beneath the *kitrely* platform. No stone is available in the forest; hearths are nothing more than low heaps of ash and cinders. Inside the house, normally to the left as one enters, is another hearth used for heat at night. The adjacent portion of the front wall often slopes outward at the bottom, so the smoke rises and finds its way through the cracks in the wall and roof. Archaeologically speaking, the hearth appears to lie on the wall between the doorway and the corner post.

The doorways of these houses almost always face north, and settlements are normally a linear north-south scatter of houses. In addition, the rear (south) wall of the house is sometimes built more solidly than the others, with reeds or planks if available. During the dry season there is a stiff, cold wind from the south; the better-constructed south wall keeps this wind from entering the house. Having the doorway (which may or may not have a working door) face north also prevents the wind from blowing hot sparks around that might cause the house to burn. (Doorways in villages are more likely to face in a variety of directions. Doors and windows are often left open during the hot, wet season, so that the evening west or north wind helps to cool the house; no fire is burning at this very hot time of the year, so sparks are not a concern.)

Forest hamlets may also contain smaller tent-like structures made of grass or bark. We have seen lean-tos, three-sided boxes, A-frames (small structures like pup tents), and Quonsets (Figure 4.3). People use such structures when they first move into or establish a hamlet until they have built their more permanent

Figure 4.3. Expedient shelters found in hamlets and camps. (a) A-frame, (b) lean-to, (c) Quonset (photo courtesy of James Yount), (d) three-sided box.

house. The tent-like structures are then turned into children's houses. By the age of about 10, children of both sexes sleep outside their parents' house.

Other structures include maize-threshing bins, animal pens, and troughs (hollowed-out logs) placed beneath the eaves of houses to catch rainfall for drinking water in the wet season. If fenced enclosures are present, they are usually ceremonial enclosures (as noted above).

Trash in forest hamlets tends to be deposited in an arc some 3 to 9 m away from the outside hearth. Sweeping debris out of the living space around houses and into the "trash ring" is a weekly housekeeping chore. Trash piles contain large amounts of maize husks and other debris that are periodically burned (although clean corncobs are kept for use as toilet paper). Vegetation seems to play a role in the distance to the trash. Recently established settlements have a considerable growth of bushes near the houses. Trash is deposited there (or is blown and caught there), because no one walks through the bushes. Settlements that have been occupied longer have less vegetation, which slowly becomes trampled, eaten by goats, or pulled up by kids to be used as brooms or in games. Trash disposal then moves farther away from the house. But even where there is no vegetation, trash is deposited some 8–9 m from the house's doorway, often in an arc in front of it. Houses are placed far enough apart that this 8–9 m arc does not fall within a neighbor's space.

Seasonal Hamlets

Seasonal hamlets are forest hamlets that are less permanently occupied. The occupants spend most of the year in other places, particularly villages, but live in these hamlets while tending their *hatsake* or pasturing livestock. Seasonal hamlets have fewer houses, which are often (though not always) made in a more expedient fashion, as lean-tos, three-sided boxes, A-frames, or Quonsets. These houses have interior and exterior hearths, but fewer *kitrely* platforms. The houses may have the same floorspace as those in forest hamlets but tend to have lower roofs. They may be used for more than one season but are not thought of as a central residence, their primary purpose being to keep rain off the occupants. These houses can be built in a few days and last about a season.

Many if not all forest hamlets began life as seasonal hamlets. Several of the communities that were seasonal hamlets when the data in this chapter were collected in 1993–1994 had become forest hamlets in 1997–1999. A settlement was classed as a seasonal hamlet depending on whether anyone was living there when it was visited in the dry season.

Foraging Camps

Foraging effort is often intensified in June through August, particularly in years when the maize harvest is inadequate. Households may decide that it is more difficult to transport wild foods to families living in hamlets than to relocate their families closer to the wild food patches, shifting from logistical to residential mobility, in Binford's (1980) terms. Most Mikea households practice pure nomadic foraging (*mihemotse*) for at least a few weeks in some years. In the dry season there is virtually no rainfall and no surface water in the forest outside the Namonte Basin except for a few enhanced wells. Most water comes from *babo* (a tuber). Hence adults sometimes leave their children in others' care in hamlets or villages to increase their mobility and decrease drinking water needs. More often an individual household or pair of households occupies a foraging camp. We have also encountered foraging camps composed of young unmarried adults from different families.

Foraging camps are frequently not far from hamlets, and people may go back and forth between the two. Other camps are remote and solitary and contain no buildings, except simple lean-tos or box-like shade structures in some cases (Figure 4.3). These structures can be built in one or two days and last two to four weeks. But foraging camps often have no structures, consisting of widely spaced hearths nestled in the tangled brush. Mats are laid on the ground around the hearth, and the few possessions (such as baskets, buckets, and tools) are hung on surrounding trees. Sometimes people must bend over and nearly crawl through tunnels in the brush to move between hearths, even where more open savanna space is located only a few meters away. Informants said that the main reason for shunning the open space is that the brush provides a windbreak against the dry season's cool night wind and shade against the noon sun. When modest structures are built in foraging camps, they are constructed where people choose to station themselves more in the open and hence need protection from wind and sun (rain being extremely rare in the dry season).

The little trash produced in foraging camps that is not eaten by dogs (only once did Kelly and Poyer see a bone lying in a camp) is tossed not more than 1–2 m from the hearth, normally in the bushes.[3] The hearths are also periodically emptied of their ashes, which are unceremoniously tossed aside in the nearby vegetation.

ANALYSES

The people who live in and near the Mikea Forest implement a variety of settlement options that differ in terms of people's perception of their permanence. How could an archaeologist sort out these settlements, indicating different lengths of stay?

First, we need to consider the effects of differential availability of raw materials. The wattle-and-daub houses are partly a function of the presence of clay along the eastern edge of the forest but not within the forest itself or in the Namonte Basin. Likewise, houses that use burnt shell as mortar obviously can only be built on the coast. Wood is widespread, however, and the numbers or types of poles used in construction are not affected by the availability of trees. All people in the settlements in this study had access to similar types and sizes of wood in the immediate vicinity of their settlement or near their *hatsake* fields, which they regularly visited.

A number of ethnoarchaeological studies point to several variables that reflect differences in mobility (for example, Diehl 1992; Kent 1991, 1992, 1993; Kent and Vierich 1989; see also Kelly 1992). These include site size, investment in house construction, artifact and feature diversity, and distance to areas of trash disposal. Permanent habitations tend to contain more people, are thus larger, and leave more physical traces behind. It is often difficult, however, for the archaeologist to determine whether site size and density are due to length of occupation or to multiple occupations. Other factors may indicate more subtle differences in mobility. As Kent (1992) points out, the length of time that people anticipate remaining at a location strongly conditions how much effort they invest in their housing. In making a permanent house they may be more selective about the wood species, choosing those that resist insects. Alternatively, they may overbuild a structure so that it will last longer. Likewise, if people intend to remain in one place for a long period they might invest time in constructing facilities designed to fulfill particular purposes: cook houses, storage racks or sheds, threshing bins, and the like. Other studies suggest that disposal of trash will take place farther from where it was produced and/or in pits as the length of occupation increases (for example, Hitchcock 1987).

House Size

People who intend to remain in one place for a longer period might be expected to invest in larger houses. Kent (1992; Kent and Vierich 1989) found that population size did not account for differences in house size (although it did account for the number of houses) and that a measure of anticipated mobility was the best predictor of size. In our sample, mean house size is largest in the permanent villages. This is mainly a function of the wattle-and-daub houses, although the pole-and-thatch structures in the villages are slightly larger than houses in the forest and seasonal hamlets (Tables 4.1 and 4.2). We found no significant difference, however, between the size of houses in forest hamlets and in seasonal hamlets. We cannot make adequate comparisons between populations in part because the seasonal hamlets are not occupied in the dry season, when many of the data here were gathered. Kent (1992: 640) suggests that house size increases with greater anticipated length of stay because people in such communities have more belongings. Although we have no systematic measure, our impression is that people in the villages did indeed have more belongings. As noted below,

Table 4.1. Summary of Main and Secondary Support Post Size by Settlement Type

Settlement type	Mean	Standard deviation	N	Coefficient of variation
Main support posts				
Forest hamlets	6.7	1.2	69	.18
Seasonal hamlets	6.1	1.5	85	.25
Villages	7.9	1.6	97	.20
Villages without Vorehe thatch houses	8.4	1.5	54	.17
Secondary support posts				
Forest hamlets	4.5	.89	171	.20
Seasonal hamlets	4.1	.94	167	.23
Villages	5.4	1.4	359	.26
Villages without Vorehe thatch houses	6.1	1.2	212	.20

Table 4.2. Data from Settlements

Settlement	Mean house size (m²)	Main posts (n)	Secondary posts (n)	Main post C.V.	Secondary post C.V.	Mean main post diameter (cm)/s.d.	Mean secondary post diameter (cm)/s.d.	Mean secondary posts per wall (n)	Mean secondary post distance (cm)
Villages									
Vorehe: wattle & daub	24	12	58	.06	.14	8.4/.5	6.7/0.9		
Vorehe: thatch	8.3	42	145	.20	.25	7.1/1.4	4.6/1.2	5.6	41
Ankililaly	5.7	19	68	.13	.15	9.6/1.2	6.7/1.0	8	57
Bevondro	6	23	86	.16	.16	7.4/1.2	6.5/1.0	5	56
Forest hamlets									
Montobe	5.2	13	32	.18	.16	6.4/1.2	4.9/.8	4.5	60
Bedo	4.9	30	48	.16	.16	4.3/.7	4.9/.8	3.5	65
Behisatse	5.5	22	69	.16	.20	6.6/1.1	4.6/.9	4.4	54
Seasonal hamlets									
Besy	5.7	17	52	.15	.17	4.5/.7	5.7/.9	3.5	72
Antsoñiobe	4.9	29	40	.25	.25	3.8/1.0	4.9/1.2	3.5	51
Antrañovoroke	5.2	10	22	.29	.27	5.4/1.6	4.3/1.2	3	
Ampanañira	4.9	29	53	.24	.21	6.7/1.6	3.9/.8		

larger houses may result not only from having "more stuff" to store but also from an increased need to store these things out of sight.

Feature Diversity

We tabulated 31 features found in 29 settlements (some of these are the same settlements but in different years). The list included 11 different kinds of habitations, ranging from lean-tos to tin-roofed cement houses (very rare), that included *kitrely* platforms, fenced housing compounds, animal corrals, other fenced areas, wash areas, cook houses, public troughs, drying racks/miscellaneous posts, churches, schools, stores, wells, ceremonial enclosures, maize threshers, "guest houses," clinics, bellows, wash houses, outside hearths, and storage bins.

As might be expected, we found a positive relationship between the overall number of features and their diversity (after converting the number of features to a log scale, $r = 0.85$; $df = 27$; $p < 0.001$). Because settlements contain more features, they also contain a greater diversity of features. Villages have a greater diversity of features than the other types of settlements; forest hamlets and seasonal hamlets have about the same diversity, and foraging camps have the least (Figure 4.4).

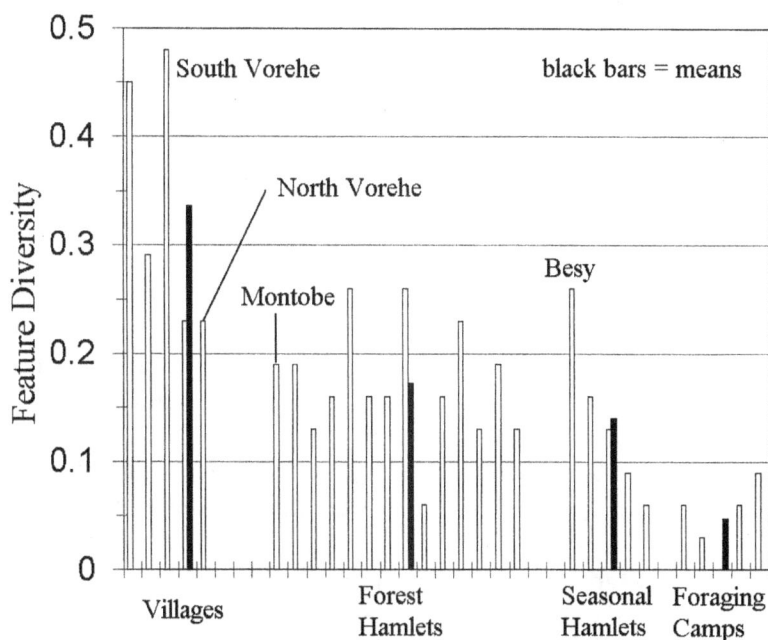

Figure 4.4. Histogram showing feature diversity with means for the villages, forest hamlets, seasonal hamlets, and foraging camps.

This is not surprising, for other studies have shown that feature diversity and frequency mirror a settlement's longevity (reviewed in Kelly 1992). We can see this in action for the forest hamlets of Bedo and Behisatse. Bedo was established between the dry seasons of 1993 and 1994. It had a feature diversity of 0.13 in 1994; this increased to 0.16 in 1995. By 1996 additional facilities had been added to the hamlet, including a larger fenced enclosure, and feature diversity rose to 0.26. The same pattern holds at Behisatse. This hamlet contains two linked patrilineal household clusters, one of which was established when the hamlet was first visited in 1993; members of the other lineage had apparently just arrived. The highest-ranking man in this cluster at the time was living in a small A-frame structure that had been replaced by a more substantial house by 1994. Feature diversity registered as 0.16 in both 1993 and 1994. By 1996 more houses had been added to both clusters, and feature diversity had increased to 0.26. Our Malagasy colleague Tsiazonera (personal communication, 2002) informs us that the inhabitants of Behisatse have recently upgraded their homes from *trañ oholits'hazo* to the more time-consuming (because the reeds must be brought from the Namonte Basin and/or purchased) *trañ ovondro*. This is an indication that the settlement is becoming more permanent.

The types of features that appear in the different settlements are also worth noting. The rare features that appear in villages are schools, churches, stores, clinics, and "guest houses" (unoccupied houses set aside for visitors; this was only seen in South Vorehe, where a mission hosts foreign visitors). All of these are features that result from and permit increased interaction with the wider world: the Malagasy government, medical and religious missions, and tourists. In addition, villages have a larger diversity of house types, including a few rare houses that are made of milled lumber or cement block or that have tin roofs. These houses reflect the greater wealth that a few individuals are able to muster, probably by tapping into the wider Malagasy social, economic, and political networks. Thus the increased diversity of features is not simply a product of the increased length of stay but a function of the role that a larger, more permanent settlement plays in the wider political, economic, and social world.

In this respect, it is interesting to note that North Vorehe has a much lower feature diversity than South Vorehe. North Vorehe is a section of the village also known as the "Mikea Quarter." The people who live here appear to have closer ties to the forest. (For example, a dwarf who lives there was once portrayed in a traveling carnival as an original Mikea, a "man of the forest." We had first met several of South Vorehe's inhabitants in foraging camps.) There is a clear physical separation between this part of Vorehe and the rest of the village, and a social separation as well. In 1995, for instance, a celebration in North Vorehe was attended by few people from the rest of Vorehe; and a clinic in South Vorehe did

Table 4.3. House Type by Vorehe Neighborhood, 1998

Neighborhood	Wattle-and-daub houses	Reed houses	Grass houses	Stone houses	Total
Iaborao	33 (63.4)	17 (32.6)	2 (3.8)	0 (0)	52
Ankiliraiky	23 (85.1)	0 (0)	4 (14.8)	0 (0)	27
Lutheran Station	43 (87.7)	3 (6.1)	0 (0)	3 (6.1)	49
South Vorehe	114 (72.2)	41 (25.9)	3 (1.9)	0 (0)	158
North Vorehe	44 (41.5)	26 (24.5)	36 (33.9)	0 (0)	106
Total	257 (65.5)	87 (22.1)	45 (11.4)	3 (0.01)	392

Note: Row-wise percentages in parentheses.

not acknowledge the existence of the clinic in North Vorehe only a few hundred meters away. The lower feature diversity here suggests that inhabitants of this part of the village may see themselves as somewhat more temporary inhabitants, less committed to a "permanent" life in the village than are the people who live in South Vorehe. A 1998 tabulation of different house types in the various neighborhoods of Vorehe indicates that over one-third of houses in North Vorehe were thatched with grass, a thatch type that is very rare in the rest of the village (Table 4.3). Grass-thatch houses are the typical form found in the forest habitations nearby. We return to the difference between North and South Vorehe below.

Besy has a higher feature diversity than other settlements classed as seasonal hamlets. In many ways it appears to be a forest hamlet or even a village. Unlike other seasonal hamlets, Besy had several corrals and animal pens, as well as a relatively high number of *kitrely* platforms. We classified it as a seasonal hamlet because no one was living there in the dry seasons of 1994 and 1995. Only one house, in fact, had any personal belongings in it—a pattern seen at other seasonal hamlets. The investment in this settlement suggests that it was intended as more than a temporary seasonal settlement (as discussed below).

It is also worth noting that *kitrely* platforms are rare in villages. In fact, one of the foraging camps contained the greatest number of them (six). The purpose of these platforms was to move belongings off the ground so that they were not destroyed, scattered, or polluted by dogs or wild guinea fowl (*Numida meleagris*). These platforms were all small and made with very thin poles, however, reflecting the few belongings that needed storage in this temporary foraging camp. *Kitrely* platforms are more common (relative to the number of houses) in forest hamlets and seasonal hamlets than in large villages. Villages have one *kitrely* platform per three houses, seasonal hamlets about one for every two houses, and forest hamlets nearly one per house.

Kitrely platforms in forest hamlets are used to store dried maize in the husk.

This means that everyone is able to see how much maize every other family in the settlement has stored. Tools (such as axes, blowguns, spears, and digging sticks) are also often stored on top of the *kitrely* platforms, where they may be borrowed by other people (as discussed below).

Trash

As in other ethnoarchaeological studies, we see that the distance between the place where trash was created and where it is deposited increases as the length of stay increases. Trash is tossed only 1–2 m from hearths in foraging camps, some 2–3 m away from the outside hearth in seasonal hamlets, 4–9 m away in forest hamlets, and 10–40 m away in the villages, depending on how close a house is to the edge of the settlement. As noted above, in the villages trash is also deposited in pits near houses that were originally excavated to obtain clay for house construction (these pits are also accidental trash deposits: they catch anything blowing across the hard flat clay surface that typifies most villages).

Investment in Housing

Differences in house posts may reflect investment in housing. Binford (1990) notes that more permanent houses usually have roofs fashioned from materials different from those used for the walls. This is the case in our study: wattle-and-daub houses have walls of post and mud but roofs of thatch. Walls and roofs in forest settlements are most often made of the same material.

Here we focus on the archaeological manifestation of houses. Many of these houses would leave few remains: interior hearths, a slightly raised, packed clay floor in the case of wattle-and-daub and pole-and-thatch houses in villages, and post molds. We focus on this last feature, which is commonly recorded in archaeological sites. We collected data on post diameter by measuring the posts with a caliper at ground level (we included only posts for which we could obtain accurate measurements, avoiding those that had drying, peeling bark). In addition, we also counted the number of posts along the long sides of the houses and measured the distances between them. We do not have a truly random sample, because our schedule did not permit us to take measurements at all settlements and we were not given permission to measure all houses.

Informants stated that if they intended to remain in a house for a long time they were more selective in their choice of the species of wood for posts and in using a more standardized diameter for particular tasks. Thus in villages we often saw poles of approximately the same diameter bundled together (with each bundle destined to be used as primary supports, secondary supports, roof beams, and so forth). But informants were clear that in building more temporary structures they would use whatever poles could easily be obtained in the forest nearby or scavenged from abandoned dwellings. Post selectivity therefore could be reflected in the amount of variation, as measured by the coefficient of variation (CV), in houses' post diameters.

Houses have six main supports, four at the corners and two posts holding up the central roof beam. They also have some secondary support posts along the four walls. Surprisingly, mean house size (by settlement) is not significantly correlated with mean main post diameter ($r = 0.43$, $df = 9$, $p > 0.10$), though it is correlated with mean secondary post diameter ($r = 0.64$, $df = 9$, $p < 0.05$). The sample of house sizes is skewed to smaller houses, however; as noted above, the large houses are wattle-and-daub and are considerably larger than others in the sample. Without the wattle-and-daub house sample, the correlation between mean secondary post diameter and house size is not significant ($r = 0.2$, $df = 8$, $p > 0.1$). While large houses require large-diameter posts, small houses can use both small- and large-diameter posts. The mean diameters of main and secondary posts by settlement are correlated ($r = 0.87$, $df = 9$, $p < 0.001$), as are their CVs ($r = 0.85$, $df = 9$, $p < 0.001$). We have different sample sizes of the different post types; but this is not a significant factor in analysis, because CV is not correlated with sample size (for main posts, $r = 0.21$, $df = 9$, $p > 0.10$; for secondary posts, $r = 0.07$, $df = 9$, $p > 0.10$).

If we look simply at the mean post diameter for a settlement type, there is a significant difference (using the Feltz and Miller 1996 statistic) in the coefficients of variation among main posts in villages, forest hamlets, and seasonal hamlets (D'AD = 8.53, $df = 2$, $p < 0.025$). The highest CV is in the seasonal hamlet sample, as expected. There is also a significant difference among the coefficients of variation of secondary posts (D'AD = 14.67, $df = 2$, p < 0.001); but the greatest coefficient of variation is found, contrary to expectations, in the village sample. This village sample contains both wattle-and-daub and Vorehe's thatch houses. If we remove Vorehe's thatch houses from the village secondary post sample, the seasonal hamlets have the highest CV, as expected, but there is no significant difference among the samples (D'AD = 4.25, $df = 2$, $p < 0.25$). Hence the prediction that less selectivity in building materials would be reflected in greater coefficients of variation in the forest and seasonal hamlets holds true only for the main support posts.

It may be misleading to look at post variation simply by settlement type. Further inspection of the data with knowledge of the particular characteristics of some of the settlements is instructive. Figure 4.5 and Table 4.2 show the data by settlement within the settlement types. The least variation is found in the village of Vorehe's wattle-and-daub houses. Our sample of these houses is small, because we can only measure post diameters in houses that are under construction (once a house is completed the posts are encased in mud). Variation in post diameter of

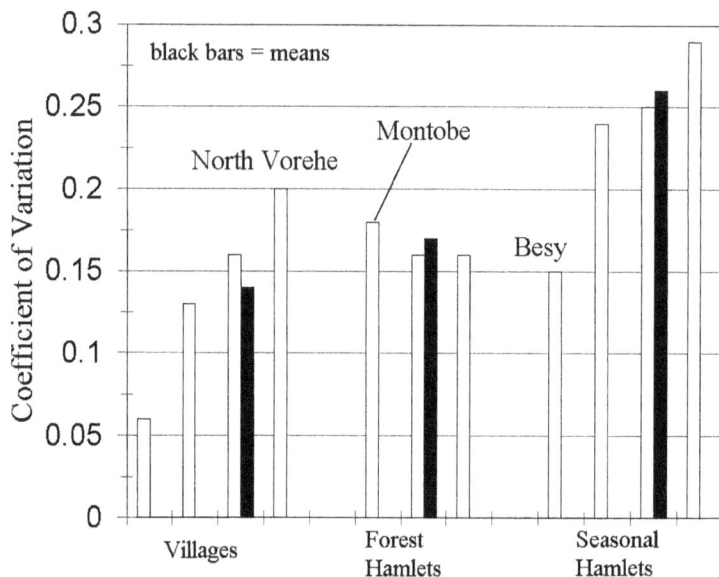

Figure 4.5. Histogram showing coefficients of variation with means for villages, forest hamlets, and seasonal hamlets.

wattle-and-daub houses, however, may be not be low simply because the builders were more selective. Large houses, especially those with heavy mud walls and thick thatch roofs, require large posts, so there is a minimum size for support posts. Because the effort to cut down a tree increases exponentially relative to its diameter, people have little incentive to cut down a tree much larger than the minimum required size. Thus they may cut down the smallest trees that will do the job and hence tend to cut trees that are of nearly the same diameter.

The highest coefficients of variation in the village sample are found in the pole-and-thatch structures of Vorehe. One possible explanation is that most of the data here were collected by Kelly in 1994 and 1995, while the Vorehe pole-and-thatch measurements were made by Tucker in 1996 when Kelly was not present. We believe that we measured the posts in the same manner, however, and that any errors of this type are minimal. (The Vorehe sample is also the largest; but random sampling of the sample produces the same CV, so the high CV is not simply a function of sample size.) One alternative explanation is that the house builders obtained their raw materials in different locations and thus did not have the same range of building materials from which to choose. Only sparse clumps of trees are found in the immediate environs of the village; clearing for cultivation has left the village surrounded by anthropogenic savanna. The builders of these houses probably harvested their materials from their *hatsake* fields, which were 3 to 20 km away and in different parts of the forest. At present we cannot evaluate this possible source of the variation.

Another possible explanation for the high coefficients of variation in post diameter in Vorehe is that most of the pole-and-thatch structures measured were in North Vorehe. As noted above, North Vorehe is spatially, socially, and historically distinct from the rest of the village in that its inhabitants tend to have closer ties to the forest. It is possible that the people of North Vorehe think of their settlement as being more like a forest hamlet. One-third of these houses are grass-thatched, a type of construction that is rare in other village contexts but common in the forest. The small grass-thatch houses are potentially mobile: the roof and four walls can be untied from one another, the wall supports excavated, and the house moved as four separate walls and roof. Once an entire cluster of houses was packed up and moved due to sickness and death among the inhabitants (the land was said to be *tany mafana*: literally "hot land," signifying "unhealthy land"). Hence the high CV in the posts perhaps is caused by the use of more scavenged posts and the use of less than optimal materials from the depauperate nearby forests. But it may also reflect people's unwillingness to commit themselves permanently to Vorehe. This interpretation is consistent with the lower feature diversity mentioned above.

We should note that we originally classified Montobe in the village category but shifted it to the forest hamlet category. Our informant there, the village's eldest man and founder, stated that he had grown tired of the quarreling in Vorehe and so had emigrated (a few kilometers away) and founded Montobe as a new village. Montobe was present in 1994 and had grown larger by 1995. By 1996 the settlement had failed, however, and most people had left. Apparently the founder was considered a sorcerer and may have been forced to leave Vorehe. In making their houses there, many of Montobe's inhabitants may never have considered the settlement to be as permanent as the founder had claimed to us. Neither did the founder, perhaps, for he maintained a wattle-and-daub house in Vorehe. Thus this site was classified as a forest hamlet. In their day-to-day activities people may have treated the settlement as if it were to be permanent: trash was deposited at the outskirts of this settlement rather than in rings or arcs around the houses (Figure 4.2e).

The lowest coefficient of variation in the seasonal hamlet sample is found in Besy. As noted in the discussion of feature diversity above, Besy appears to be less like the other seasonal hamlets and more like a village. Other data also reflect this. Trash was more commonly located in an arc some 8–9 m from the fronts of houses rather than 2–3 m away (Figure 4.2c). Additionally, more houses at Besy were made of reeds rather than of bark; some of the houses were still low in height, however, as they are in the other settlements in the seasonal hamlet sample. If we were to reclassify Besy as a forest hamlet (again leaving the Vorehe grass and reed houses out of the village sample) there is still a significant difference in the main posts' coefficients of variation (D'AD = 14.631, $df = 2$, $p < 0.005$). But now there is also a significant difference in the secondary posts' coefficients of variation (D'AD = 8.92, $df = 2$, $p < 0.025$). For each sample, the highest CV is in the seasonal hamlet sample (0.26 for the main posts, as opposed to 0.18 and 0.17 for the forest hamlet and village samples, respectively; 0.24 for the secondary posts, as opposed to 0.19 and 0.20 for the forest hamlet and village samples). With this reclassification, the CV of both main and secondary supports helps to differentiate seasonal hamlets from forest hamlets/villages.

A higher investment in housing across the settlement categories is also indicated by an increased number of secondary posts per wall in village houses as opposed to those in forest and seasonal hamlets. By excluding the large wattle-and-daub houses, we effectively hold house size and wall length constant. The data are few, but the distance between secondary posts becomes smaller from seasonal hamlets to forest hamlets to villages (see Table 4.2). If we add some cases not shown in Table 4.2 (and for which we only gathered data on secondary post counts) the average number of posts in village (not wattle-and-daub) houses is 6.2; in forest hamlets, 4.1; and in seasonal hamlets, 3.25. House size is roughly constant, so the number of posts declines and by necessity the distance between them increases, as we might expect. We should add that in villages posts are more likely to have had their bark removed before being placed in the ground (the bark tends to break away as it dries and may loosen the bindings between posts), corner posts are more likely to have secondary supports placed right next to them, and main supports may be adzed to a square cross section (Figure 4.2b).

It is worth noting, however, that these data still include Besy as a seasonal hamlet and the Vorehe thatch houses as part of the village sample. Rather than being contradictory, these data may show some ambivalence of the occupants as to whether they are permanent residents or not or may reflect a change over time in the degree of permanence of a settlement. If Besy began as a seasonal hamlet, then people might very well have built some houses with poles left over from construction in their home villages (which might have not been very variable in diameter because they were originally selected for more permanent homes). But they may have used fewer of those poles in constructing houses at Besy. Likewise, the people of North Vorehe (with thatch houses) may have been less selective in the poles they used from the forest and thus incorporated posts of variable diameters into their houses. But perhaps they incorporated more poles into their structures because poles are fairly easy to come by in the forest or from their *hatsake* fields and because they were building slightly larger houses.

This ambivalence in a settlement's intended permanency is partly a function of the fact that virtually every adult male who has a house in a forest or seasonal hamlet also simultaneously owns a house in a village. People may claim that the houses in villages are their permanent houses and those in hamlets are temporary. Yet they sometimes spend more time in their "temporary" hamlet houses than in the "permanent" houses in villages. The eldest male at Behisatse, for example, spent most of his time in his *trañ oholits'hazo* house in this forest hamlet than in his *trañ ofotake* equipped with table and chairs in Namonte. The house in Namonte is maintained partly in order to have a place to stay during ceremonies and partly because a man must maintain a residence in his *tanindraza* (the place of origin of his lineage) to demonstrate that he belongs. Archaeological studies must take into account that different houses types could be constructed and used by the same individuals and that these houses may reflect not only mobility but also social functions. In this case, a house maintained for social reasons but occupied infrequently is more elaborate and substantial than a "temporary" house maintained for economic reasons and used for much of the year.

DISCUSSION

Looking at house size, post diameter variation, feature diversity, and distance to trash from outside hearths, we see that the different settlement types can be differentiated (Table 4.4). In general, these differences meet the expectations outlined by Kent: the anticipated length of stay is related to the degree of effort placed into house construction, reflected by the size of the house, the number of posts used, the standardization of those posts, and the distance to trash.

While Kent's predictions are based on settlement permanence, we consider how distance to trash, house size, and feature diversity may be related to the social environments of settlements that differ by size. In small settlements occupants tend to be close kin, while larger settlements contain more distant kin and nonkin. Foraging camps are often single nuclear families. Hamlets contain clusters of patrilineally related households (an elder man and his sons) or dual clusters related to each other through a matrilineal or marital link, living in spatially separated areas. (This is another instance in which social relations are mapped out spatially: see Whitelaw 1991.) Villages are composed of many of these household clusters packed together. The social functions of houses and other architecture are related to the extent to which people feel comfortable sharing their belongings and their lives with their neighbors.

Food sharing is generally thought to be ubiquitous among foragers (Sahlins 1972; Ingold 1988; Bird-David 1992; Lee and Daly 1999). Nevertheless, analysis of food transfers at the forest hamlet of Behisatse suggests that the Mikea are

Table 4.4. Summary of Observations

Site Type	House size	Fenced compounds	Post variability	Secondary posts	*Kitrely* platforms	Distance to trash (m)	Featur diversit
Villages	Various, but can be large	Present	Low	Many, closely spaced	Rare	10–40+	High
Forest hamlets	Small	Rare (normally ceremonial)	Low-medium	Fewer, farther apart	Present to common	4–9	Mediur
Seasonal hamlets	Small	Absent	High	Fewer, farther apart	Rare/Absent	3–4	Medium-.
Foraging camps	Lean-tos or "boxes," if present at all	Absent	N.A.	N.A.	Rare/Absent	1–2	Low

an exception (Tucker n.d.). Food transfers at Behisatse are largely consistent with the predictions of kin selection: people provision their own households and rarely give food away. Tucker (n.d.) proposes that the general absence of food sharing at Behisatse is best explained from the perspective of tolerated theft (Blurton Jones 1984, 1987; Winterhalder 1996). Tolerated theft considers food sharing as a solution to the conflict that occurs when one person has a food item that another does not have. Because food usually delivers diminishing marginal utility, a resource "holder" is likely to devalue a food item subjectively if he or she holds a large enough quantity of it. The person without food, the "scrounger," has a greater need for it. Because the scrounger values the food more than the holder does, the scrounger will work harder to take it from the holder than the holder will work to defend it. If the holder finds that the costs of defending the food are greater than its diminished marginal utility, the holder will prefer to avoid the conflict altogether by giving portions of the food to the scrounger.

At Behisatse carbohydrate staples—maize, manioc, wild tubers, and wild cucurbits—are not shared because everyone has similar access to these foods. Indeed, it may be easier to dig a wild tuber than to scrounge it from a neighbor. The only food product that is widely shared is meat from slaughtered livestock. Like meat from large game animals, which is the most commonly shared food in many foraging populations (Wiessner 1982: 67; Kaplan and Hill 1985; Ba-huchet 1993; Lu 1999: 178–189; Hawkes et al. 2001), slaughtered livestock fit tolerated theft's conditions for a resource transfer. Animals are slaughtered by only one household at a time, leaving those in other households in the scrounger role. The food package is large enough that the holder's marginal utility diminishes, so that the holder does not mind giving excess portions of meat to needy neighbors. There is also a good probability of reciprocity, for livestock are usually slaughtered for ceremonies that all households are required to hold eventually.

Small game and honey offer the greatest potential for resource conflicts. These are acquired asynchronously, so that some households have them when others do not, providing a strong incentive to scrounge. People have little incentive to donate them, however. Individual prey animals are too small for holders to experience diminishing marginal utility. The largest prey item is the feral cat, averaging 2 kg—hardly enough meat to feed a family. People collect estivating animals, particularly the small *tambotrike* tenrec (*Echinops telfairi*), in baskets over time. Buckets of honey weighing 7 or 8 kg and baskets of torpid tenrecs constitute medium-sized food packages. Nevertheless, the holder's utility for these foods remains constant because they can be preserved. Tenrecs are stored alive in torpor and will not rot or spoil for months. In addition, tenrecs and honey have a high market value: each marginal unit can be exchanged for equal cash value, which can then be turned into goods such as tobacco, soap, and clothing. For small game and honey, the conflict between holder and scrounger is not easily resolved. Holders actively defend their foods by hiding them within their houses (while tubers and maize are frequently cooked outside houses, meat is almost always cooked inside a house). In a few cases scroungers stole honey outright, but the more common form of scrounging was demand-sharing—the application of social pressure to cajole people into generosity (Peterson 1993).

We observed somewhat greater generosity when it came to sharing tools. People appeared to pick up their neighbors' spades, buckets, and axes with little negotiation, although they only used them for short durations before returning them to their owners. It was our sense, however, that such sharing was limited to individuals within a patrilineal cluster.

The degree to which people are willing to be generous with their food and tools influences the spacing and size of their houses and the degree to which they store their belongings inside or outside their homes.

Following Kent's logic, the distance of trash from the hearth represents the degree of labor invested in camp maintenance and thus permanence. It is also correlated with settlement size and the spacing of houses. Trash in larger settlements must be removed farther, because houses are more closely packed together. Throwing trash a few meters away would mean throwing it in someone else's space, with obvious social repercussions. The trash ring in forest hamlets and seasonal hamlets is far from houses, because they are distant from each other (often 10 to 20 m). Our informants claimed that houses were spaced far apart to reduce the risks of fire spreading from one house to the next. It also seems likely that houses are widely spaced to enable people to hide things they do not wish to share. It appears to be cross-culturally true that if people cannot see something (or if they are not supposed to have seen it) then they cannot ask for it (Cribb 1991; Hitchcock 1987; Layne 1987; Wilson 1988). Our genealogical data and settlement maps show that kinship is reflected in house spacing. The closer people are socially to another, the closer the houses are. (A house located a significant distance to the south of the cluster was always, in our experience, the house of a daughter and son-in-law, reflecting the tense relations that often exist between fathers and sons-in-law in patrilineal/patrilocal societies.) Tucker (n.d.) demonstrates considerably more sharing of prepared meals among people who are close kin.

The size and location of the archaeologically recoverable trash ring may be an indicator of the spatial limits of residential units that share food in hamlets. In Mikea foraging camps (where each hearth represents a nuclear family household) foods are widely shared within the hearth group, and trash is distributed relative to each hearth. In forest hamlets trash is most heavily distributed around kinship clusters; a line of trash usually delineates one patrilineal group from another.

The linear north-south orientation of most hamlets may function to reduce visibility and enable resource hiding. The Kalahari Bushmen and the pygmies in the Ituri Forest both share food (especially meat) widely. Their camps are built as rings of houses with all the doorways facing inward, toward public space (Tanaka 1980; Fisher and Strickland 1989, 1991). It is difficult to hide foods with such a site layout, for the interiors of all houses are equally visible to everyone. By contrast, in a forest hamlet a person can only see into someone's house when standing to the north of it (and then only if its door is open). It may be significant that the elder's house is almost always the most northerly house, providing him with a preferential view of the belongings and activities of his neighbors (his sons). Likewise, hearths in foraging camps are widely spaced (> 10 m) in thick brush, making it is impossible to see from one to another. Several experiences in these camps made it clear that one reason for the seclusion was to prevent neighbors from seeing if someone had gathered any game during the day. Thus the arrangement of houses/hearths may mirror the degree of food sharing in predictable ways.

House size and the presence of *kitrely* platforms appear to be a function of the degree to which households store their products outdoors (where they can be targeted by scroungers) or inside (where they may be more easily defended).

Villages have very few *kitrely*, because people store maize and manioc in bins or gunnysacks inside their houses. In maize-growing hamlets, however, the maize is stored outdoors atop *kitrely* platforms. Households in the same hamlet are likely to experience the same maize payoff; people can store their maize outdoors, in public view, because they know that their neighbors are unlikely to scrounge. It is much easier to erect a *kitrely* with just four to six upright posts than to build a house large enough to store the maize indoors. (It may also be significant that most households have a *kitrely* or two hidden in the forest, so that much of their harvested maize remains hidden although physically unprotected.) The *kitrely* platforms also provide shady workspaces that household clusters use collectively. Beneath these platforms they conduct maize-threshing and meal-preparation tasks in full view of others and can leave tools out for others to use.

Things are different in villages. Households have unequal access to agricultural foodstuffs because their fields, which are in scattered locations, perform differently. This inequity means that scrounging is likely. Huge piles of maize left outdoors are difficult to defend from thieves as well as from demand-sharers. It is hard to say no to a request; in places with a large population, it is more likely that the person asking to borrow a tool or insisting on food sharing does not have a close kin tie to the person with the tools or food. With such ties, and without the frequent face-to-face interaction of small residential groups, it is harder to ensure reciprocity. People are more likely to be free-riders (some men in the villages of Vorehe, Ankililaly, and Namonte are notorious for this). By moving things inside houses or inside fenced enclosures, people not only prevent theft but remove things from the realm of demand-sharing.

Other structures are also built for increased privacy and security. On our first visit in 1993, a weekly market had recently been established in the village of Vorehe that was attracting people from many other villages and hamlets. Initially this market was small; we noted no fenced house compounds on our first visit. During the next two years the market became quite large, and several houses (notably those of people living near the market, who were also politically important in the village) had fences built around their house compounds. Informants told us that these enclosures were built simply because there were "too many strangers" in town on market days.

Houses are obviously built as places to stay dry (or somewhat dry) in the wet season and as places to keep out the cold nightly wind and dew in the dry season. Equally important, however, is the function of houses as a private space to conduct activities and store belongings that people do not want to share with the community. More sharing occurs in small settlements where people are closely related to their neighbors; houses are smaller and closer together, with more external storage on *kitrely* platforms. Houses in large settlements frequented by distant kin and nonkin are built larger to create more internal private space, for hiding possessions and foodstuffs not intended for public use and consumption. Archaeological data on structures and site layout may reflect the degree to which kin co-reside, items are shared, and space is privatized. The archaeological transition to sedentism most likely entails an increase in the privatization of space that is related to sharing and that affects site structure as much as simple mobility does.

Conclusion

When combined with existing research on mobility and houses, the case described in this chapter helps to provide archaeology with some ways to interpret patterns in house form and site structure. Increasing distance to areas of trash disposal, the use of more uniform building materials and evidence of "overbuilding" (as reflected, for example, in the number of secondary posts), evidence of more function-specific materials (such as the use of different materials for the roof and the walls), larger houses, and using a variety of house types and a greater variety of more function-specific features all point to lower levels of residential mobility.

Archaeologists have long used architecture as a way to infer various aspects of prehistoric behavior. Margaret Nelson (1999), for example, used patterns in architecture to infer that seasonal farmsteads during the Classic Mimbres period in southwest New Mexico became more permanent residences in the twelfth century through the mid-thirteenth century. This is evidenced in part by a change in construction from light materials and open-sided construction to more enclosed structures built of more durable materials, with additional posts. Michael Diehl (1997) also uses an increase in number of posts/m² of floor space in a sample of Mogollon pithouses ranging in age from AD 200 to 1000 to argue for slowly decreasing residential mobility over this period. The data presented here support these inferences.

We must be cautious, however, in applying any ethnoarchaeological lesson to an archaeological case. It is not enough simply to seek correlations between material culture (attributes of houses) and behavior (residential mobility). We must consider those correlations within a theoretical framework (O'Connell 1995), such as the existing research on sharing behavior presented in this chapter. This process lets us see that the changes in mobility are concomitant with changes in social relations but that the variables affecting those social relations could be independent and unrelated. This would make it harder to recognize the part played by mobility alone in shaping the archaeological record. This caveat should be considered in any application of the insights in this chapter.

ACKNOWLEDGMENTS

A condensed version of this chapter appeared in the September 2005 issue of *American Anthropologist*.

We thank, first and foremost, the many Mikea who welcomed us into their *trañio* and permitted us to do utterly bizarre things like measure their house posts; we are especially grateful to those Mikea who defended us against accusations of sorcery. We wish to thank our Malagasy colleagues Jean-François Rabedimy, Jaovola Tombo, Zaramody, Tsiazonera, and Veve Tantely and to thank James Yount for his assistance in 1995 and 1996. The notable Mr. Resariny, resident of North Vorehe, directly facilitated the measurement of houses there. Kelly and Poyer are grateful for support from the Leakey Foundation, Wenner-Gren Foundation, and the National Geographic Society. Tucker's research was supported by grants from Fulbright IIE, the National Science Foundation, Sigma Xi, and a Travel Grant from the University Center for International Studies, University of North Carolina, Chapel Hill.

NOTES

1. Kent's sample (Kent and Vierich 1989) consists of 30 settlements classified as to whether inhabitants intended for them to be of short (< 2.9 months), medium (3–5.9 months), or long (> 6 months) duration and whether they actually were of short, medium, or long duration. Of the 6 settlements with "anticipated short" duration, 4 were actually short, 1 medium, and 1 long. Of the 7 settlements with "anticipated medium" duration, 3 were actually short, 2 medium, and 2 long. Of the 17 settlements with "anticipated long" duration, 2 were actually short, 1 medium, and 14 long. Cell frequencies in a 3×3 chi-square table are too small for a realistic test of association between anticipated and actual mobility. But the major difference lies between long durations on the one hand and short/medium durations on the other, so the data can be collapsed into a 2×2 table. In this case a significant association exists between short-medium actual and anticipated settlements and long actual and anticipated settlements (Fisher's Exact Test, $p = 0.00162$). Thus, for archaeologists, the distinction between anticipated and actual mobility is perhaps not so critical.

2. "Behisatse" is a pseudonym, used to protect the identities of our key informants.

3. The game that is routinely caught is so small that it is eaten in its entirety: flesh, bones, and entrails.

REFERENCES

Ames, K.
1991   Sedentism: A Temporal Shift or a Transitional Change in Hunter-Gatherer Mobility Patterns? In *Between Bands and States*, edited by S. Gregg, pp. 108–134.

Center for Archaeological Investigations Occasional Papers No. 9. Southern Illinois University Press, Carbondale.

Astuti, R.

1995a *People of the Sea: Identity and Descent among the Vezo of Madagascar*. Cambridge University Press, Cambridge.

1995b "The Vezo Are Not a Kind of People": Identity, Difference, and "Ethnicity" among a Fishing People of Western Madagascar. *American Ethnologist* 22: 464–482.

Bahuchet, S.

1993 Food Supply Uncertainty among the Aka Pygmies (Lobaye, Central African Republic). In *Peoples of the Ituri*, edited by M. Pulford, pp. 171–200. Harcourt Brace College Publishers, Fort Worth.

Binford, L. R.

1980 Willow Smoke and Dogs' Tails: Hunter-Gatherer Settlement Systems and Archaeological Site Formation. *American Antiquity* 45: 4–20.

1990 Mobility, Housing, and Environment: A Comparative Study. *Journal of Anthropological Research* 46: 119–152.

Bird-David, N.

1992 Beyond "The Original Affluent Society": A Culturalist Reformulation. *Current Anthropology* 33: 25–47.

Blurton Jones, N. G.

1984 A Selfish Origin for Human Food Sharing: Tolerated Theft. *Ethology and Sociobiology* 5: 1–3.

1987 Tolerated Theft: Suggestions about the Ecology and Evolution of Sharing, Hoarding and Scrounging. *Social Science Information* 26: 31–54.

Cribb, R.

1991 *Nomads in Archaeology*. Cambridge University Press, Cambridge.

Diehl, M.

1992 Architecture as a Material Correlate of Mobility Strategies: Some Implications for Archaeological Interpretation. *Behavior Science Research* 26: 1–35.

1997 Changes in Architecture and Land Use Strategies in the American Southwest: Upland Mogollon Pithouse Dwellers, A.C. 200–1000. *Journal of Field Archaeology* 24: 179–194.

Dina, J., and J. M. Hoerner

1976 Étude sur les populations Mikea du Sud-Ouest de Madagascar. *Omaly Sy Anio* 3–4: 269–286.

Eder, J.

1984 The Impact of Subsistence Change on Mobility and Settlement Pattern in a Tropical Forest Foraging Economy: Some Implications for Archaeology. *American Anthropologist* 86: 837–853.

Fanony, F.

1986 À propos des Mikea. In *Madagascar: Society and History*, edited by C. P. Kottak,

J. A. Rakotaorisoa, A. Southall, and P. Vérin, pp. 133–142. Caroline Academic Press, Durham, N.C.

Feltz, C. J., and G. E. Miller
1996   An Asymptotic Test for the Equality of Coefficients of Variation from K Popula-
        tions. *Statistics in Medicine* 15: 647–658.

Fisher, J. W., and H. C. Strickland
1989   Ethnoarchaeology among the Efe Pygmies, Zaire: Spatial Organization of Camp-
        sites. *American Journal of Physical Anthropology* 78: 473–484.
1991   Dwellings and Fireplaces: Keys to Efe Pygmy Campsite Structure. In *Ethnoar-*
        chaeological Approaches to Mobile Campsites: Hunter-Gatherer and Pastoralist Case
        Studies, edited by C. S. Gamble and W. A. Boismier, pp. 215–236. International
        Monographs in Prehistory, Ann Arbor.

FTM (Foiben-Taosarintanin'I Madagasikara; Institut National de Géodesie et Car-
        tographique)
1968a   *Carte de Madagasikara, 1:500,000.* Toliara. FTM, Antananarivo.
1968b   *Carte topographique au 1:100,000.* Feuille B-55, Befandefa. FTM, Antananarivo.

Hawkes, K., J. F. O'Connell, and N. G. Blurton Jones
2001   Hadza Meat Sharing. *Evolution and Human Behavior* 22: 113–142.

Hitchcock, R. K.
1987   Sedentism and Site Structure: Organizational Change in Kalahari Basarwa Resi-
        dential Locations. In *Method and Theory for Activity Area Research*, edited by S.
        Kent, pp. 374–423. Columbia University Press, New York.

Ingold, T.
1988   Notes on the Foraging Mode of Production. In *Hunters and Gatherers, Volume*
        1: History, Evolution, and Social Change, edited by T. Ingold, D. Riches, and J.
        Woodburn, pp. 269–286. Berg, Oxford.

Kaplan, H., and K.Hill
1985   Food Sharing among Ache Foragers: Tests of Explanatory Hypotheses. *Current*
        Anthropology 26: 223–246.

Kelly, R. L.
1992   Mobility/Sedentism: Concepts, Archaeological Measures, and Effects. *Annual*
        Review of Anthropology 21: 43–66.

Kelly, R. L., J.-F. Rabedimy, and L. Poyer
1999   The Mikea of Southwestern Madagascar. In *The Cambridge Encyclopedia of Hunter-*
        Gatherers, edited by R. B. Lee and R. Daly, pp. 215–219. Cambridge University
        Press, Cambridge.

Kent, S.
1991   The Relationship between Mobility Strategies and Site Structure. In *The Interpre-*
        tation of Spatial Patterning with Stone Age Archaeological Sites, edited by E. Kroll
        and T. D. Price, pp. 33–59. Plenum, New York.
1992   Studying Variability in the Archaeological Record: An Ethnoarchaeological
        Model for Distinguishing Mobility Patterns. *American Antiquity* 57: 635–659.

1993   Models of Abandonment and Material Culture Frequencies. In *Abandonment of Settlements and Regions: Ethnoarchaeological and Archaeological Approaches*, edited by C. Cameron and S. Tomka, pp. 54–73. Cambridge University Press, Cambridge.

Kent, S., and H. Vierich

1989   The Myth of Ecological Determinism: Anticipated Mobility and Site Spatial Organization. In *Farmers as Hunters*, edited by S. Kent, pp. 96–130. Cambridge University Press, Cambridge.

Layne, L.

1987   Village-Bedouin: Patterns of Change from Mobility to Sedentism in Jordan. In *Method and Theory for Activity Area Research*, edited by S. Kent, pp. 345–373. Columbia University Press, New York.

Lee, R. B., and R. Daly

1999   Introduction: Foragers and Others. In *The Cambridge Encyclopedia of Hunters and Gatherers*, edited by R. B. Lee and R. Daly, pp. 1–19. Cambridge University Press, Cambridge.

Lu, F. E.-S.

1999   Changes in Subsistence Patterns and Resource Use of the Huaorani Indians in the Ecuadorian Amazon. Ph.D. dissertation. Curriculum in Ecology, University of North Carolina at Chapel Hill.

Madsen, D. B., and S. R. Simms

1998   The Fremont Complex: A Behavioral Perspective. *Journal of World Prehistory* 12: 255–336.

Molet, L.

1958   Aperçu sur un groupe nomade de la forêt épineuse des Mikea. *Bulletin de l'Académie Malgache* 36: 241–243.

1966   Les Mikea de Madagascar: Ou vivre sans boire. *Revue de Madagascar* 36: 11–16.

Nelson, M.

1999   *Mimbres during the Twelfth Century*. University of Arizona Press, Tucson.

O'Connell, J.

1995   Ethnoarchaeology Needs a General Theory of Behavior. *Journal of Archaeological Research* 3: 205–255.

Peterson, N.

1993   Demand Sharing: Reciprocity and the Pressure for Generosity among Foragers. *American Anthropologist* 95: 860–874.

Poyer, L., and R. L. Kelly

2000   Mystification of the Mikea: Constructions of Foraging Identity in Southwest Madagascar. *Journal of Anthropological Research* 56: 163–185.

Sahlins, M.

1972   *Stone Age Economics*. Aldine, Chicago.

Schlanger, S.

1991   On Manos, Metates, and the History of Site Occupation. *American Antiquity* 56: 460–474.

Seddon, N., J. Tobias, J. W. Yount, J. R. Ramanampamonjy, S. Butchart, and H. Randrianizahana

2000   Conservation Issues and Priorities in the Mikea Forest of South-west Madagascar. *Oryx* 34(4): 287–304.

Stiles, D.

1991   Tubers and Tenrecs: The Mikea of Southwestern Madagascar. *Ethnology* 30: 251–263.

Tanaka, J.

1980   *The San Hunter-Gatherers of the Kalahari: A Study in Ecological Anthropology.* University of Tokyo Press, Tokyo.

Tucker, B.

2001   The Behavioral Ecology and Economics of Variation, Risk, and Diversification among Mikea Forager-Farmers of Madagascar. Ph.D. dissertation. Department of Anthropology, University of North Carolina, Chapel Hill.

2003   Mikea Origins: Relics or Refugees? *Michigan Discussions in Anthropology* 14: 193–215.

n.d.   Giving, Scrounging, Hiding, and Selling: Minimal Food Transfers among Mikea Forager-Farmers of Madagascar. In preparation.

Whitelaw, T.

1991   Some Dimensions of Variability in the Social Organization of Community Space among Foragers. In *Ethnoarchaeological Approaches to Mobile Campsites: Hunter-Gatherer and Pastoralist Case Studies,* edited by C. S. Gamble and W. A. Boismier, pp. 139–188. International Monographs in Prehistory, Ann Arbor.

Wiessner, P.

1982   Risk, Reciprocity and Social Influences on !Kung San Economics. In *Politics and History in Band Societies,* edited by E. Leacock and R. B. Lee, pp. 61–84. Cambridge University Press, Cambridge.

Wilson, P. J.

1988   *The Domestication of the Human Species.* Yale University Press, New Haven.

Winterhalder, B.

1996   A Marginal Model of Tolerated Theft. *Ethology and Sociobiology* 17: 37–53.

Yount, J., Tsiazonera, and B. Tucker

2001   Constructing Mikea Identity: Past or Present Links to Forest and Foraging. *Ethnohistory* 48: 257–291.

5

How Much Land Do the Wana Use?

MICHAEL S. ALVARD

This chapter reviews home range data for the Wana, a group of traditional slash and burn horticulturalists living in the uplands of central Sulawesi, Indonesia. The data were initially collected to examine the sustainability of Wana hunting but are used here to examine the general issue of home range (see Alvard 2000a, 2000b, where some of these data are presented). I used GPS (Global Positioning System) technology to calculate home range estimates for 22 adult males, based on house, garden, and trap sites. I also present data on the size area that the Wana require to harvest in a sustainable way their most important prey species, the Celebes warty pig (*Sus celebensis*). I compare the Wana ranging data to the home ranges of other horticulturalists, hunter-gatherers, and other mammals by examining the allometric relationship of body mass, home range, and subsistence pattern.

THE WANA

The study population is a group of Wana living in the Posangke region of Morowali Nature Reserve in central Sulawesi, Indonesia. This group is a sample of a larger population of Wana living in the highlands of the central eastern arm of Sulawesi (Figure 5.1). The Wana are ethnographically known from a study of shamanism by Jane Atkinson (1989) in the late 1970s and a general ethnography of the Posangke Wana written by the Dutch missionary A. Kruyt (1930). They adapt to interior mountainous rain-forest environments, practice swidden dry rice and manioc horticulture, and hunt and trap wild game. According to a World Wildlife Fund report (WWF 1980) approximately five thousand Wana live in and around Morowali Nature Reserve. The Wana's historic isolation in upland valleys insulates them from the often politically tumultuous coast, although warfare was not uncommon in the interior (Atkinson 1989). Until the early part of the twentieth century, and through much of the early Dutch period, the coastal areas were under the influence of various marine-based polities. There is evidence that the Wana paid nominal tribute to the sultanates of Ternate and Bungku (Kruyt 1930). The Wana, however, have maintained relative isolation from much of the outside world. Culturally intact, most adults within the study sample speak no or very little Indonesian, have had little or no interaction with

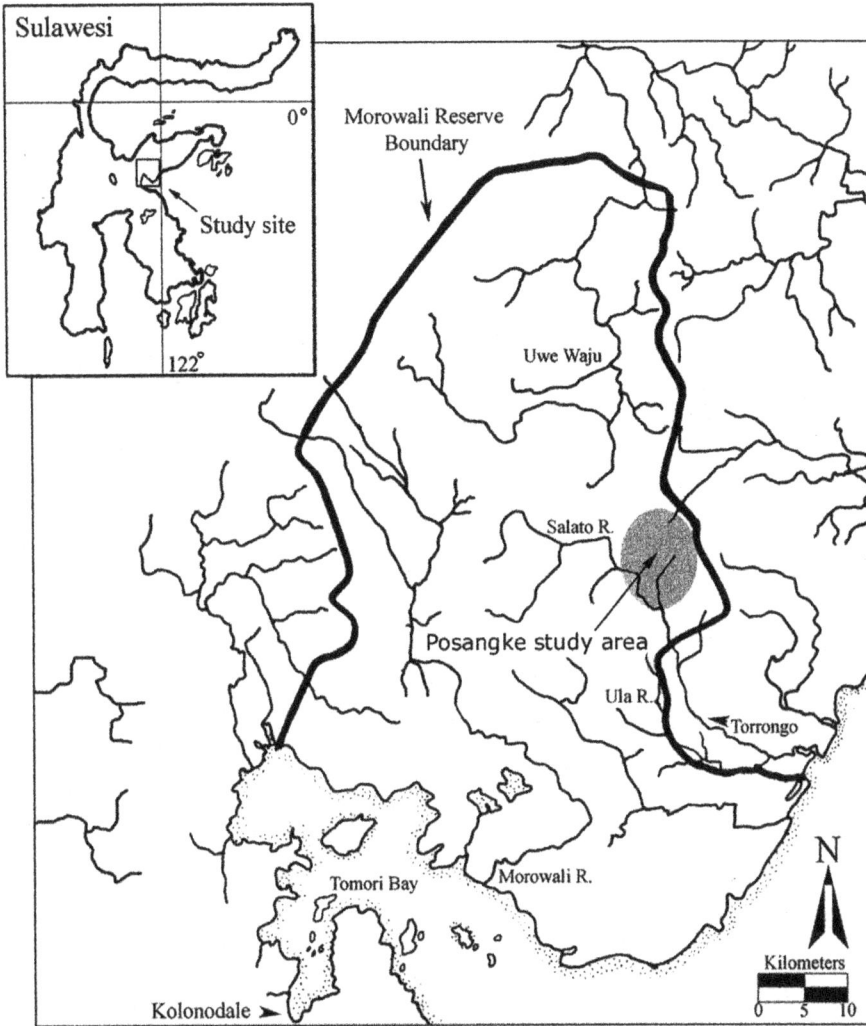

Figure 5.1. Map of Morowali Nature Reserve, Sulawesi, Indonesia.

the cash economy, and maintain a traditional religious belief system (Alvard 2000b). It is unclear how long the Wana have inhabited the area.

This study focuses on a Wana population located in Morowali Nature Reserve along one segment of the upland course of the Salato River valley and its tributaries in an area called Posangke. The tributary streams come from Mt. Tokala to the west, which forms a high ridge along the western side of the study area. The peaks reach a height of 2,600 m. The area includes all of one watershed (drained by the Sumi'i River into the Salato River) and the upper part of the next watershed (drained by the Uwe Kiumo River to the east). The entire population of Posangke consists of approximately 367 individuals living in a number of watersheds and valley systems that all feed the Salato (Alvard 2000b).

ENVIRONMENT

Morowali Nature Reserve was established in 1980 by the Indonesian government and encompasses approximately 225,000 hectares of upland and coastal areas. The terrain in the interior of Morowali Reserve is mountainous and extremely rugged. The reserve has three peaks over 2,200 m in altitude. Mt. Tokala rises to 2,593 m immediately to the east of the Posangke study site. Wana inhabit the northeastern and eastern portions of the reserve. The rest of Morowali is uninhabited, with the exception of a small number of Wana settlements to the west and a number of fishing villages along the coast to the south. Vegetation includes mangroves and lowland alluvial plain forests along the southern coastlines and lowland and montane rainforests in the higher elevations where most of the Wana live (Schweithelm et al. 1992; Whitten et al. 1987; WWF 1980).

The habitat surrounding Wana settlements is a very diverse and anthropogenic one. Gallery forests are found along the watercourses. Along the bottom of the basins, *alang alang* (*Imperata cylindrica*) meadows are interspersed among active Wana fields, old fields, secondary forests, and bamboo groves. Primary forest is found along the low slopes, ridges, and nearby mountains, up to the summits of the peaks. Additional active and fallow Wana gardens are located along numerous lower ridges. For reasons probably related to distance, difficult travel, and slope, few fields are found above 1,200 m.

POSANGKE WANA SETTLEMENTS

The Wana of the Posangke region inhabit a habitat ranging in altitude from about 300 m to about 1,100 m, although most of the houses in the sample were located between 300 and 600 m. Unlike the groups of Wana studied by Atkinson (1989: 3) in areas north of Morowali, who live in centralized villages, the Posangke Wana live in scattered households of 5 to 20 people. The Wana keep two types of dwellings or *banuah*. *Banuah mpu'u* are the main dwellings (see Figure 5.2). Each family also maintains secondary houses (*banuah kendepe*), often located at garden sites. Depending on the season, a family may actually spend more time at their *kendepe* than at the *mpu'u*. The number of *banuah mpu'u* for the sample at the start of the study was 22. Residential sites are moved, minimally, on a yearly basis as field sites change. During the course of observations, all families moved their *mpu'u* sites at least once, and many more than once. The Wana are not, however, nomadic in any sense. House sites are moved frequently

Figure 5.2. Typical Wana house site with two visible house structures surrounded by fallow rice fields.

but often only a short distance and always to a new site within the Posangke area. Rarely are they moved even 1 km. In addition to building new *kendepe* (field) houses, main dwellings are often moved after a death (for reasons related to hygiene) or if floods or landslides threaten.

A typical household consists of one or two couples, usually connected by kinship, and their children. The marriage residential pattern is marginally matrilocal. There are many widowers and widows, and a number of orphans float from household to household. Houses are located at distances ranging from several meters to several hundred meters away from their nearest neighbors. The community shows a pattern of fluid mobility. Families sometimes share houses for varying amounts of time. Children can be found sleeping for extended periods at houses other than their parents' house.

A typical house might be 4 m by 5 m. It has a bamboo slated floor 1 m off the ground (often much higher), supported by bamboo or tree trunk posts. The roof is thatched rattan or sago leaves. Walls are minimal, sometimes made from the same thatch as the roof, from woven bamboo, or from the flattened bark of trees. The hearth is in the house, sunken at a level approximately 0.5 m below the main floor and placed on packed earth. Cooking is done at and around the hearth. There are no partitioned sleeping areas. People sleep in their sarongs on woven mats placed on the floor. Houses are not heated, except by the warmth that the hearth can provide. People living at altitudes higher than 800 m often complain of being cold. Dogs, cats, and especially chickens roam with semifreedom throughout the house.

WANA SUBSISTENCE

The Wana have adapted to an upland rainforest habitat that does not allow irrigated rice agriculture (*sawah*) and the social complexity often associated with that subsistence pattern. The Wana obtain most of their carbohydrates from rainforest slash and burn horticulture (*ladang*) for a variety of crops, primarily nonirrigated rice and manioc. Sago is also cultivated. The Wana follow a common pattern of slash and burn. Fields are cut in August, September, and October then burned and planted with rice in October and November.

The Wana obtain most of their animal protein and fat from hunting and trapping and have a broad diet (Alvard 2000a, 2000b; Alvard and Winarni 1999). Game includes the smallest mice and birds, bats, reptiles, and amphibians as well as the larger ungulates and primates. Trapping for both small and large game is common and accounts for most of the meat consumed on a day-to-day basis. Rodents are frequently trapped, but pigs (*Sus celebensis*) and *anoa* (dwarf buffalo, *Bubalus* spp.) provide most of the meat (Alvard 2000a). Children frequently kill birds using blowguns. Deer (*Cervus timorensis*) and monkeys (*Macaca tonkeana*) are occasionally taken as well. Blowpipes and darts, sometimes poisoned with the resin of a tree (*Antiaris toxicaria*—the Wana call the poison *impo*), are used primarily for bats and birds but also occasionally for primates and other small mammals. For the largest terrestrial game such as the *anoa*, they hunt with spears and dogs. Although river organisms such as fish, eels, and crayfish are taken in the dry season, the rivers seem to be relatively unproductive.

The Wana use two techniques to capture large game: active hunting (with spears and dogs) and trapping. Trapping is done throughout the year and is the method that provides the greatest number of large game kills, primarily pigs (Alvard 2000a). The Wana have a great number of trap types, but the *kaiyoro* and the *saiya* are the two major types used for larger animals. *Kaiyoro* are common snare traps and accounted for 40% of the pig kills during the field session. The hunter bows over a sapling or segment of bamboo with a length of cord tied on the end. The other end of the cord is tied into a loop placed on the ground and attached to a trigger device that keeps the trap in place and set. When the leg of an animal disturbs the trigger, the sapling quickly straightens, tightening the loop around the leg of the prey and trapping it. The animal is dispatched with a spear. The Wana also occasionally trap *anoa* and deer with *kaiyoro*, although none of these traps were reportedly set during the field session with these prey types in mind. One man using spear and dogs killed all nine of the *anoa* killed during the field session.

Saiya are traps that disable and kill by causing the animal to impale itself on a picket of bamboo staves. Of the harvested pigs, 44% were killed with *saiya* traps. The trap employs a trip line activated by the passage of an animal that causes a log to fall. The falling log frightens the animal, causing it to leap forward to its death, transfixed on the staves. Hunters placed the majority of *saiya* along pig trails, often on steep slopes to facilitate the pig's journey toward the stakes. Hunters frequently set both types of traps near the previous year's fields, which are thick with undergrowth and still contain a substantial amount of unharvested manioc that attracts the pigs. Hunters also killed a number of pigs with dogs and spears (12%), by hand, or with other trap types (4%).

METHODS

The study population for the project consisted of 153 individuals and 35 nuclear families. Data were collected from March 1995 through July 1996 from a subsample of 22 adult Wana males. I used house, garden, and trap locations to delineate the extent of ranging. Adult males are responsible for field placement and clearance, trap setting and checking, and house construction and can be clearly associated with each site. For this reason, range data are associated with males. No time allocation data were collected, but it was clear that individuals spend the majority of their time within the bounds of the area defined by these points, engaged in activities associated with these sites. Other activities infrequently took them beyond the area. A common reason to travel was social visiting. This usually occurred within Posangke. Some men occasionally visited places like Torrongo (13 km to the south) or Uwe Waju (a day's walk to the north). Some rare individuals also visited places like Kolonodale (a town about 60 km to the southwest). Another common reason that men and women traveled beyond the immediate area was to harvest *damar* (*Agathis* spp.), a tree resin that is sold to middlemen in Torrongo. A small proportion of the sample accomplished this activity seasonally.

At the start of the project, all active *mpu'u* and *kendepe* were located for each male. I noted all subsequent residential moves and measured new locations ($n = 117$). The locations of the men's *saiya* and *kaiyoro* traps ($n = 342$) were determined during each of five sampling periods. With the assistance of each man, I visited all active trap sites during these periods. Finally, I measured all fields ($n = 77$) that were cut by the men in 1994 and 1995 and noted the locations.

Locations were measured using GPS, a technology originally developed in the United States as an aid for military positioning and navigation applications (Hofmann-Wellenhof et al. 1993). The technology is now widely available for commercial and scientific uses. GPS allows an accurate determination of the absolute position of any point on earth. Satellite transmissions are used to trilaterate and calculate the latitude, longitude, and altitude of the receiver. At the time of the study, differential GPS (DGPS) was used to increase data accuracy. DGPS requires at least two simultaneously activated receivers.[1] The base station receiver must be placed on a known locality within 300 km of a roving receiver. Both units see the same satellites, so both produce the same error. The direction and magnitude of the error can be calculated relative to the known location of the base station. The error data are then used to correct the roving unit. Data that are differentially corrected in this manner are associated with an accuracy of 1 m to 5 m. For this project, a Trimble Pathfinder Basic Plus receiver was used as a roving unit. The Trimble Community Base Station was used to provide differential correction. The base station was located in Kolonodale, a town approximately 60 km southeast of the study site.

Data collection involved placing the receiver on the site for which a location datum was required. When the unit was activated, approximately 300 data points were recorded. Radio contact with the base station in Kolonodale assured that the base station collected correction data concurrently. The data were subsequently corrected using postprocessing Trimble Pfinder software. The points were then averaged to provide the final datum. The data were imported into GIS (Geographic Information System) software (Intergraph Microstation MGE). All the fields, houses, and traps were plotted for each man, and the area was calculated using the minimum area polygon method (White and Garrott 1990). The minimum area polygon was constructed by connecting the outer locations to form a convex polygon. The area within the polygon was calculated in the MGE Terrain Analysis module using the MEASURE AREA command.

Results

To produce the ranging estimates for the sample of 22 men, 117 house sites were mapped with GPS. The average man had 5.3 house sites (range = 1–11). For the five rounds of trap data collection, 342 large game traps (*saiya* and *kaiyoro*) were mapped, for an average of 15.5 traps per man. The 22 men cut a total 77 fields or 1.75 fields per year. In 1994, 53 of the fields that were located were also measured. The average size was 0.375 ha, and the fields ranged from 0.043 ha to 1.107 ha. All were rice and manioc fields.

In spite of relatively high residential mobility, the absolute area of land used by each man was relatively small. Figure 5.3 presents data on the ranges for the

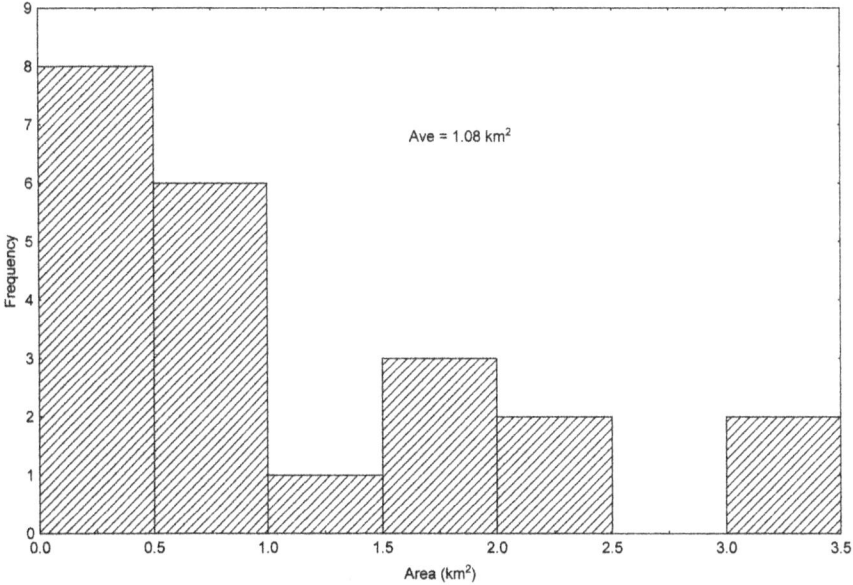

Figure 5.3. Frequency distribution of ranges for the sample of 22 men. The mean size of the area encompassing all of one man's fields, traps, and house locations is 1.08 km².

22 individuals in the sample. The mean size of the area encompassing all of one man's fields, traps, and house locations was only 1.08 km². This is equivalent to a circular area with a radius of 0.58 km. The home range values ranged from 0.63 km² to 3.03 km². All the ranges overlapped with one or more ranges of other men. The area that encompassed all of the houses, fields, and traps for the entire sample of 22 men was only 18.1 km². Population density within this area is 8.45 individuals/km². It should be noted that this is an ecological density (Haila 1988) defined to include only the habitat in which the Wana of Posangke habitually travel.

While these figures are accurate to the extent that they describe the areas over which males range, it is important to note that they are not the areas from which all important resources are culled. I have reported elsewhere (Alvard 2000b) that Wana trappers harvest game from an area much larger than the area in which the hunters physically travel. As mentioned, all the houses, fields, and traps for the entire sample fall within an area of 18.1 km². At least 88% (50 of 57) of the pigs that were killed during the field session were harvested within this area. A harvest of that size, however, is unsustainable from an area that small.

To determine the area required to sustain a harvest of that size, I calculated the potential harvest (Alvard 2000b). The potential harvest is the number of animals or equivalent kilograms of biomass that can be taken per year per square kilometer without not depleting the standing population (Robinson and Redford 1991). Pigs have a high potential harvest, although the actual values depend on the density values. Density values at Morowali were calculated from data collected using line transect methods at one hunted and two unhunted, uninhabited areas (Alvard 2000a). Comparisons of encounter rates for the hunted and unhunted areas show that pigs were equally common in all three areas, with no statistically significant differences in encounter rates between sites. The density estimate from the pooled sample of the three sites was 1.54 pigs per km^2 (Alvard 2000a). Given this density value, the potential harvest per km^2 is 0.16 pigs per year (Alvard 2000b). (For details on the methods for calculating the potential harvest, see Robinson and Redford 1991; also Alvard et al. 1997; Fitzgibbon et al. 1995.)

The Wana remove pigs at an unsustainable 15 times the estimated potential harvest if we assume that they are extracting pigs from an area of only 18.1 km^2. The transect data reported above, however, are strong evidence that the Posangke Wana pig harvest *is* sustainable at current levels. The three sites have no significant differences in the rates of encounter. In spite of hunting, pig numbers are just as great at the Posangke site as they are at the two other unhunted sites (Alvard 2000a).

This analysis of the transect data suggests that the Wana must be harvesting from pig production occurring *outside* the community's home range area of 18.1 km^2 if we assume that the pig harvest is indeed sustainable. I determined the size of the minimum catchment area that would allow the observed harvest at Posangke to be sustainable by dividing the number of pigs harvested annually (46 pigs) by the potential harvest of 0.16 pigs per km^2 per year (Alvard 2000b). A catchment area of approximately 290 km^2 is required to provide a potential harvest equal to the actual annual pig harvest of the Posangke Wana. In another words, for the pig harvest to be sustainable at the current rate, the Wana must be harvesting from a prey population in an area greater than 290 km^2—a circular area with a radius of 9.6 km. Thus, while the Wana home range narrowly defined is 18.1 km^2, the area from which they harvest their resources is much larger (Alvard 2000b).

DISCUSSION

Although no detailed observations on the day-to-day activities of the Wana were made, I infer ranging from the location of important subsistence activities. The definition of home range in this chapter is similar to definitions offered by students of animal ecology. J. L. Gittleman and P. H. Harvey (1982) define home range as the total area used by a group (or individual in solitary species) during normal activities. This definition is ambiguous: it does not indicate the time scale during which observations are made or define what constitutes "normal activities." Others (Harestad and Bunnell 1979) define home range as "the area normally traversed by a individual animal or group of animals during activities associated with feeding, resting, reproduction and shelter seeking." As for the Wana data, A. S. Harestad and F. L. Bunnell do not include infrequent movements outside this area as part of the home range. These definitions are analogous to what Lewis R. Binford (1983) calls the "annual range" for hunter-gatherers: an area used during the course of a year for residential and subsistence purposes.

I compare the Wana results to data from other horticulturalists, hunting and gathering groups, nonhuman primates, and other mammals to show patterns that clarify a number of issues important to archaeologists (such as the relevance of comparative home range to the question of the transition from foraging to food production, discussed below).

I plotted home range data as an allometric function of species body mass (see Figure 5.4). Many morphological, physiological, life history, and ecological traits, including home range size, vary with body mass among species according to the relationship $X = AW^b$ (Peters 1983). W is body mass, A is the intercept, b is the slope of the line for the linear relationship of the log-transformed variables $\log(X) = \log(A) + b[\log(W)]$. The best-known allometric relationship is described by Kleiber's rule, which states that metabolic rate is a function of body mass where $b = 0.75$ power (Kleiber 1961). Recent work has shown that these relationships are generated from the properties of hierarchical resource transport networks in the vascular systems of mammals and plants (West et al. 2002). Such allometric relationships provide useful tools for interspecific comparisons because they control for the effects of body mass.

Figure 5.4 presents home range data for 171 mammals, including 46 non-human primates. It also includes ranging data for a sample of 69 hunting and gathering groups (compiled by Kelly 1996); two other tropical horticultural groups, the Yanomamö (Good 1989) and Highland New Guinea (Pataki-Schweizer 1980); one agricultural group, the Koyfar (Stone 1996); and the Wana (this study).

It is reasonable to assume that the size of an animal's home range is determined fundamentally by its energy requirements. Thus larger animals are expected to have larger ranges because their metabolic requirements are greater. For example, Harestad and Bunnell (1979) report that the North American brown bear (*Ursus arctos*) weighs 204 kg and has a range of 9,283 ha, while the gray squirrel (*Sciurus carolinensis*) has a range of 0.95 ha and a body mass of 0.5 kg. All other things being equal, home range size should scale according to

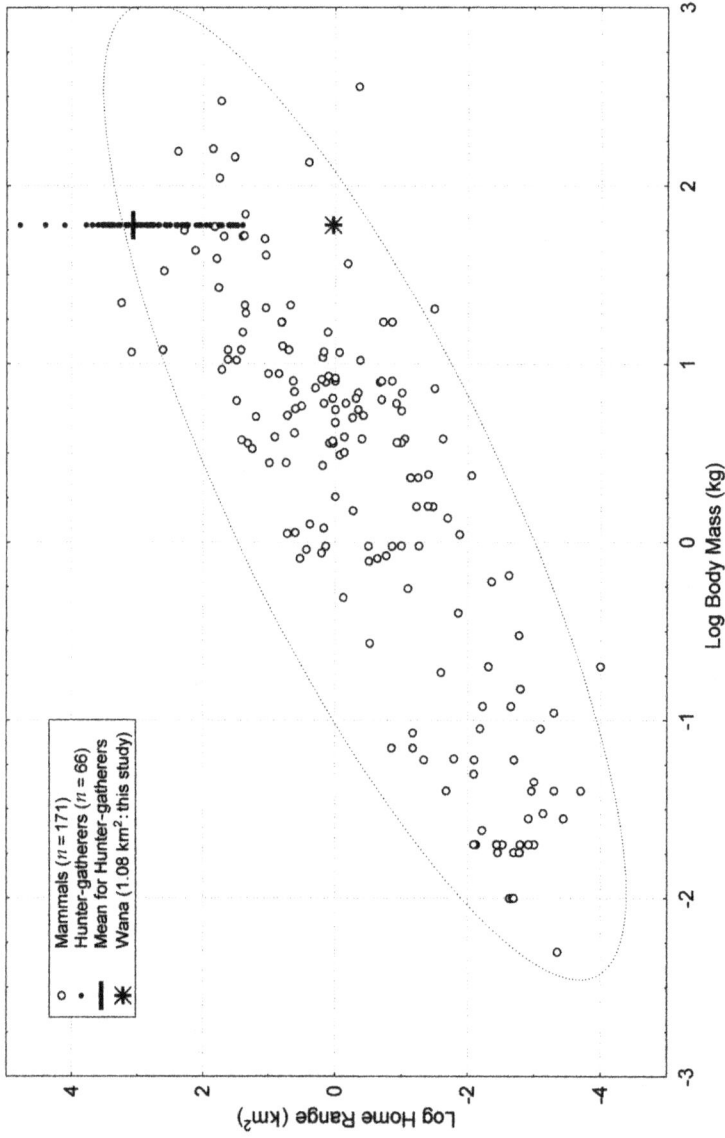

Figure 5.4. Plot of Log Body Mass and Log Home Range. The data show a strong allometric effect—animals with large body sizes have larger home ranges. The ellipse represents the 95% confidence ellipse based on the sample of 171 mammals. Body mass for humans is set at 60 kg. The hunter-gatherer data are from Kelly (1996). The mammal sample was compiled from Milton and May (1976), McNab (1963), Lindstedt et al. (1986), Gittleman and Harvey (1982), and Swihart et al. (1988).

body mass following Kleiber's 0.75 exponent for metabolic rate. It is therefore interesting to find that larger animals generally need more space than predicted from their metabolic needs. That is, the scaling factor that relates body mass to home range size is greater than Kleiber's exponent of $b = 0.75$. For example, Robert Swihart et al. (1988) found $b = 1.42$ for a sample of $n = 23$ mammals. The combined sample of $n = 171$ mammals from Figure 5.4 has a scaling factor of $b = 1.2$.

One explanation of this observation may be that density scales negatively to body size with a coefficient of -0.75 (large animals live at lower density), but biomass (mass × number of animals) scales positively, with a coefficient of 0.25. This means the biomass of a large species per unit area is greater than that of a small-bodied species (see Alvard and Kuznar 2001: Figure 3). In addition, the relationship between density and home range is negative; that is, large animals live at lower density and have larger home ranges. With respect to body mass, however, density does not go down as fast as home range goes up. While the number of individuals per unit area is fewer for larger animals, they tend to share their home ranges with more individuals than do smaller species (Damuth 1981). Individuals of larger-bodied species may have to range farther than expected compared to individuals of smaller species, because of greater intraspecific competition (see also Nunn and Barton 2000).

How do humans fit into this picture? Hunter-gathers have large ranges. The ranges in Kelly's (1996) sample of hunter-gatherers vary from 62,000 km² for the Crow, to 25 km² for the coastal Andamanese; 95% of groups in the sample (66 of 69), however, range between 25 km² and 6,000 km². The mean range for the entire sample of 69 societies is 2,571 km². If the three large outliers are removed, the mean home range for hunter-gatherers drops to 1,177 km²—an area with a radius of 19.4 km. This mean value is within the 95% confidence ellipse of the regression (for $n = 171$ mammals) in Figure 5.4 and is thus consistent with the home range predicted strictly from body mass.

Much of the work to explain the variability in home ranges independent of body mass has focused on diet. B. K. McNab (1963) presented data to show that what he called "hunters" (carnivores, insectivores, frugivores, and granivores) have larger home ranges than do "croppers" (grazers and browser). T. W. Schoener (1968) hypothesized that this was because of the presumed greater density of herbivore food. Carnivores must travel greater distances to obtain equivalent amounts of resource. Harestad and Bunnell (1979), Schoener (1968), and Swihart et al. (1988) provided additional data on mammals and birds. They concluded that, for animals of the same body mass, home range is larger for species that have a greater proportion of animal prey in their diet (see also Kelt and van Vuren 2001). The data from Kelly (1996) suggest that the same is true for

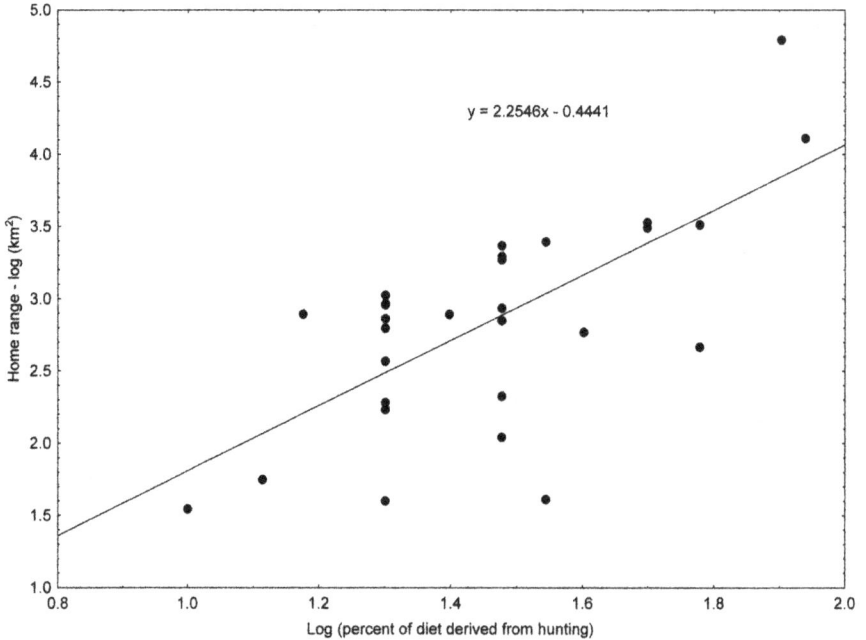

Figure 5.5. Plot of Log Home Range and Log Percent of Diet Derived from Hunting for hunter-gatherers (*n* = 29). Data from Kelly (1996).

human hunter-gatherers. The mean home range value for foragers falls above the regression line in Figure 5.4, consistent with foragers' omnivorous diet. Diet also explains variance *within* the sample of human foraging groups. For a sample 29 societies for which data were available, Kelly (1996) found a significant positive relationship between the proportion of the diet obtained from hunting and the size of the group's home range (see Figure 5.5). It makes sense that predators need larger ranges, because meat is a scarce resource relative to fruit and foliage (Gittleman and Harvey 1982; Kelly 1996). The conclusion is consistent with the observation that *all* the hunter-gathers in Figure 5.4 fall above the line and tend to have larger home ranges than expected, given their body mass. In fact, foragers may be pushing the limit of possible range size for animals of their mass. D. A. Kelt and D. H. van Vuren (2001) argue that home range area is constrained both by the need to obtain resources sufficient for survival and by decreasing gains relative to costs associated with larger home ranges. These costs include travel costs, the lack of familiarity with a large range, and the simple physiological limitations of travel.

Fewer data exist for horticultural groups, but as expected their home ranges are smaller. Some of the best data come from K. J. Pataki-Schweizer (1980), on highland New Guinea groups. He reports a mean home range of 12.5 km² for a sample of 92 groups. An estimate of Yanomamö home range (based on Figure 13 in Good 1989) is 126 km². These values may be higher than those presented above for the Wana because they are for entire villages rather than for individuals (see also below). Glenn Stone (1996: Figure 9.2) reports that for the Koyfar (moderately intensive farmers located in Nigeria) the number of food production trips decreases sharply above distances of 700 m. Approximately 73% of all trips traversed less than 700 m from the house. An area with a radius of 700 m is 1.5 km².

The data show that home range on average is greater for hunter-gatherers than for horticulturalists. While reliance on meat explains much of the variance in home range for hunter-gatherers, however, the difference in home range between hunter-gatherers and horticulturalists is likely related to both plant and animal foods. Gardens are localized and concentrated carbohydrate patches. This is one great proximate advantage of agriculture over gathering. In agricultural systems the travel distance to reliable food patches is minimized. One reason for the relatively large range of the Yanomamö is that they spend a portion of their seasonal round on treks searching for forest plant resources (Good 1989). As reliance on agricultural produce increases, the data suggest that a shrinking home range is required to sustain a given population results.

From the comparative data presented here it is unclear how much of the difference between foragers and horticulturalists is due to the proportion of the diet obtained from hunting. The Wana hunt primarily by trapping animals that are attracted to fields; active mobile pursuit is relatively rare. Such trapping concentrates game acquisition activities into a much smaller area, although (as noted above) the effective catchment area can be much larger than the physical area traveled during the course of trapping. The Posangke Wana have very few domesticated animals. Most families have chickens; only three domesticated pigs were observed. More intensive animal husbandry, as practiced by many agriculturalists, would have the same effect as trapping, by limiting the need for individuals to travel long distances in search for prey.

Understanding the patterns of variability in the area of land that humans use is important for a wide range of issues important to anthropologists. For example, some of the most interesting changes in landscape use have occurred during technological revolutions. One of the most notable is the Neolithic transition from foraging to food production that first occurred at the end of the Pleistocene. Associated with the transition were a number of well-documented factors, including sedentism (Brown 1985; Bar-Yosef and Belfer-Cohen 1989; Harris 1977; Lieberman 1993), increased population density (Cohen 1977; Harris 1977), and the development of political, social, and economic complexity (Brown 1985; Byrd 1994; Harris 1977; Price and Brown 1985; Wright 1978).

Given the assumption that population pressures increased in the late Pleistocene, foragers could have responded by increasing their range. As noted above, however, many hunter-gatherers may be near their maximum range, given the costs of maintaining large ranges. In addition, a solution that involves increased range in response to population pressures at the end of the Pleistocene may not have been viable if all groups adapted such a strategy. In a context of greater density, increasing range results in each group simply sharing a larger range with more people.

In contrast, the changes in subsistence associated with the Neolithic transition resulted in the development of more productive means to acquire resources in a much smaller range. Smaller ranges allow more people to live in a given area, increasing sedentism and carrying capacity while accommodating increasing population density. According to the allometric analysis above, the range of 1.08 km^2 exhibited by the Wana is what is predicted for a mammal the weighs 6.4 kg with a density of 24 individuals per km^2. This predicted density is greater than the 8.45 individuals/km^2 estimated for the density of the Wana but is within the correct order of magnitude. Although hunter-gatherer density varies, it is generally at least an order of magnitude below what we usually see with horticultural groups. Binford (2001) reports mean hunter-gatherer densities for a sample of 339 ethnographic cases living in 24 vegetative classes. Density within vegetative class ranges from 0.6 to 0.006 persons/km^2 with a mean of 0.18 persons/km^2. The Neolithic transition to food production allowed a constricting home range that accommodated a much higher population density than predicted for an animal with a human body mass.

Whether or not this shift was related to the proportion of meat in the diet— as suggested by Kelly's (1996) hunter-gatherer data—is unknown. It does seem very likely that the farmers' smaller range was related to their exploitation of localized food resources in the form of fields and husbanded animals. Whatever the cause, watershed changes in ranging behavior had a major impact on carrying capacity and subsequent social development. To my knowledge, no systematic study relating diet to ranging in farming groups has been done. To understand such changes, more high-quality data on ranging in farming groups and a cross-species comparative perspective are required.

ACKNOWLEDGMENTS

This research was funded by the Nature Conservancy, the National Geographic Society, the Wenner-Gren Foundation, the Indonesian Field Office of the Nature Conservancy, and the Faculty of Arts and Sciences at the State University of New York at Buffalo. I would like to thank Fred Rawski, Martarinza, Nurul

Winarni, Sharon Gursky, and Jabar Lahadgi for assistance in the field. I also thank Soeprapto Hadi Pranoto and Iksan Tengkow at the Indonesian Forestry Department (PHPA); Jatna Supriatna at the University of Indonesia; Marty Fujita, Nengah Wirawan, Mulyanto, Duncan Neville, and Totok Hartono of the Indonesian Field Office of the Nature Conservancy; and Dewi Soenariyadi at the Indonesian Institute of Science (LIPI). I am grateful to David Nolin for assistance with data management and analysis. I offer special thanks to the Wana of Posangke.

NOTE

1. Selective availability, a U.S. government scheme to degrade the GPS satellite signals for reasons of national security, was recently disabled by the government (May 1, 2000). Data before that date that were not differentially corrected were significantly compromised.

REFERENCES

Alvard, M.

2000a The Impact of Traditional Subsistence Hunting and Trapping on Prey Populations: Data from the Wana of Upland Central Sulawesi, Indonesia. In *Hunting for Sustainability in Tropical Forests*, edited by J. Robinson and Elizabeth Bennett, pp. 214–230. Columbia Press, New York.

2000b The Potential for Sustainable Harvests by Traditional Wana Hunters in Morowali Nature Reserve, Central Sulawesi, Indonesia. *Human Organization* 59: 428–440.

Alvard, M., and L. Kuznar

2001 The Transition from Hunting to Animal Husbandry: Prey Choice When Harvests Are Deferred. *American Anthropologist* 103(2): 295–311.

Alvard, M., J. Robinson, K. Redford, and H. Kaplan

1997 The Sustainability of Subsistence Hunting in the Neotropics. *Conservation Biology* 11: 977–982.

Alvard, M., and N. Winarni

1999 Human Disturbance and Avian Biodiversity in Morowali Nature Reserve, Sulawesi, Indonesia. *Tropical Biodiversity* 6: 59–74.

Atkinson, J.

1989 *The Art and Politics of Wana Shamanism*. University of California Press, Berkeley.

Bar-Yosef, O., and A. Belfer-Cohen

1989 Origins of Sedentism and Farming Communities in the Levant. *Journal of World Prehistory* 3: 447–498.

Binford, L. R.

1983   Long-term Land-use Patterns: Some Implications for Archaeology. In *Lulu Linear Punctuated: Essays in Honor of George Irving Quimby*, edited by R. Dunnell and F. Grayson, pp 27–53. Museum of Anthropology Anthropological Papers No. 72. University of Michigan, Ann Arbor.

2001   *Constructing Frames of Reference*. University of California Press, Berkeley.

Brown, J.

1985   Long-term Trends to Sedentism and the Emergence of Complexity in the American Midwest. In *Prehistoric Hunter-Gatherers: The Emergence of Cultural Complexity*, edited by T. Price and J. Brown, pp. 201–231. Academic Press, New York.

Byrd, B.

1994   Public and Private, Domestic and Corporate: The Emergence of the Southwest Asian village. *American Antiquity* 59: 34.

Cohen, M.

1977   *The Food Crisis in Prehistory: Overpopulation and the Origins of Agriculture*. Yale University Press, New Haven.

Damuth, J.

1981   Home Range, Home Range Overlap, and Species Energy Use among Herbivorous Mammals. *Biological Journal of the Linnean Society* 15: 183–193.

Fitzgibbon, C., H. Mogaka, and J. Fanshawe

1995   Subsistence Hunting in Arabuko-Sokoke Forest, Kenya, and Its Effects on Mammal Populations. *Conservation Biology* 9: 1116–1126.

Gittleman, J., and P. Harvey

1982   Carnivore Home Range Size, Metabolic Needs and Ecology. *Behavioral Ecology and Sociobiology* 10: 57–63.

Good, K.

1989   Yanomami Hunting Patterns: Trekking and Garden Relocation as an Adaptation to Game Availability in Amazonia, Venezuela. Ph.D. dissertation. University of Florida.

Haila, Y.

1988   Calculating and Miscalculating Density: The Role of Habitat Geometry. *Ornis Scandinavica* 19: 88–92.

Harestad, A., and F. Bunnell

1979   Home Range and Body Weight—A Reevaluation. *Ecology* 60: 389–402.

Harris, D.

1977   Settling Down: An Evolutionary Model for the Transformation of Mobile Bands into Sedentary Communities. In *The Evolution of Social Systems*, edited by J. Friedman and M. Rowlands, pp. 401–417. Duckworth, London.

Hofmann-Wellenhof, B., H. Lichtenegger, and J. Collins

1993   *GPS Theory and Practice*. 2nd ed. Springer-Verlag, Vienna.

Kelly, R.

1996   *The Foraging Spectrum*. Smithsonian Institution Press, Washington, D.C.

Kelt, D. A., and D. H. van Vuren

2001 The Ecology and Macroecology of Mammalian Home Range Area. *American Naturalist* 157: 637–645.

Kleiber, M.

1961 *The Fire of Life*. John Wiley and Sons, New York.

Kruyt, A.

1930 De To Wana op Ost-Celebes. *Tijdschrift Voor Indische Taal-, Land-, en Volkenkunde* 70: 398–625.

Lieberman, D.

1993 The Rise and Fall of Seasonal Mobility among Hunter-Gatherers: The Case of the Southern Levant. *Current Anthropology* 34: 599–632.

Lindstedt, S., B. Miller, and S. Buskirk

1986 Home Range, Time, and Body Size in Mammals. *Ecology* 66: 418–423.

McNab, B.

1963 Bioenergetics and the Determination of Home Range Size. *American Naturalist* 97: 133–140.

Milton, K., and M. May

1976 Body Weight, Diet and Home Range Area in Primates. *Nature* 259: 459–462.

Nunn, C., and R. Barton

2000 Allometric Slopes and Independent Contrasts: A Comparative Test of Kleiber's Law in Primate Ranging Patterns. *American Naturalist* 156: 519–533.

Pataki-Schweizer, K.

1980 *A New Guinea Landscape*. University of Washington Press, Seattle.

Peters, R.

1983 *The Ecological Implications of Body Size*. Cambridge University Press, Cambridge.

Price, T., and J. Brown

1985 Aspects of Hunter-Gatherer Complexity. In *Prehistoric Hunter-Gatherers: The Emergence of Complexity*, edited by T. D. Price and J. A. Brown, pp. 3–20. Academic Press, New York.

Reiss, M.

1988 Body Size and Home Range Area. *Trends in Ecology and Evolution* 3: 85–86.

Robinson, J., and K. Redford

1991 Sustainable Harvest of Neotropical Animals. In *Neotropical Wildlife Use and Conservation*, edited by J. Robinson and K. Redford, pp. 415–429. University of Chicago Press, Chicago.

Schoener, T. W.

1968 Sizes of Feeding Territories among Birds. *Ecology* 49: 123–141.

Schweithelm, J., N. Wirawan, J. Elliot, and A. Khan

1992 *Sulawesi Parks Program Land Use and Socio-economic Survey Lore Lindu National Park and Morowali Nature Reserve*. Report commissioned by the Nature Conservancy, Jakarta, Indonesia.

Stone, G.
1996   *Settlement Ecology: The Social and Spatial Organization of Kofyar Agriculture*. University of Arizona Press, Tucson.
Swihart, R., N. Slade, and B. Bergstrom
1988   Relating Body Size to the Rate of Home-Range Use in Mammals. *Ecology* 69: 393–399.
West, G. B., W. H. Woodruff, and J. H. Brown
2002   Allometric Scaling of Metabolic Rate from Molecules and Mitochondria to Cells and Mammals. *Proceedings of the National Academy of Sciences* 99: 2473–2478.
White, G., and R. Garrott
1990   *Analysis of Wildlife Radio-Tracking Data*. Academic Press, New York.
Whitten, A., M. Mustafa, and G. Henderson
1987   *The Ecology of Sulawesi*. Gadja Mada University Press, Bulaksumur, Indonesia.
Wright, G.
1978   Social Differentiation in the Early Natufian. In *Social Archaeology: Beyond Subsistence and Dating*, edited by C. Redman, pp. 201–223. Academic Press, New York.
WWF (World Wildlife Fund)
1980   *Cagar Alam Morowali: Suatu rencana pelestarian*. Report for Indonesian Forestry Department, Bogor, Indonesia.

6

Forager Landscape Use and Residential Organization

RUSSELL D. GREAVES

The archaeological study of settlement patterns involves unique interpretive difficulties. Survey provides regional data with coarse-grained information regarding individual component sites. Excavation data from sites rarely can be securely linked to other sites. Inferences about the systemic relationships between identified archaeological sites rely on assumptions about the meanings of variability in an array of remains. These different classes of materials may carry unique taphonomic histories that do not intuitively implicate residential systems. Additionally, the archaeological view of settlement is a nonintuitive abstraction at a scale ambiguously related to ethnographic observations.

These complexities are challenges to the development of methods that can link archaeological sites and regions into informative views of past land-use patterns. Recent improvements in survey methods have addressed the use of sampling strategies, remote sensing (Conyers and Goodman 1997: 137–147), and increasing appreciation of the intricacies of regional dynamics of settlement (Schiffer 1987: 100–120). Especially critical is the recognition that the landscape record represents accretions of cultural material that may not be analytically equivalent to recognizing discrete human occupations (Camilli and Ebert 1992:114; Dewar 1986; Thomas 1975).

Ethnoarchaeological studies of settlement have provided vital clues about the complexities of settlement adaptations in different environments. They have demonstrated that simple analogies from observable behaviors will not sum to archaeological patterns (Fisher 1995: 188–191, 196–200; Yellen 1977:103). Seasonal variations in reoccupation trajectories do not produce equal archaeological visibility of annual residential components (Brooks and Yellen 1987: 86–88). Even in systems with highly redundant site use, diagnostic inferences about site roles may come from intersite comparisons, not intrasite assemblage analyses (Graham 1993: 29–39). Systemic views of past regional adaptations can productively combine ethnoarchaeological observations of relevant variability with archaeological inquiry (Binford 1980, 1982, 1983). Ethnoarchaeology provides opportunities not only to examine the range of realized human settlement responses but to explore relationships between strategies that result in identifiably contrasting systemic archaeological signatures.

This study of Pumé settlement and subsistence mobility provides observational data on dramatically variable residential, food-gathering, and support activities within a single system. Twenty-four months of observation offer a longitudinal view of seasonal responses of Pumé foragers. Characteristics of settlements, food remains, in-camp processing, and manufacture activities provide potential links with the archaeological record. Some aspects of these data are case specific; however, the relationships examined have direct relevance to existing archaeological analytical methods and procedures for inferring past human behaviors.

STUDY LOCATION

The Pumé (Yaruro) are a foraging people living in the savannas (llanos) of southwestern Venezuela (Figure 6.1; Besnerais 1948, 1954, 1962; Leeds 1960, 1961, 1964; Mitrani 1975, 1988; Petrullo 1939). A population of 5,400 Pumé (OCEI 1995a: 32) live in the state of Apure primarily along the Capanaparo, Riecito, and Cinaruco Rivers and in the savanna between these drainages (Mitrani 1988:

Figure 6.1. Location of Pumé population in Venezuela.

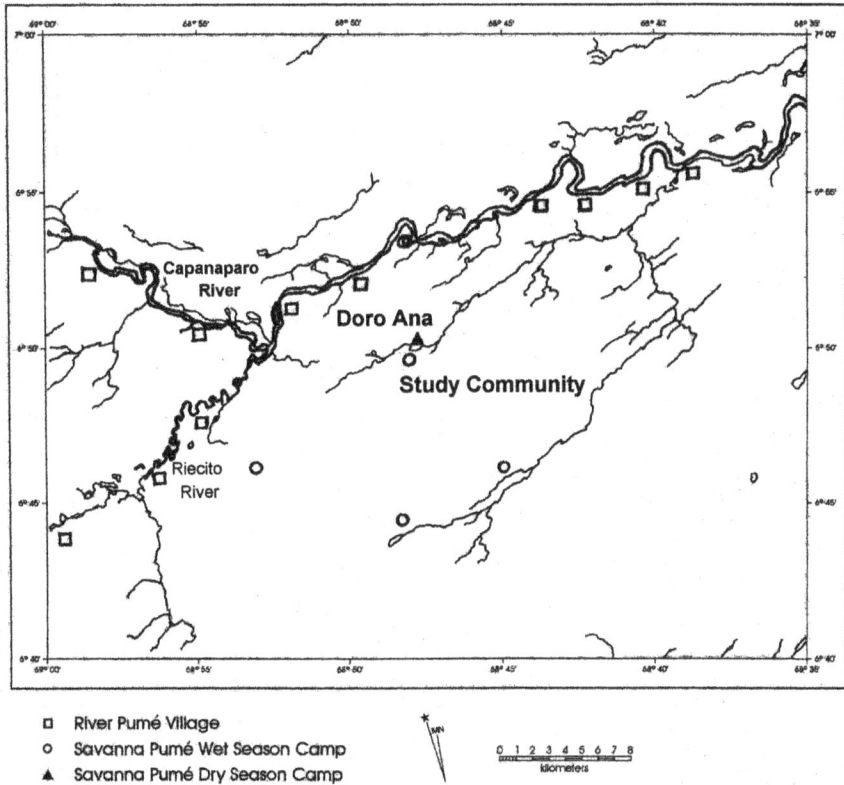

Figure 6.2. Location of Doro Aná study area and neighboring Pumé communities.

Map 1). Pumé subsistence economy is based on hunting, fishing, manioc horti-culture, and wild root collection. Communities adjacent to major drainages are large permanent villages that are reliant on agriculture, wage labor, and animal husbandry in addition to wild food resources. Savanna Pumé communities are smaller, more dispersed, and most dependent on hunting, fishing, and wild root collection (Figure 6.2).

Ethnoarchaeological research was conducted during 1990 and 1992–1993 in the savanna Pumé community of Doro Aná, located south of the Capanaparo River (Gragson 1989; Greaves 1997). In 1992–1993 the village contained ap-proximately 22 adult women, 23 adult men, and 22 children. Investigations fo-cused on collecting time-allocation data on mobile activities. Travel distances, resource encounters, tool use, and other data were recorded routinely as part of these behavior samples. Subsistence data for mobile foraging trips by adults and teenagers are presented in Table 6.1.

Table 6.1. Female and Male Subsistence Trips, 1992–1993

Resource trip	Sample size	Mean distances (m)	Mean food returns (kg)	Mean total trip time (min)
Female				
mango (D)	27	16,158	29.15	311
wild fruits (D)	17	3,640	2.20	114
small roots (W)	21	1,569	3.33	156
large roots (W)	25	5,647	11.75	293
dry season	44	11,321	18.74	235
wet season	50	3,934	8.21	236
all trips	94	7,391	13.14	236
Male				
bow and arrow fishing (D)	26	5,168	1.93	203
hook and line fishing (D)	17	5,739	2.58	288
total fishing (D)	43	5,394	2.19	237
hunting (W)	25	11,732	2.18	230
all trips	68	7,724	2.18	235

Note: D indicates dry-season activity; W indicates wet-season activity.

The Venezuelan llanos (plains) are a seasonally flooded grassland savanna that is part of the depression of the Apure Basin. The llanos range from 40 to 80 m above sea level. These grasslands are maintained by edaphic controls, fire, and hyperseasonal variation in rainfall (Foldats and Rutkis 1965; Sánchez et al. 1985; Sarmiento and Monasterio 1975: 239; Silva and Sarmiento 1976). Mean daily temperatures vary little between the six-month wet and dry seasons. Rainfall data collected at the study site indicate that 78% (1,404 mm) of the total annual precipitation (1,805 mm) occurred during the wet-season months of May through October (Figure 6.3).

This is a hyperseasonal savanna characterized by extreme variation in rainfall (Blydenstein 1967:1–2; Bulla et al. 1980; Ramia 1959:21; Sarmiento 1984; Troth 1979). River levels change markedly in relation to the pronounced seasonality of rain (Novoa 1989: Figure 2; Zinck 1982: Figure 8). Plant and animal responses to seasonal fluctuation in rainfall establish a biseasonal pattern of subsistence responses by the Pumé. Fishing, as in many parts of South America, is a dry-season activity for the Pumé, when fish are concentrated in restricted pools and small segments of streams (Gragson 1989, 1992a, 1992b; Leeds 1961; Mitrani 1988). Women collect roots during both the dry and wet seasons. Wild roots are critical resources throughout the year (Gragson 1997: 382, Tables 1–2; Greaves 1997: Table 6.1). Some fruits are gathered during the dry season. Except for mangos, available primarily near the Capanaparo River, fruit does not constitute

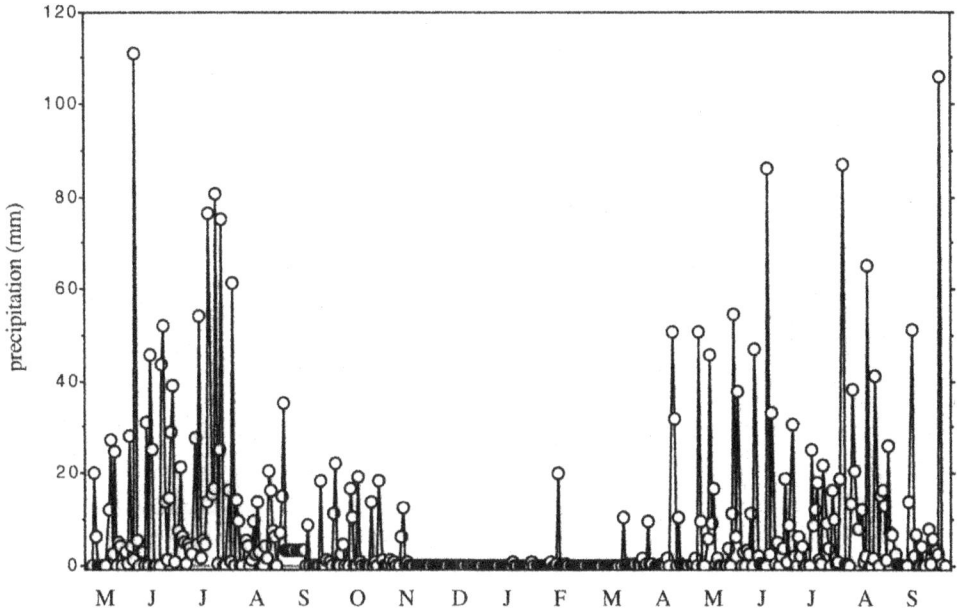

Figure 6.3. Precipitation at Doro Aná study site, 1992–1993.

significant dietary bulk or energetic effort. Hunting is performed during the wet season (Gragson 1989; Mitrani 1988). Agricultural work is done by both sexes. Garden preparation can be a significant component of time allocation in the late dry season. The majority of agricultural labor and harvesting takes place during the wet season. Manioc (*Manihot esculenta*) is the only important crop. The highest wild root returns also were collected during the agricultural season.

RESIDENTIAL MOBILITY AND SETTLEMENT PATTERN

The Pumé practice several kinds of residential mobility. The extreme seasonality difference in the llanos creates marked contrasts in the plant and animal dynamics. This seasonality is associated with distinct shifts in Pumé subsistence activities and is responsible for a biseasonal settlement pattern (Gragson 1989: 287–288; Sanoja 1961). Separate villages are occupied for approximately six months as central-place bases during the dry and wet seasons. Dry-season camps are located adjacent to streams to take advantage of water and proximity to fishing locations (Besnerais 1948:14; Chaffanjon 1986:159; Gragson 1989:299–301, Mitrani 1988: Photo 10; Petrullo 1939: 211, Plates 12:2, 14, 15:1, 15:2). Architecture at dry-season camps (Figure 6.4) consists of brush sun shades and small circular or rectangular thatched structures that protect against the infrequent

Figure 6.4. Doro Aná dry-season architecture, 1993.

rain (Besnerais 1948: 14; Chaffanjon 1986: 159; Gragson 1989: 299–301; Mitrani 1988: Photo 10; Petrullo 1939: 211, Plates 12-2, 14, 15-1, 15-2). Dry-season habitations usually contain only nuclear family units.

Dry-season camps were situated within 20 m of this stream on its floodplain. While stream flow is intermittent in this caño, or drainage, during the dry season, isolated pools persist even during the periods of minimal rainfall (December–January). The proximity to fishing locations has been considered the most important reason for placing dry-season camps near streams. However, camps near streams take advantage of water that is unavailable away from these ephemeral sources. The timing of moves to dry-season camps is dictated by women's decisions regarding the decreasing availability of water at wet-season camps and when it is procurable near potential dry-season camp locations. The locations of dry-season base camps change every year.

Wet-season camps were situated away from these streams. They were located on higher ground above the maximum levels of savanna flooding. Wet-season camps were located approximately 750–1,000 m from the Caño el Morichal. Wet-season houses are large, sturdy, rectangular, thatched habitations that often contain extended families (Figure 6.5; Besnerais 1948; Gragson 1989: 291–296, 332–335; Grossa 1966: 81; Mitrani 1988: Photo 8). The Doro Aná Pumé move their camps from areas that are above annual flooding of the savanna to locations adjacent to the ephemeral stream (Caño el Morichal) as the water table drops so that well water is not available near to camp. Wet-season camps are reoccupied

Figure 6.5. Doro Aná wet-season camp, 1990.

for two or more years, probably because of the greater architectural investment. These campsites also were reoccupied after short abandonments (one to two years). Reoccupations of camps used the previous year always were associated with rebuilding and reorientation of at least one, and usually several, of the houses already existing at that camp.

Many short (a few days to three weeks) or long-term (up to two months) residential moves are made from these seasonal central-place base camps. All of these residence locations may have multiple occupations within one or more seasons. At any time of the year small groups of individuals or whole families visit nearby and distant Pumé communities from one day up to two weeks.

Wet-Season Camps

In 1990 three different wet-season base camps were occupied. These can be characterized as main base camps and garden camps. Architecture and site structure are identical at both of these locations. The main camp is where most of the wet season is spent, and garden camps are occupied for one to two months. The main camp was abandoned by the entire community for approximately one month to allow regular agricultural work in gardens 2 km away. Two separate garden camps were occupied in 1990. Once back in the main camp, segments of the population made periodic visits back to one of these garden camps for additional garden work. Similar uses of different base and garden camps were observed in 1992–1993.

A single wet-season main camp was occupied in 1992 and 1993. This camp location was the 1991 main wet-season camp but not the 1989–1990 wet-season base camp. A one-month visit was made to the garden camp (1990 wet-season

garden camp) for additional garden work. Figure 6.2 indicates the position of
the 1992–1993 main wet-season camp.

From 1990 until 1993 four different wet-season camps were used. Two of these
were short-term garden camps with at least four separate occupations. The two
main wet-season camps had at least five occupations. Additionally, in 1990 another village community moved nearby and built two wet-season camps within
700 m of the Doro Aná villages and one dry-season camp approximately 1 km
from the Doro Aná dry-season camp. That wet-season camp was abandoned in

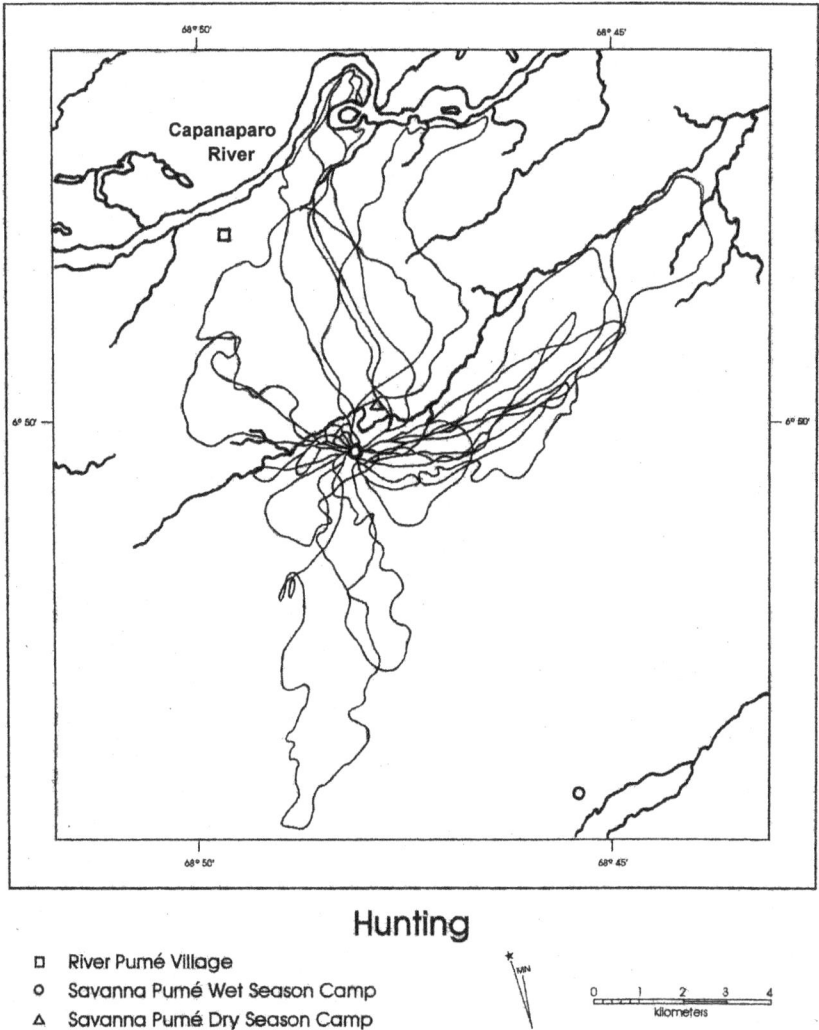

Hunting

□ River Pumé Village
○ Savanna Pumé Wet Season Camp
△ Savanna Pumé Dry Season Camp

Figure 6.6. Hunting trips by Pumé men, 1992–1993.

1991. Due to a dry-season fire, many of the architectural elements were never borrowed or "robbed."

Dry-Season Camps

For the period from January 1990 to September 1993 the Doro Aná Pumé occupied four different dry-season base camps, with one occupation each. Separate locations were used as dry-season base camps for each year from 1990 to 1993. Temporary camps were established for up to two weeks to perform intensive

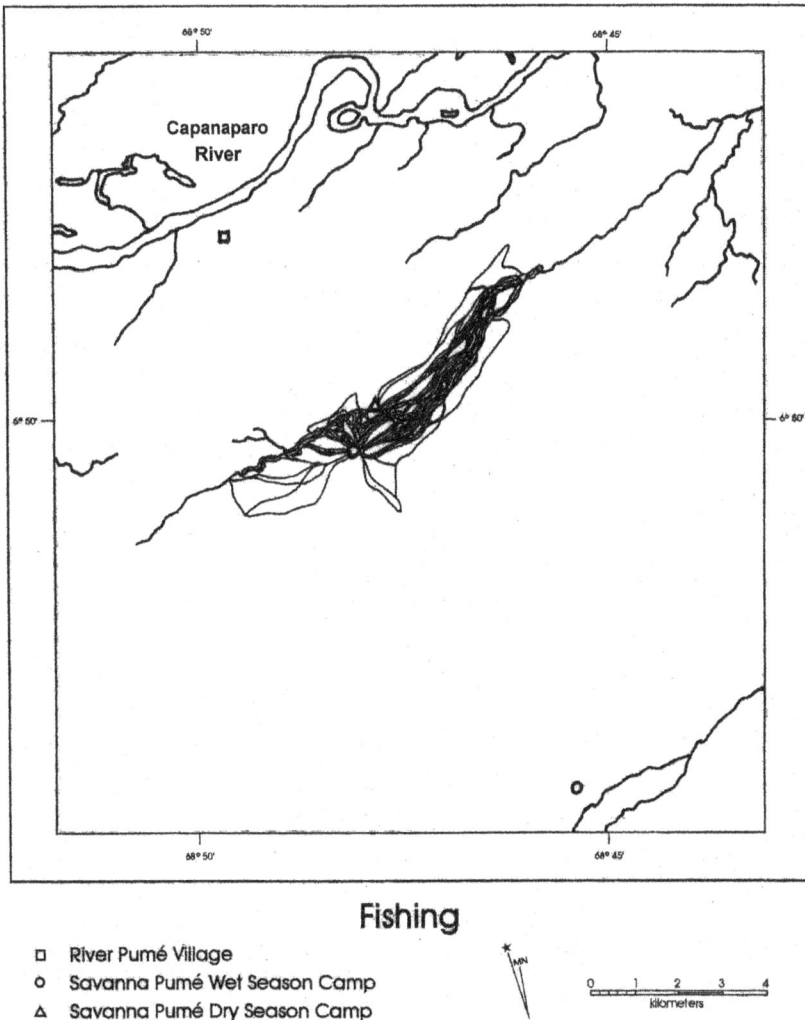

Fishing

□ River Pumé Village
○ Savanna Pumé Wet Season Camp
△ Savanna Pumé Dry Season Camp

Figure 6.7. Fishing trips by Pumé men, 1992–1993.

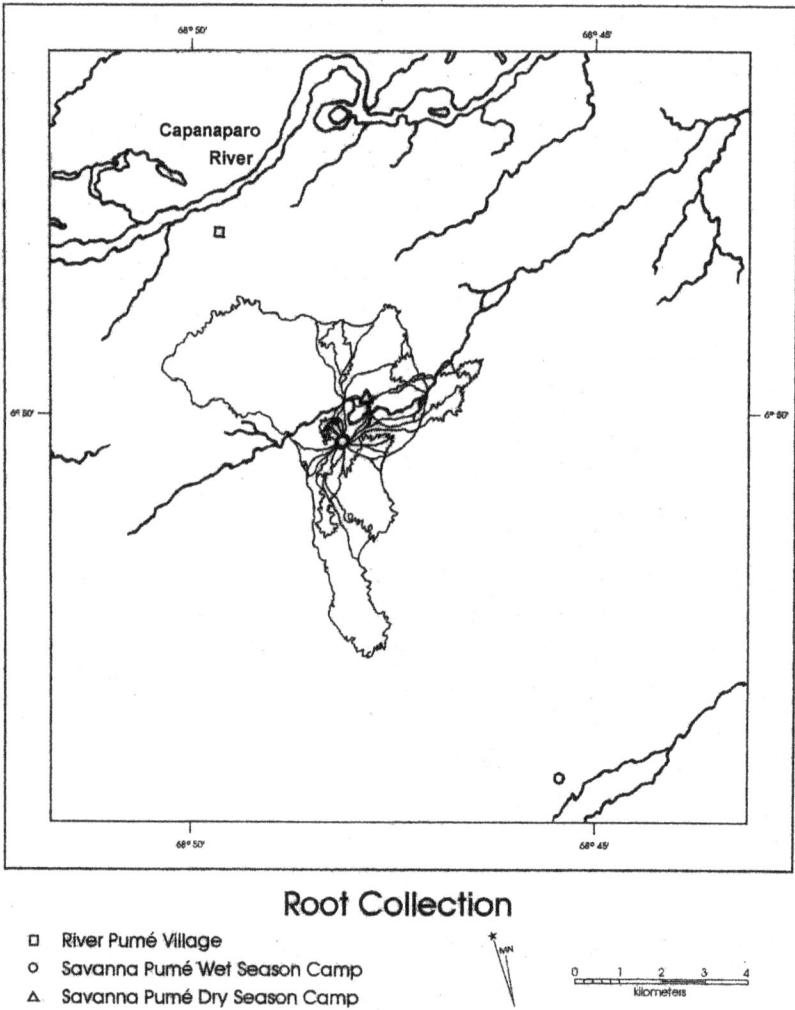

Root Collection

□ River Pumé Village
○ Savanna Pumé Wet Season Camp
△ Savanna Pumé Dry Season Camp

Figure 6.8. Wild root collection by Pumé women, 1992–1993.

fishing. Unlike the wet-season garden camp occupations, no events were witnessed where the entire population abandoned the dry-season main camp. The location of the 1989–1990 dry-season camp was observed, but no dry-season camp was observed before fieldwork ended in December 1990. In 1992 the dry-season camp was situated at a different location from those used in 1989, 1990, and 1991. The 1992–1993 dry-season camp was a location that had not been used during previous fieldwork or in 1991. Figures 6.6–6.8 indicate the locations of three dry-season camps used from 1991 to 1993.

Residential mobility within the habitual range of the Doro Aná Pumé was observed in 1990 and 1992–1993. Reliable informant data can be combined with data on the same community from Theodore L. Gragson (1989: Table 7.1, appendix B) to reconstruct habitation locations from the wet season of 1986 to the wet season of 1993. During these seven years three different main wet-season camps were used. Six different dry-season base camps were occupied during that same period. Additionally, a village community usually living approximately 12 km southwest of Doro Aná occupied one main wet-season and dry-season camp within the habitual use area of the study group during 1990. Figure 6.7 shows the location of three dry-season temporary fishing camps observed in 1992–1993. One or two temporary wet-season garden camps and dry-season fishing camps were occupied during each of those years. Most of these temporary camps experienced two or more occupations. A few dry-season camps were used for single evenings or for less than three days. I observed only one overnight camp, occupied during an overnight bird and caiman hunting trip.

SUBSISTENCE MOBILITY

Mobility associated with subsistence activities can be described as primarily central-place foraging. Few trips took more than one day. During the dry season, residential mobility is associated with periodic intensive fishing bouts focused on one lagoon and with palm raw material extraction. The young leaf shoots of moriche palm (*Mauritia flexuosa*) are the most important raw material collected during the dry season. Men may also collect and tiller new palm wood bow staves at this time of the year, primarily from *Astrocaryum jauari*. Wet-season residential mobility was for the purpose of occupying temporary garden camps.

Hunting

Pumé men hunt during the wet season (Gragson 1989; Greaves 1997; Mitrani 1988). The switch away from aquatic resources appears to be due to dispersal of fish in the flooded savanna. Although the characterization of the Pumé as a fishing culture (Kirchoff 1948: 456) is a simplification of their biseasonal subsistence, it is essentially accurate. Fishing is the most productive male protein-acquisition activity and is not practiced during the wet season only because of the impracticability of pursuing the dispersed fish. Terrestrial game in the llanos is not abundant (August 1984; Comerma and Luque 1971; Emmons 1984; Medina 1980: 311; O'Connell 1989; Ojasti 1983: 431). When available, fish provide a more reliable food source. Terrestrial hunting involves much greater search efforts but similar risk compared with fishing: 7 of 25 hunting trips in

1992–1993 provided no food returns, while 8 of 26 fishing trips in 1992–1993 had no catches.

The mean round-trip travel distance for hunting trips was 11,409 m. The total hunting area used during focal observation follows was 432 m² (Figure 6.6). This is the largest area used by the Pumé to search and obtain any class of subsistence foods. High search mobility is related to a very dispersed terrestrial biomass. During the 24-month investigation, only one capybara, two deer, six anteaters, and seven caimans were captured. Pumé hunting targets small animals, approximately 0.5–1.5 kg. Armadillos, tegu lizards, and small teiid lizards weighing less than 100 gm represent 79% of captured game for this sample of 1992–1993 hunting trips (a total of 47 kills on 25 trips). Small-animal hunting results in very low mean return weights for Pumé hunters. Men averaged 2.18 kg per trip, ranging from 0 to 11.25 kg. Only one trip's hunting return exceeded 6.27 kg.

Pumé terrestrial hunting emphasizes relatively high search mobility because of the low encounter rates with game. Search efforts are positively correlated with the variety and amount of game returned to camp (Greaves 1997: 301, Figure 3). Travel distances for hunting trips and the area exploited (Figure 6.6) are significantly higher than for fishing or wild plant food collection. Hunting use areas are larger and exploit more different kinds of habitats than any other subsistence trips.

Fishing

Fishing is the main source of protein during the dry season (Gragson 1989, 1992b; Mitrani 1988). Pumé take a wide variety of fish and other aquatic fauna using bows and arrows, hooks and lines, piscicides, and dams (Gragson 1992b: 117–122; Leeds 1962: 600–601). Pumé men practice bow and arrow intercept fishing from trees and platform facilities. Mean fishing returns are statistically undifferentiated from hunting returns. For 26 bow and arrow fishing trips, the mean food return was 1.93 kg. The mean round-trip distance was 5,168 m. For a sample of 265 catches the mean individual captured fish weight was 188 gm. Hook and line fishing is now commonly practiced. Hook and line mean returns were 2.58 kg. Although these individual catches are undifferentiated from returns, more men in camp were engaged in fishing each day, and they went fishing almost every day. Hook and line travel distances show a bimodal distribution, with a mean of 3,088 m as the round-trip distance of short trips and 9,526 m for longer trips. The mean fish size for 440 hook and line catches was 100 gm. Temporary fishing camps (Figure 6.7) were occupied for two to three weeks during the dry season. Cooperating individuals formed camps adjacent to fishing locations to take advantage of temporal blooms in fish availability and to reduce travel time to these periodically productive patches. The total area used during focal fishing observations was 70 m². This is the smallest resource use area for any food source other than gardening. Although fishing has longer mean trip distances than root gathering, the exploitation area is smaller because these resources are restricted to the stream and portions of the flooded savanna.

Travel distance is not important in searches for fish, involving nonmobile search time spent exploiting resource patches (Figure 6.7). Time spent at these patches does not correlate with numbers of fish caught or with fish weight. Productivity of fishing locations is dependent on the variable trajectories of fish life cycles and migration, not on time spent at fishing patches. Men fishing cannot increase search time and assure successful returns in proportion to that investment. Exploitation of fish was characterized by the use of more alternative technologies than in wild root collection. Fishing sampled and abandoned unproductive patches more frequently than women did on root collection trips.

Wild Root Gathering

Collection of wild roots provides major portions of the Pumé diet. Wild roots were a very important food even during the agricultural season. Wild root collection involved significantly lower total round-trip distances (Figure 6.8) than male resource acquisition trips. The intensive physical activities involved in excavation and carrying of wild root resources, however, represent a considerable energetic effort even under conditions of lower mobility. The mean round-trip distance of trips to gather small roots was 1,636 m. These roots were most frequently collected during the dry season. Gathering of large roots involved significantly higher mean round-trip distances: 5,602 m. Large roots were gathered from patches found in open, flooded savanna, during the wet season. For 20 person trips that targeted the collection of small roots the mean total weight of roots collected per trip was 5.25 kg. The mean return weight of large roots was 11.83 kg for 28 trips. The total area used during all root collection focal follows was 127 m² (Figure 6.8). Although the average individual trip distances indicated lower individual trip mobility for women's root collection than for hunting or fishing, gathering trips investigate and exploit a larger total resource area than fishing does.

Root trips, with shorter travel distances than hunting or fishing trips, concentrated on a small spectrum of food species, using one exploitation technique to access those resources. Pumé women's knowledge of local conditions and seasonal variation in root condition allowed them to select particular patches with a higher frequency of successful returns. Even if the productivity of those patches was below expectations, time spent in such a patch still yielded proportional returns on search and collection time. Women did not engage in alternative exploitation techniques when they encountered root patches that produced fewer roots than anticipated. Unlike both hunting and fishing, no root collection trips registered zero returns. In comparison with the relatively low returns from hunting and fishing, this high predictability of root returns may explain the persistence of wild root collection in Pumé subsistence.

ARCHAEOLOGICAL IMPLICATIONS OF SETTLEMENT AND SUBSISTENCE MOBILITY

The repetitive reoccupation of a limited number of appropriate wet-season camp locations is dictated by available areas above the level of maximum savanna flooding. These locations also are selected for proximity to patches of forest appropriate for swidden agriculture. The camps are associated with the highest distance subsistence mobility by men and women. Hunting search areas are the largest territorial use during any season. Women's wet-season root-gathering trips involved longer distances and a greater total area exploited than in dry-season root collection. Dry-season camps located at different areas of stream banks each year are associated with much lower overall territorial mobility. Areas exploited for fishing are restricted almost exclusively to the corridor along the creek and around one lagoon.

These patterns result in an association of highest architectural investment with greatest subsistence mobility. More abbreviated architecture and greater camp mobility are associated with lower total territorial use. This is counterintuitive to many archaeological inferences regarding the relationships between structural investment and subsistence behavior (Hunter-Anderson 1977: 290–291). It emphasizes the behavioral independence of different classes of material remains commonly used in archaeology as overall systemic indicators. These co-associations are not presented as warning tales. The subsistence implications of seasonal territorial use would be available archaeologically from faunal and ethnobotanical remains. Robust wet-season architecture is a response to climate, not systemic investment in a more sedentary system. Paleoethnobotany, stable isotope analyses, insect remains, gastropods, soil analysis, and geomorphological data from archaeological sites are commonly used for climatic reconstructions. In combination with architectural remains, features, subsistence residues, and technological debris, climate and seasonality information provides a strong inferential framework for the interpretation of archaeological deposits.

The dynamics of site reoccupation can be read in the patterns of site placement and the taphonomic record of spatial reorganization. The resulting archaeological landscape shows a redundancy in the Pumé use of appropriate wet-season camp locations. Robust evidence of camp reuse is available from posthole patterns that document multiple reconstruction events of households and other camp features. Dry-season camps do not redundantly locate at the same places each year. More dry-season sites are created that contain lower-density archaeological debris. Dry-season camps are restricted in their occurrence to a narrow corridor associated with stream locations and periodic use of one lagoon. Many of these dry-season camps will show limited evidence of reoccupation and restructuring of camp space. In combination with subsistence and environmental data, variations in the site locational associations and taphonomic dynamics are important clues about organizational responses to seasonality.

These camp areas also have functional uses during the seasons when they are not inhabited. Caching is one common response to mobility where unoccupied sites are still active parts of a settlement system. Pumé wet-season camps obviously offer greater caching opportunities, because they contain more robust houses and are maintained as base camps for two to three years at a time. They represent a "bank" of architectural elements that may be robbed, especially at the end of the dry season, as infrequent rains necessitate construction of some rainproof shelters. Archaeologically, recycled architectural wood use has been identified in contexts with good organic preservation (Dean 1969: 74–77) and in the old wood "problem" in radiocarbon dating (Bowman 1990: 51). If reoccupation of wet-season camps was anticipated, then most of the building elements and thatch were left in place. Dry-season camp architecture was more thoroughly robbed by the middle of the wet season.

Some tools are kept in wet-season camps throughout much of the dry season. Most of the dry-season camps are inundated by 50–100 cm of water during much of the wet season and have lower potential as cache locations. They do serve as caching locations as water levels rise during the early wet season. Most storage of gear in old houses takes place in the more substantial wet-season houses. Commonly used tools such as cooking pots, grinding stones (whetstones), hammers, and bottles were frequently left in camps for up to two months after camp moves. Most of these, except bottles, were retrieved within the first month after camp relocation. Cached tools left in houses for the full six months of abandonment were smaller, more replaceable tools. The most common cached items were fire drill sets. Dry-season camps had more limited and specialized gear caches. Two abandoned dry-season camps that were above most inundation were observed to serve as caching locations during two wet seasons. One old man kept rare wooden-tipped arrows as a backup gear cache in the event of situational encounter with birds. Caching adjacent to old dry-season camps was commonly noted for caiman-hunting paraphernalia. A wooden harpoon shaft was cached at an old dry-season camp (Greaves 1997: 298), and moriche palm raft segments also were left near such camps.

All camp locations represent landscape points that are used for a variety of activities during mobile foraging. Abandoned camps are frequently located on trails and serve as convenient stopping places for rest, readying of tools, or eating. Old wet-season camps are frequently visited during hunting trips to obtain melons as field snacks. These grow from volunteer seeds in middens or from informal cultivation of midden gardens during wet-season occupation. None of the midden gardens are tended during the dry season. Men occasionally select old camps as locations for field snacking on a portion of hunting returns. Field snack debris could readily be identified archaeologically as a different taphonomic disposal of bones than in habitation faunal remains, even in the absence of seasonal indications. Low hunting returns and emphasis on small animals result in dispersal even of small game at base camps from extensive sharing. Field snacking results in the localized discard of all skeletal elements. Even destruction of some of these bones by dogs produces a nonhabitation signature of relative completeness. Field snacks exhibit burning of many elements from roasting. Possibly in relation to conserving fat from low protein inputs, habitation cooking emphasizes boiling of fauna and fish. Roasting is much rarer at base camps. Old camps often served as stopping points during mobile subsistence trips. Women did occasionally visit old camps, but field snacking during gathering trips was not observed.

The archaeological result of the biseasonal Pumé residential mobility is a robust record of wet-season occupations in a limited number of locations. Two camps used twenty years ago are still visible, and one of these locations is habitually visited as a stopping point along a major trail. Dry-season camps are practically invisible after only five years. These camps are located all along the Caño El Morichal and represent a more spatially extensive record. Infrequent use of these camps does occur during the dry season. Visitation during nonresidential seasons adds some faunal remains, removes architectural elements, and moves cached tools and raw materials. Identification of seasonality from faunal remains is possible with taphonomic approaches to sampling and interpretation. Architectural and locational differences indicate tactical contrasts in regional use. The links between these residential manifestations and their systemic roles are not encoded in any single class of archaeological data. Inferences from multiple evidence such as faunal analysis, seasonality, "old wood" juxtaposition, architectural investment, and climatic reconstruction are required. Combined patterning from different classes of archaeological remains is necessary to provide clues about landscape use strategies.

CONCLUSION

The Pumé are a group of subsistence foragers living in a hyperseasonal savanna in southwestern Venezuela. They practice central-place foraging from wet-season and dry-season camps that are occupied for six months. Their subsistence is based on hunting of armadillos and teiid lizards, manioc horticulture, and the collection of wild roots during the wet season. Dry-season subsistence is dependent on fishing, wild fruit gathering, and mango collection. These different subsistence activities are associated with separate village locations during each of these seasons. Architecture and some aspects of camp structure are very different in each of these seasonal residences. These dramatically variable seasonal responses make Pumé adaptations an ideal ethnoarchaeological study group.

Data on subsistence mobility in relation to each seasonal occupation provide quantified views of areal use that can be identified through archaeological techniques. Recovery and analytic methods addressing food remains are among the best-developed aspects of prehistoric research. This makes subsistence resources the most useful ethnoarchaeological anchoring point for linking behaviors in other, less well understood systems. Terrestrial fauna, fish, and wild roots are all classes of remains that can be identified from archaeological sites. Especially notable are innovations in analyses of phytoliths (Kealhofer et al. 1999; Lalueza Fox et al. 1996; Piperno 1988; Piperno and Pearsall 1998), starches (Cortella and Pochettino 1994; Piperno and Holst 1998), and residues (Evershed et al. 1991; Hastorf and DeNiro 1985; Kobayashi 1994) and use-wear (Jahren et al. 1997; Jensen 1994; Kealhofer et al. 1999) on tools. Taphonomic approaches to fish (Casteel 1976; Lyman 1994: 435–445; Stewart 1989: 78–98; Weigelt 1989: 137–147; Wheeler and Jones 1989: 62–78) and root foods (Ford 1988: 217; Minnis 1978; Mulholland 1989: 508; Pearsall 1988; Piperno 1988: 131–167; Rhode and Madsen 1998: 1203–1205) are currently less developed than those for terrestrial fauna. This research provides data on plant food collection and processing that are relevant to paleoethnobotanical taphonomic studies. Pumé fishing data also provide behavioral information on exploitation of aquatic resources.

Inferences about past subsistence mobility can integrate identification of food remains with their past distribution in paleo-landscapes. The importance of food acquisition and robust analytic knowledge of the taphonomic implications of archaeological assemblages make this a significant comparative anchor. Many archaeological patterns are not unambiguously related to documented subsistence differences. Nonetheless, this strategy can link data from well-studied classes of archaeological remains to the development of better inferences about more obstinate interpretive problems.

Pumé settlement mobility is directly responsive to seasonality, but it is not brokered by simple relocations relative to food resources. Dry-season habitations are located near aquatic resources, due to the need for access to water. That important result of this study has not received adequate archaeological attention. Modern technology (plastic buckets and aluminum pots) for hauling water makes no difference to women who consider distances of 700 m too great for their daily needs. The importance of water in tethering mobility (Taylor 1964) is a critical component of this tropical settlement system. Pumé well features are present at both dry-season and wet-season camps. Wells can be readily visible archaeologically (Meltzer 1991). Spatial associations combined with sedimentological and soil manifestations of water table dynamics make wells critically important features in understanding regional environmental and settlement dynamics (Meltzer 1991). Similarly, wet-season camps are located primarily in relation to areas that are above savanna flooding, not in relation to proximity to food targets. Because Pumé seasonal camps are moved only short distances, it should be apparent that site location is not dramatically changing access to most necessary resources. Seasonal residential mobility must be recognized as potentially responsive to tactical decisions about a variety of contrasting opportunities.

Architectural investment is not directly related to overall subsistence mobility. Pumé dry-season camps contain brush shelters and shades. This is directly determined by climatic conditions, not mobility. Subsistence mobility during the dry season is much lower than both women's and men's mobility during the wet season. Sturdy wet-season architecture must withstand dramatic thunderstorms and nearly daily showers. The highest subsistence mobility is associated with hunting and distant root collection during this season. Although the Pumé have a very evenly divided annual residential pattern, the archaeological effects are not simple. Wet-season campsites are restricted to a limited number of appropriate geomorphic settings. Dry-season camps have some reoccupation overlap but are located in different places each year. Leaving aside problems of archaeological visibility for the moment, survey would identify more dry-season than wet-season camps. Excavation, however, could readily identify dramatic differences in the amount of site structural modification. Postholes in wet-season camps show extensive reconfiguration of houses and dance plaza facilities across decades. Because dry-season camps are established at a greater number of locations, evidence of changing site structure from more punctuated reoccupation would be less robust.

The relationships in these examples of Pumé subsistence mobility, camp location, residential mobility, and settlement characteristics are not archaeologically obscure. They are interpretable when it is recognized that behavior is dynamic and highly variable (Wandsnider 1992:286). There is no *a priori* reason why our initial expectations regarding which classes of archaeological data are most informative about research goals are relevant even if they seem logical. Compelling analytical relationships must be sought for inferences about past settlement systems from archaeological data. Locational data do not inherently provide a view of past settlement landscapes. Relationships in one archaeological record may not identify residential organization in another. It is currently well appreciated that artifact debris can have complex taphonomic histories that make simple and direct interpretation problematic. The taphonomy of landscape records can be much more complex. In conjunction with advances in material studies, survey, and excavation methods, ethnoarchaeology offers two important contributions. Observations of modern peoples provide data on behavioral variability that are obscure in the ethnographic literature and often tautological in archaeological interpretation. Ethnoarchaeology also seeks links between multiple classes of data that are available from the archaeological record and their relationships with significant systemic dynamics.

Ethnoarchaeological data do not distill classes of human mobility that can be generally applied in archaeology. The scale of ethnographic experience is not readily applicable to projections of archaeological effects. Ethnoarchaeology seeks to discover a nexus of links that can be useful to interpretation of the myriad events that preserve individual archaeological records. As a tool for learning how to unravel some of the complexities of the taphonomy of residence, this is not a quest for normative descriptions of settlement "types." Observations of living systems are opportunities to test our assumptions about human organization and learn more of the strategic complexity involved in campsite and regional dynamics. Productive ethnoarchaeological work examines relationships between different behavioral systems that potentially have archaeological signatures.

References

August, Peter V.
1984   Population Ecology of Small Mammals in the Llanos of Venezuela. *Special Publications, Museum of Texas Tech University* 22: 71–104.
Besnerais, Henry le
1948   Algunos aspectos del río Capanaparo y de sus indios Yaruros. *Memoria de la Sociedad de Ciencias Naturales La Salle* 8(21): 9–20.
1954   Contribution à l'étude des Indiens Yaruro (Vénézuéla): Quelques observations sur le territoire, l'habitat et la population. *Journal de la Société des Américanistes* 43: 109–122.
1962   Contribution à l'étude des Indiens Yaruro et Otomaco, suite II (1). *Bulletin de la Société Suisse des Américanistes* 24: 7–25.

Binford, Lewis R.

1980   Willow Smoke and Dogs' Tails: Hunter-Gatherer Settlement Systems and Ar-
        chaeological Site Formation. *American Antiquity* 45(1): 4–20.

1982   The Archaeology of Place. *Journal of Anthropological Archaeology* 1(1): 5–31.

1983   Long-term Land-use Patterns: Some Implications for Archaeology. In *Lulu Lin-
        ear Punctated: Essays in Honor of George Irving Quimby*, edited by R. C. Dun-
        nell and D. K. Grayson, pp. 27–54. Anthropological Papers No. 27. University of
        Michigan, Ann Arbor.

Blydenstein, John

1967   Tropical Savanna Vegetation of the Llanos of Colombia. *Ecology* 48 (1): 1–15.

Bowman, Sheridan

1990   *Radiocarbon Dating: Interpreting the Past.* University of California Press, Berke-
        ley.

Brooks, Alison, and John Yellen

1987   The Perspective of Activity Areas in the Archaeological Record: Ethnographical
        and Archaeological Work in Northwest Ngamiland, Botswana. In *Method and
        Theory for Activity Area Research: An Ethnoarchaeological Approach*, edited by S.
        Kent, pp. 63–106. Columbia University Press, New York.

Bulla, Luis, Rafael Miranda, and Jesús Pacheco

1980   Producción, decomposición, flujo de materia orgánica y diversidad en una sabana
        de banco del Módulo Experimental de Mantecal (Estado Apure, Venezuela). *Acta
        Científica Venezolana* 31: 331–338.

Camilli, Eileen L., and James I. Ebert

1992   Artifact Reuse and Recycling in Continuous Surface Distributions and Implica-
        tions for Interpreting Land Use Patterns. In *Space, Time, and Archaeological Land-
        scapes*, edited by J. Rossignol and L. Wandsnider, pp. 113–136. Plenum Press, New
        York.

Casteel, Richard W.

1976   *Fish Remains in Archaeology and Paleo-Environmental Studies.* Academic Press,
        New York.

Chaffanjon, Jean

1986   *El Orinoco y el Caura: Relación de viajes realizados en 1886 y 1887.* Translated by
        Joelle Lecoin. La Fundación Cultural Orinoco. Editorial Croquis, Caracas.

Comerma , J. A., and O. Luque

1971   Los principios suelos y paisajes del Estado Apure. *Agronomía Tropical* 21: 379–
        396.

Conyers, Lawrence B., and Dean Goodman

1997   *Ground Penetrating Radar: An Introduction for Archaeologists.* Altamira Press,
        Walnut Creek, Calif.

Cortella, A. R., and M. L. Pochettino

1994 Starch Grain Analysis as a Microscopic Diagnostic Feature in the Identification of Plant Material. *Economic Botany* 48: 171–181.

Dean, Jeffrey S.

1969 *Chronological Analysis of Tsegi Phase Sites in Northeastern Arizona.* Papers of the Laboratory of Tree-Ring Research No. 3. University of Arizona Press, Tucson.

Dewar, Robert E.

1986 Discovering Settlement Systems of the Past in New England Site Distributions. *Man in the Northeast* 31: 77–88.

Emmons, Louise H.

1984 Geographic Variation in Densities and Diversities of Non-flying Mammals in Amazonia. *Biotropica* 16(3): 210–222.

Evershed, R. P., C. Heron, and L. J. Goad

1991 Epicuticular Wax Components Preserved in Potsherds as Chemical Indicators of Leafy Vegetables in Ancient Diets. *Antiquity* 65: 540–544.

Fisher, John W.

1995 Shadows in the Forest: Ethhnoarchaeology among the Efe Pygmies. Ph.D. dissertation. Department of Anthropology, University of California, Berkeley.

Foldats, E., and E. Rutkis

1965 Influencia mecánica del suelo sobre la fisionomía de algunas sabanas del llano venezolano. *Boletín de la Sociedad Venezolana de Ciencias Naturales* 25: 355–392.

Ford, Richard I.

1988 Commentary: Little Things Mean a Lot—Quantification and Qualification in Paleoethnobotany. In *Current Paleoethnobotany: Analytical Methods and Cultural Implications of Archaeological Plant Remains*, edited by C. A. Hastorf and V. S. Popper, pp. 215–222. University of Chicago Press, Chicago.

Gragson, Theodore L.

1989 Allocation of Time to Subsistence and Settlement in a Ciri Khonome Pumé Village of the Llanos of Apure, Venezuela. Ph.D. dissertation. Department of Anthropology, Pennsylvania State University.

1992a Fishing the Waters of Amazonia: Native Subsistence Economies in a Tropical Rainforest. *American Anthropologist* 94(3): 428–440.

1992b Strategic Procurement of Fish by the Pumé: A South American "Fishing Culture." *Human Ecology* 20(1): 109–130.

1997 The Use of Underground Plant Organs and Its Use in Relation to Habitat Selection among the Pumé Indians of Venezuela. *Economic Botany* 5(14): 377–384.

Graham, Martha

1993 Settlement Organization and Residential Mobility among the Rarámuri. In *Abandonment and Settlement of Regions: Ethnoarchaeological and Archaeological Approaches*, edited by C. M. Cameron and S. A. Tomka, pp. 24–42. Cambridge University Press, Cambridge.

Greaves, Russell. D.

1997   Hunting and Multifunctional Use of Bows and Arrows: Ethnoarchaeology of Technological Organization among Pumé Hunters of Venezuela. In *Projectile Technology*, edited by H. Knecht, pp. 287–320. Plenum Press, New York.

Grossa, Dino J.

1966   Una visita a los indios Yaruros de Riecito. *Boletín Indigenista Venezolano* 10(1–4): 67–81.

Hastorf, C. A., and M. J. DeNiro

1985   Reconstruction of Prehistoric Plant Production and Cooking Practices by a New Isotopic Method. *Nature* 315: 489–491.

Hunter-Anderson, Rosalind L.

1977   A Theoretical Approach to the Study of House Form. In *For Theory Building in Archaeology: Essays on Faunal Remains, Aquatic Resources, Spatial Analysis, and Systematic Modeling*, edited by L. R. Binford, pp. 287–315. Studies in Archaeology. Academic Press, New York.

Jahren, A. H., N. Toth, K. Schick, J. D. Clark, and R. G. Amundson

1997   Determining Stone Tool Use: Chemical and Morphological Analyses of Residues on Experimentally Manufactured Stone Tools. *Journal of Archaeological Science* 24: 245–250.

Jensen, Helle Juel

1994   *Flint Tools and Plant Working: Hidden Traces of Stone Age Technology.* Aarhus University Press, Aarhus.

Kealhofer, Lisa, Robin Torrence, and Richard Fullager

1999   Integrating Phytoliths within Use-Wear/Residue Studies of Stone Tools. *Journal of Archaeological Science* 26(5): 527–546.

Kirchoff, Paul

1948   The Yaruro. In *Handbook of South American Indians, Volume 4, The Circum-Caribbean Tribes*, edited by J. H. Steward, pp. 456–463. Smithsonian Institution, Bureau of American Ethnology Bulletin 143. U.S. Government Printing Office, Washington, D.C.

Kobayashi, Masashi

1994   Use-Alteration Analysis of Kalinga Pottery. In *Kalinga Ethnoarchaeology: Expanding Archaeological Method and Theory*, edited by W. A. Longacre and J. M. Skibo, pp. 127–168. Smithsonian Institution Press, Washington, D.C.

Lalueza Fox, Carles, Jordi Juan, and Rosa M. Albert

1996   Phytolith Analysis on Dental Calculus, Enamel Surface and Burial Soil: Information about Diet and Paleoenvironment. *American Journal of Physical Anthropology* 101: 101–113.

Leeds, Anthony

1960   The Ideology of the Yaruro Indians in Relation to Socio-economic Organization. *Antropológica* 9: 1–8.

1961   The Yaruro Incipient Tropical Forest Horticulture: Possibilities and Limits. In *The Evolution of Horticultural Systems in Native South America: Causes and Consequences*, edited by J. Wilbert, pp. 13–46. Antropológica Supplement No. 2. Editorial Sucre, Caracas.

1962   Ecological Determinants of Chieftainship among the Yaruro Indians of Venezuela. *Akten des 34 Internationalen Amerikanistenkongresses*, 597–608. Verlag Ferdinand Berger, Horn, Austria.

1964   Some Problems of Yaruro Ethnohistory. *Actas y Memorias del 25 Congreso Internacional de Americanistas* 2: 157–175.

Lyman, R. Lee
1994   *Vertebrate Taphonomy*. Cambridge University Press, New York.

Medina, F.
1980 Ecology of Tropical American Savanna: An Ecophysiological Approach. In *Human Ecology in Savanna Environments*, edited by D. R. Harris, pp. 297–319. Academic Press, London.

Meltzer, David J.
1991   Altithermal Archaeology and Paleoecology at Mustang Springs, on the Southern Plains of Texas. *American Antiquity* 56(2): 236–266.

Minnis, Paul
1978   Paleoethnobotanical Indicators of Prehistoric Environmental Disturbance: A Case Study. In *The Nature and Status of Ethnobotany*, edited by R. Ford, M. F. Brown, M. Hodge, and W. L. Merrill, pp. 347–366. Anthropological Papers No. 67. University of Michigan, Ann Arbor.

Mitrani, Philippe
1975   Remarques sur l'organisation sociale, la parenté et l'alliance des Yaruro de l'Apure. *Antropológica* 40: 2–24.

1988   Los Pumé (Yaruro). In *Los Aborígenes de Venezuela*, Vol. 3, *Etnología Contemporánea II*, edited by J. Lizot, pp. 147–213. Fundación La Salle de Ciencias Naturales, Caracas.

Mulholland, Susan C.
1989   Phytolith Shape Frequencies in North Dakota Grasses: A Comparison to General Patterns. *Journal of Archaeological Science* 16: 489–511.

Novoa, Daniel F.
1989   The Multispecies Fisheries of the Orinoco River: Development, Present Status, and Management Strategies. In *Proceedings of the International Large River Symposium (LARS)*, edited by D. P. Dodge, pp. 422–428. Canadian Special Publication of Fisheries and Aquatic Sciences 106. Department of Fisheries and Oceans, Ottawa.

OCEI (Oficina Central de Estadística e Informática)
1995a   *Censo Indígena de Venezuela 1992, Tomo I*. República de Venezuela, Presidencia de la República, Oficina Central de Estadísticas e Informática. Taller Gráfico de la OCEI, Caracas.

1995b   *Censo Indígena de Venezuela 1992: Nomenclador de Asentamientos, Tomo II.*
República de Venezuela, Presidencia de la República, Oficina Central de Estadísticas e Informática. Taller Gráfico de la OCEI, Caracas.

O'Connell, Margaret A.

1989   Population Dynamics of Neotropical Small Mammals in Seasonal Habitats. *Journal of Mammalogy* 70(3): 532–548.

Ojasti, Juhani

1983   Ungulates and Large Rodents of South America. In *Ecosystems of the World 13:
Tropical Savannas*, edited by F. Bourlière, pp. 427–439. Elsevier Scientific Publishing Company, New York.

Pearsall, Deborah M.

1988   Interpreting the Meaning of Macroremain Abundance: The Impact of Source and
Context. In *Current Paleoethnobotany: Analytical Methods and Cultural Implications of Archaeological Plant Remains*, edited by C. A. Hastorf and V. S. Popper,
pp. 97–118. University of Chicago Press, Chicago.

Petrullo, Vincenzo

1939   The Yaruros of the Capanaparo River, Venezuela. Anthropological Papers 11, Bureau of American Ethnology Bulletin 123, pp. 161–290. Smithsonian Institution,
Washington, D.C.

Piperno, Dolores R.

1988   *Phytolith Analysis: An Archaeological and Geological Perspective.* Academic Press,
New York.

Piperno, Dolores R., and Irene Holst

1998   The Presence of Starch Grains on Prehistoric Stone Tools from the Humid Neotropics: Indications of Early Tuber Use and Agriculture in Panama. *Journal of
Archaeological Science* 25(8): 765–776.

Piperno, Dolores R., and Deborah M. Pearsall

1998   *The Silica Bodies of Tropical American Grasses: Morphology, Taxonomy, and Implications for Grass Systematics and Fossil Phytolith Identification.* Smithsonian Contributions to Botany No. 85. Smithsonian Institution Press, Washington, D.C.

Ramia, Mauricio

1959   *Las Sabanas de Apure.* Ministerio de Agricultura y Cría, Dirección de Recursos
Naturales Renovables, División de Investigaciones, Sección de Sabanas, Caracas.

Rhode, David, and David B. Madsen

1998   Pine Nut Use in the Early Holocene and Beyond: The Danger Cave Archaeobotanical Record. *Journal of Archaeological Science* 25: 1199–1210.

Sánchez, Pedro Vicente, J. J. San José, and Jorge Paolina

1985   Efectos del fuego sobre la bioproducción de pastizales: La adición de cenizas y la
eliminación de la hojarasca en una sabana de *Trachypogon*. *Sociedad de Ciencias
Naturales Boletín* 40: 15–29.

Sanoja, Mario

1961   La vivienda de los Yaruros. *Revista Venezolana de Geografía* 1(3): 241–253.

Sarmiento, Guillermo

1984   *The Ecology of Neotropical Savannas.* Translated by O. Solbrig. Harvard University
        Press, Cambridge, Mass.

Sarmiento, Guillermo, and M. Monasterio

1975   A Critical Consideration of the Environmental Conditions Associated with the
        Occurrence of Savanna Ecosystems in Tropical America. In *Tropical Ecological
        Systems: Trends in Terrestrial and Aquatic Research*, edited by F. B. Golley and E.
        Medina, pp. 223–250. Springer-Verlag, New York.

Schiffer, Michal B.

1987   *Formation Processes of the Archaeological Record.* University of New Mexico Press,
        Albuquerque.

Silva, Juan F., and Guillermo Sarmiento

1976   Influencia de factores edáficos en la diferenciación de las sabanas: Análisis de com-
        ponentes principales y su interpretación ecológica. *Acta Científica Venezolana* 27:
        141–147.

Stewart, Kathlyn M.

1989   *Fishing Sites of North and East Africa in the Late Pleistocene and Holocene.* Cam-
        bridge Monographs in African Archaeology 34, BAR International Series 521.
        BAR, Oxford.

Taylor, Walter W.

1964   Tethered Nomadism and Water Territoriality: An Hypothesis. In *Proceedings of
        the XXXV International Congress of Americanists, Mexico*, part 2: 197–203.

Thomas, David Hurst

1975   Nonsite Sampling in Archaeology: Up the Creek without a Site? In *Sampling in
        Archaeology*, edited by J. Mueller, pp. 66–81. University of Arizona Press, Tuc-
        son.

Troth, R. G.

1979   Vegetation Types on a Ranch in the Central Llanos of Venezuela. In *Vertebrate
        Ecology in the Northern Tropics*, edited by J. F. Eisenberg, pp. 17–30. Smithsonian
        Institution Press, Washington, D.C.

Wandsnider, Luann

1992   Archaeological Landscape Studies. In *Space, Time, and Archaeological Landscapes*,
        edited by J. Rossignol and L. Wandsnider, pp. 285–292. Plenum Press, New
        York.

Weigelt, Johannes

1989   *Recent Vertebrate Carcasses and Their Paleobiological Implications.* Translated by J.
        Schaefer. University of Chicago Press, Chicago.

Wheeler, Alwyne, and Andrew K. G. Jones

1989   *Fishes*. Cambridge Manuals in Archaeology. Cambridge University Press, New York.

Yellen, John E.

1977   *Archaeological Approaches to the Present: Models for Reconstructing the Past*. Academic Press, New York.

Zinck, Alfred

1982   *Ríos de Venezuela*, 2nd ed. Cuadernos Lagoven, Departamento de Relaciones Públicas de Lagoven, Petróleos de Venezuela. Cromotip, Caracas.

II

Archaeological Studies of Mobility

7

Changing Mobility Roles at the Advent of Agriculture

A Biobehavioral Reconstruction

MARSHA D. OGILVIE

A dramatic change in human locomotor anatomy has occurred during the Holocene, with the trend toward increasing gracilization of the lower limb (Trinkaus et al. 1994; Ruff et al. 1994; Ruff et al. 1993; see Figure 7.1). It is generally recognized that this trend is tied to one of the most important economic, social, and biological transitions in human history, the shift from food-collecting to food-producing systems (see review in Larsen 1997; Ruff 2000a; Ruff and Larsen 1990; Ruff et al. 1984). Because agriculture necessitates some degree of sedentism, this transition had a profound impact on the behavioral repertoire of highly mobile hunter-gatherer populations (Binford 2001). Behavioral adjustments facilitating a more sedentary lifeway necessarily restructured the division of labor. It is unclear how these adjustments may have impacted the juxtaposition of male and female mobility roles.

Mobility versus sedentism is difficult to distinguish conceptually. Each involves frequency, distance, and the way in which individuals organize themselves for efficient resource procurement in given environmental zones (Binford 1980, 1982, 2001; B. Huckell 1995). Traditionally foragers have been described in terms of mobility and agriculturalists in terms of sedentism. Residentially mobile farmers are reported from northern Mexico (Hard 1990), however, and sedentary foragers are noted in several studies from the southeastern United States (Brown and Vierra 1983; Buikstra 1992; Smith 1992). Such examples selected from a broad literature demonstrate the fallacy in classificatory thinking (Binford, personal communication, 2002). The potential for variability in economic organization within and between groups from diverse ecological settings is far-reaching.

BACKGROUND

Traditionally attempts to reconstruct the economic organization of past human groups from the archaeological record have relied on artifactual evidence in conjunction with ethnographic analogy. These attempts generally have not utilized

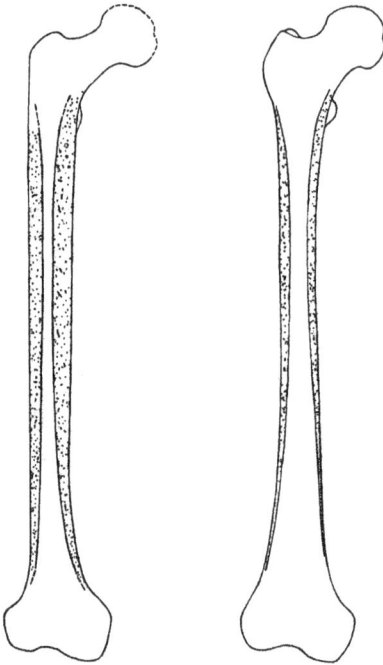

Figure 7.1. Longitudinal sections of femora illustrating differences in cortical bone thickness between an early *Homo* specimen (on left) and a contemporary specimen (on right). Modified from Ruff et al. 1993.

an important archaeologically derived resource, the human skeletal remains. Approaching this dilemma from a biological perspective can offer an innovative line of evidence to clarify issues regarding relative mobility versus sedentism in the archaeological record. Of specific interest is the expected change in locomotor behavior as the first archaeologically detectable evidence of seasonal residential stability associated with the advent of plant domestication.

BONE BIOLOGY

Bone biology research indicates a clear link between human biology and cultural behavior. The skeleton continually remodels throughout life in response to the forces or loadings imposed during habitual physical activities, such as those associated with repetitive locomotor behavior (Larsen and Kelly 1995; Ruff 1994; Ruff and Hayes 1983a, 1983b). Structural remodeling of bone results in a skeletal system better adapted to its biomechanical or functional environment, while maintaining an osseous record of its loading history (Chamay and Tschantz 1972; Carter et al. 1989; Currey 1990). Lower limb structure is the osseous template from which past locomotor behavior can be reconstructed.

THE SEXUAL DIVISION OF LABOR

An accurate assessment of past economic organization requires a clear understanding of the ethnographically documented dichotomy in male and female subsistence roles (Murdock and Provost 1973). The sexual division of labor exists in all societies studied, whether these groups are foragers, horticulturalists, or agriculturalists. The constraints associated with female reproductive biology contribute to behavioral sex differences that shape the division of labor, resulting in limited spatial range for females (Draper 1985; Draper and Cashdan 1988; Ember and Ember 1976; Hilton and Greaves 1995; Hurtado and Hill 1990; Hurtado et al. 1992). The subsistence roles assigned to females, therefore, are those that are compatible with simultaneous child care and do not require long-range travel or put children at risk (Brown 1970). Conversely, males tend to perform subsistence tasks that quite often require long-distance travel or absence from the home and increased risk to the individual (Murdock and Provost 1973; see the review in Gaulin and Hoffman 1988). These gender-specific behavioral differences in mobility roles are evident in large size and shape differences in the midshaft femur that can be used to reconstruct economic organization (Larsen 1997; Ruff 1987).

ARCHAEOLOGICAL APPLICATION

Recent archaeological investigations in southeastern Arizona yielded the first detectable signs of decreased residential mobility in the southwestern deserts (B. Huckell 1995; Mabry 1998). During this time (the late centuries BC) maize (*Zea* sp.) was incorporated into the hunting and gathering adaptive strategy (Huckell 1990; Mabry 1998; Wills and Huckell 1994). This new variant signaled the advent of plant domestication that would eventually lead to more sedentary lifeways.

This archaeologically derived evidence brings the economic organization of early farmers into question. Were they committed agriculturalists living in settled villages or seasonally mobile groups using agriculture as a buffer to wild resources?

The fortuitous recovery of skeletal remains of Late Archaic males and females from Tucson Basin sites for the first time provided the opportunity to make biological examinations of the elusive processes by which early agricultural systems developed in incipient southwestern farmers. These rare skeletal series are the oldest yet found in association with early domesticates. The presence of both males and females with intact femora can elucidate the concomitant impact of domestication on the sexual division of labor.

SAMPLE SELECTION

Three study groups were selected from the perspective of probable subsistence-related mobility, based on lines of archaeological and ethnographic evidence in each study area (see Figure 7.2). The study groups are analogs for three distinct economic stages seen in the foraging to farming transition in the American Southwest: foraging, seasonally occupied villages, and permanent agricultural settlements. Bone structure provides an independent means to track changes in subsistence behavior in groups making the transition to food production. The foragers and late agriculturalists offer a behavioral baseline of comparison for the incipient agricultural sample.

Small sample sizes involved in this study are a reality due to the paucity of preserved human remains from the Archaic period. For this reason, the chronological placement of all individuals sampled in each study group was not always contemporaneous, but they make up the "universe" for this period. All samples were confined to similar geographic areas to minimize the confounding effects of terrain and genetic variation in local populations on bone biomechanics (Bridges 1995b; Ruff 2000a, 2000b).

SKELETAL SAMPLES

The Preagriculturalists

The preagriculturalists ($n = 42$) are represented by 19 males and 23 females who traversed the arid landscape of the Trans- and Lower Pecos regions of southwest Texas, where agriculture was never prehistorically incorporated into the economy. The majority of the burials were recovered from dry rock shelters dating to the Blue Hills Phase of the Late Archaic, ca. 2300–1300 YBP (Steele and Olive 1989; Turpin et al. 1986; Collins and Labadie 1999). A drying trend at the end of the Pleistocene established arid grasslands mixed with desert succulent plant communities (Bryant 1986a, 1986b; Shafer 1986a). Hundreds of caves and overhangs in this karstic system provided shelter for the small bands of foragers who first inhabited this area.

As in all arid regions, settlement and subsistence were tempered by the location and availability of water sources (Taylor 1964; Bayham and Morris 1990; Carmichael 1990; Hard 1990; Speth 1990; Vierra 1990). The Rio Grande and Devils' and Pecos Rivers provided a reliable supply of water, critical in this otherwise arid setting. Foragers in such desert environments where there is no agriculture must move frequently to take advantage of seasonally available resources (Kelly 1983, 1995). Repeated use of sites near critical resource locations, rather than long-term occupations, attests to seasonally fluctuating hunting and

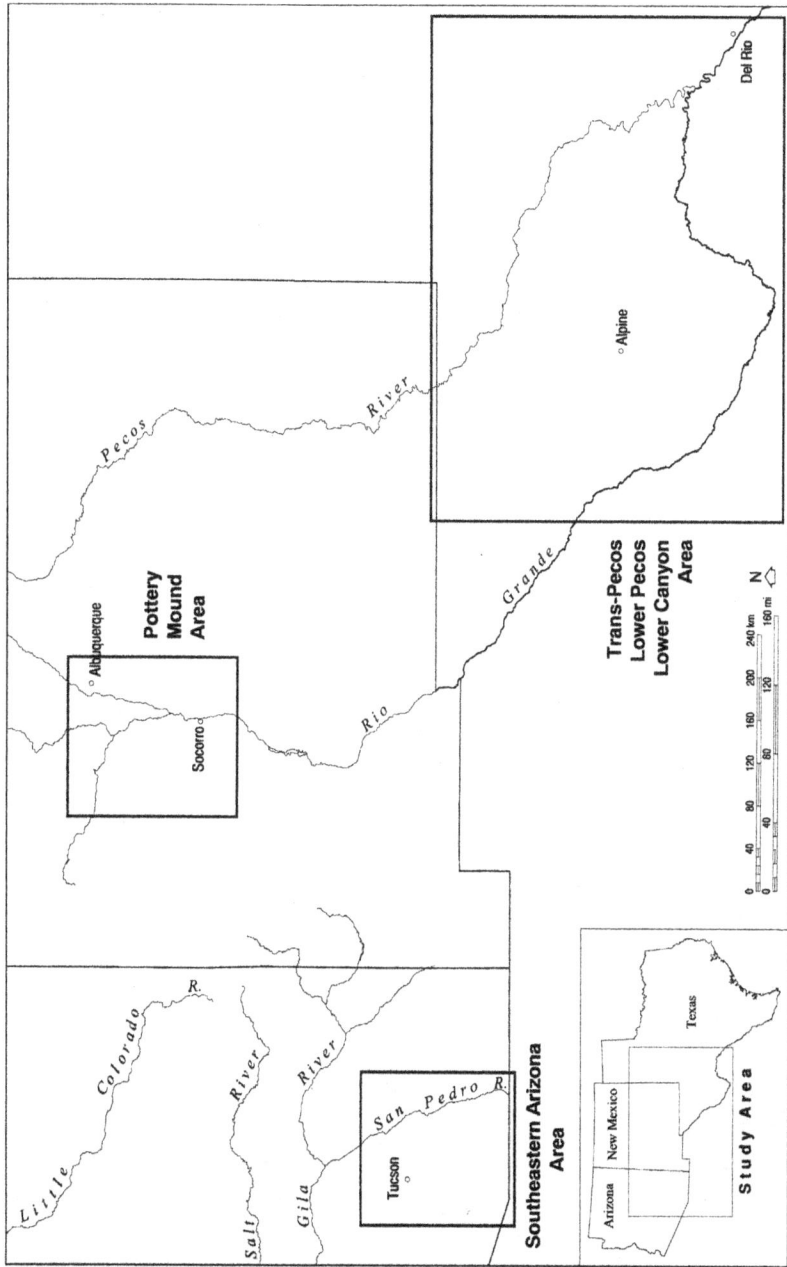

Figure 7.2. Overview map of the three study areas in the American Southwest. Boxes delineate the origin of each skeletal sample.

gathering forays for small animals and desert plants (Marmaduke 1978; Binford 1980).

Dry conditions in the Pecos preserved perishable artifacts and a wealth of plant and animal remains in the cultural deposits of the dry rock shelters (Ward 1992). Botanical and faunal studies revealed that the basic diet consisted of plants rich in Carbon 4 (^4C) and Crassulacean Acid Metabolism (CAM) and the animals that consumed them, predominantly rodents, lagomorphs, and fish (Huebner 1991, 1995a, 1995b). Plants with a C4 photosynthetic pathway include grasses and seeds, while CAM plants are primarily desert succulents. Palynological, paleofecal, and dental microwear studies indicate that sotol (*Dasylirion texanum*), agave (*Agave scabra*), prickly pear (*Opuntia phaeacantha*), and lechuguilla (*Agave lechuguilla*) were heavily exploited CAM plants, a mainstay in the Pecos diet for thousands of years (Story and Bryant 1966; Riskind 1970; Bryant 1974, 1986b; Williams-Dean 1978; Dering 1979, 1999; Stock 1983; Bryant 1986a, 1986b; Hartnady and Rose 1991; Danielson and Reinhard 1998). Sotol and lechuguilla were particularly valuable, because they represent one of the few food resources that could be harvested year-round (Huebner 1995a, 1995b).

The sexual division of labor in the Pecos region is projected from archaeological evidence, coupled with ethnographically documented cases of task specificity in male and female hunter-gatherers (Brown 1970; Murdock and Provost 1973; Shafer 1986b). Observations of desert succulent processing by present-day groups living near the south Texas/Mexican border support the cases documented above (Kludt, personal communication, 1999). Procuring and processing fibrous desert succulents is a labor-intensive task requiring the participation of both males and females (Brown 1991; Kludt, personal communication, 1999). High group mobility is implied by the necessity of travel to distant plant locations. The roasting process took several days, so all members of the group camped, foraged, and hunted near the earth-oven locations (Shafer 1986b; Dering 1999). Burned rock middens and charred remains of lechuguilla and sotol have been found by the thousands in cultural deposits in southwest Texas (Vierra 1998).

Strategic site locations in an ecotone provided hunting and foraging opportunities for males and females to exploit riverine, upland, and valley resources. This required frequent travel between these zones. The limited riparian and cienega plant communities, including seeds from a variety of grasses, were either clustered near the rivers or found in locations that necessitated strenuous travel to higher elevations (Stock 1983; Dering 1979).

Male mobility was likely tempered by logistically organized hunting forays (Binford 1980). Excellent chert sources were abundant in the local environment.

A portable tool kit geared for foraging and hunting small game was recovered during archaeological investigations at Horseshoe Cave. The kit contained multifunctional tools that could have been used to exploit any of the resources in the canyon (Hester 1983). With the exception of stylistic changes in projectile points, the basic tool kit apparently remained adequate through time for use in a foraging subsistence economy (Dibble and Prewitt 1967). Faunal remains suggest that many of the larger game species, such as white-tailed deer (*Odocoileus virginianus macrourus*), gray fox (*Vulpes* sp.), raccoon (*Procyon lotor mexicanus*), and coyote (*Canis latrans texensis*), were taken during less arid climatic phases. Most of the protein in the diet, however, came from smaller animals, such as rodents and lagomorphs, likely trapped by women and children (Murdock and Provost 1973; Huebner 1995a, 1995b).

Desert succulents, also widely available, provided the raw material for a flourishing fiber industry in the canyon (Shafer 1986a). In addition to being a dietary staple, succulent plant communities provided fiber that was used for weaving everything from technological items to clothing and children's toys (Shafer 1986b).

Resource abundance in this environment made it unnecessary to carry all burdens on routine excursions. It has been noted archaeologically that artifacts were "stashed and cached" near critical resource locations. The 9,000-year archaeological sequence, with no evidence of abandonment, clearly attests to the successful exploitation strategies employed by the inhabitants of this diverse biotic region (Lundelius 1974; Van Devender and Spaulding 1979; Van Devender et al. 1984; Reinhard et al. 1989).

The Early Agriculturalists

The early agriculturalists (*n* = 21) are represented by 14 males and 7 females from southeastern Arizona, where agriculture was thought to be in the earliest stages (B. Huckell 1988, 1995). They resided in the Tucson Basin during the Late Archaic or Early Agricultural period, approximately 3500–1850 YBP (B. Huckell 1983, 1995). Their subsistence economy was "transitional," based on lines of paleobotanical and archaeological evidence for mixed foraging and farming strategies (L. Huckell 1995; Mabry 1998).

The Tucson Basin is located on the eastern edge of the Sonoran Desert zone and is part of the Santa Cruz River drainage system. This low southern desert is surrounded by the Tucson Mountains on the west, the northern Serita Mountains on the southwest, and the southern end of the Tortolita and Santa Catalina Mountains on the north. On the east and southeast, the basin is bordered by the Rincon and Santa Rita Mountains.

Around the turn of the nineteenth century an erosional episode along Cienega

Creek and Pantano Wash in the southeastern part of the basin exposed Early Agricultural levels. During a highway expansion project in the vicinity of the Santa Cruz River additional Early Agricultural sites were subsequently encountered. These discoveries caused us to rethink previous assumptions about socioeconomic systems during the late centuries BC (Mabry 1998; Mabry and Clark 1994). Prior to this time the conventional wisdom was that domesticates had not yet been incorporated into the hunting and gathering adaptive strategy (Matson 1991; Minnis 1985, 1992; Wills 1988a, 1988b, 1992). Archaeological investigations at these recently exposed sites, however, revealed the presence of maize by at least 1000–500 BC (see references cited in B. Huckell 1995; Mabry 1998).

Cienega residential sites were strategically positioned along the perennial waterways of the Santa Cruz and its tributaries, the most critical resource for the inhabitants of the arid basin. Small special-use sites were located in three surrounding ecological zones: the riverine floodplain, the piedmont (the slopes leading from the major drainages to the mountain fronts), and the uplands bordering the basin (B. Huckell 1995). Positioning themselves in this ecotone provided the inhabitants not only with rich soil for agriculture but with a variety of microenvironments for hunting and gathering opportunities (L. Huckell 1995, 1996).

Maize fields and agricultural products at the large residential sites provide solid evidence for early agriculture (Eddy 1958; Hemmings et al. 1968; Huckell 1977; Eddy and Cooley 1983; Huckell and Haury 1988; Mabry et al. 1995; L. Huckell 1996; Mabry 1998). Dentition recovered with the skeletal remains also suggests maize consumption by site inhabitants (Minturn and Lincoln-Babb 1995; Minturn 1998). Excavations revealed large pithouse villages with thick midden deposits, extramural storage, numerous groundstone artifacts, and food-processing features (Mabry and Clark 1994; Mabry et al. 1995; Adams 1996). Wild plant species recovered from bell-shaped storage pits representing multiple seasons of the year demonstrated that the site inhabitants anticipated being in one location for extended periods (Eddy 1958; Binford 1982; Huckell and Haury 1988; L. Huckell 1995).

Because agriculture requires some degree of sedentism, it is thought that early agricultural populations were relatively more settled than highly mobile Middle Archaic groups (B. Huckell 1995). The shift toward reduced residential mobility would have required a reliance on stored resources and use of logistic strategies to exploit resources in different environmental zones most efficiently (Binford 1980; Kelly 1983, 1992, 1995; B. Huckell 1995). Mobility could decrease because stored resources were reliable and did not require daily long-distance travel. More practical targeting of economically important resources could be accomplished by logistic task groups. This is archaeologically reflected at special-use sites identified in the topographically diverse Tucson Basin. These short-term sites demonstrate that broad spectrum hunting and gathering of wild plant and animal resources was still an indispensable component of the economy (Eddy 1958; Huckell 1984a, 1984b; L. Huckell 1995, 1996).

Paleoethnobotanical studies indicate that the plant species dominating assemblages were those available in late summer and fall—maize, amaranth (*Amaranthus* sp.), purslane (*Portulaca* sp.), agave (*Agave* sp.), yucca (*Yucca elata, Yucca baccata*), rushes (*Juncus* sp.), acorns (pericarps of *Quercus* sp.), wheelscale saltbush (*Atriplex elegans*), and Arizona walnut (*Juglans major*) (L. Huckell 1996). The nutritionally important legume family (Chenopodiaceae and Amaranthaceae) thrived in the agriculturally disturbed soils along floodplains at the riverine sites (L. Huckell 1995). The limited presence of spring plants, such as stickleaf (*Mentzelia* sp.), mesquite (*Prosopis* sp.), chenopods (Chenopodiaceae), and saguaro (*Carnegiea gigantea*), suggests partial site abandonment at this time, with people moving back and forth between multiple environmental zones during seasonal rounds.

Early Agricultural groups obviously retained some degree of mobility to procure resources at varying distances from home. The rich secondary biomass in the Sonoran biotic region has implications for reduced mobility in females. Some suggest a reduction in long-range mobility during foraging forays, because wild plant resources were widely distributed and easily accessible (Fish et al. 1992). Such clustered resources would equate with a reduction in travel and search time. With foraging costs reduced, the time formerly spent in long-distance travel could be devoted to plant processing, making foraging more efficient. Moreover, this localized abundance of reliable and predictable plant communities was potentially storable, providing food sources during winter and spring months when most wild plants were not yet available (Fish et al. 1990; Fish and Fish 1991; L. Huckell 1995). In addition, many resources (such as acorns and agave) were available within relatively short distances of residences and permanent water sources (Fish et al. 1992). Based on Binford's (1980) Economic Zonation model, many resources could have been efficiently targeted by simple foraging trips near settlements.

High male mobility was likely conditioned by hunting, a commonly logistically organized activity (Binford 1978, 1980). Reaching the Sonoran desert zone and rugged uplands would have necessitated logistic mobility. Such forays are evidenced by small, extended hunting camps in the surrounding mountains and grasslands (B. Huckell 1995; Bayham et al. 1986). Mule deer (*Odocoileus hemionus canus*), white-tailed deer (*Odocoileus virginianus macrourus*), antelope (*Antilocapra americana*), bighorn sheep (*Ovis canadensis*), elk (*Cervus merriami*), bobcat (*Lynx rufus*), coyote (*Canis latrans*), and jackrabbit (*Lepus gaillardi*) constituted a significant portion of the diet (Eddy 1958; Thiel 1996). Procuring large game, more commonly targeted than smaller mammals in all biomes, required extensive travel to distant mountain ranges by male hunters, sometimes at distances of 70 km or more (B. Huckell 1995).

The topographic diversity and abundance of resources in the Sonoran Desert zone made for a rich environment for Early Agricultural populations. Our understanding of these newly investigated sites and related subsistence economies during this critical period is still incomplete.

This first biomechanical research on the remains of Cienega phase residents can help define the processes by which early agricultural systems developed and the concomitant impact on subsistence-related behaviors.

The Late Agriculturalists

The late agriculturalists ($n = 76$) are represented by 42 males and 34 females who resided at the aggregated community of Pottery Mound, New Mexico. This 500-room pueblo was occupied during the Pueblo IV period (700–500 YBP) and abandoned just prior to the Spanish *entradas* in AD 1540 (Hibben 1955, 1961, 1996). The array of ceramic types littering the site landscape, (predominantly Glaze A) reinforces the Pueblo IV designation (Brody 1964; Voll 1967).

This massive three-tiered community is located in Valencia County in central New Mexico, at the confluence of the southern edge of the Rio Grande Valley and the northern periphery of the Chihuahuan Desert Zone (Cordell n.d.; Hibben n.d.; Schorsch 1962). The site locality is bordered on the north by Mt. Taylor, on the east by the Manzano Mountains, on the west by topographically broken high canyon country, and on the south by the Magdalena and Ladron Mountains. The Rio Grande and its tributary, the Rio Puerco, course along the eastern periphery of the site.

Intensive maize agriculture with specialization of food processing was practiced at Pottery Mound (Cordell 1984, 1997; Wills et al. 1994). Major investment in multistoried adobe architecture and nonportable technology reflects the expectation of permanent residency (Wills 1988a). Ceremonial structures, storage features, flora and fauna indicative of year-round occupation, and a large burial population also demonstrate permanent ties to the land (Schorsch 1962; Wills and Huckell 1994).

This high southern desert is ecologically classified by the U.S. Department of Agriculture as the Upper Sonoran life zone (Bailie n.d.). The predominant plant species on the plain included Russian thistle (*Salsola* sp.), saltbrush (*Atriplex* sp.), sacaton grass (*Aristida* sp.), and juniper (*Juniperus* sp.) (Schorsch 1962). The ecotone within which the site was positioned provided arable soil for agriculture and high-quality wild plant and animal resources. Palynological reconstruction identified three biomes in the vicinity of Pottery Mound during Pueblo IV times: riparian, open grassland, and montane (Emslie 1981; Emslie and Hargrave 1978). The headwaters of the Rio Puerco, approximately 100 km to the north, provided a permanent source of water for the site residents.

Unique insight into the prehistoric lifeways at Pottery Mound comes from over eight hundred spectacular kiva murals, the most abundant and well preserved in the Southwest (Hibben 1960, 1966, 1967, 1975, 1996; Leviness 1959; Vivian 1961; Brody 1964, 1970; Crotty 1979). The importance of maize (*Zea* sp.) is reflected in mural depictions of corn plants and subsistence tasks associated with agricultural production. Hundreds of preserved charred cobs have been recovered from the cultural deposits (Hibben 1996). It is likely that other cultigens, beans (*Phaseolus* sp.), squash (*Cucurbita* sp.), and possibly cotton (*Gossypium* sp.) were also grown (W. H. Wills, personal communication, 1998). The grasslands adjacent to the site contained extensive agricultural fields. Additional fields were probably located farther from the main site as insurance against crop disasters (W. H. Wills, personal communication, 1998).

The economic roles of males and females documented in the ethnographic literature are clearly depicted in the kiva paintings (Brown 1970; Murdock and Provost 1973). Males were more heavily engaged in agricultural pursuits at late prehistoric communities than in earlier periods (Ember 1983). Field preparation and crop tending likely reduced the total time that males allotted for long-distance mobility in favor of more sedentary agricultural duties.

Strictly in terms of farming tasks, the proportional contribution of females was declining relative to that of males (Ember 1983). Women's total work effort, however, did not decrease. A wide variety of grinding implements at the site support ethnographic documentation that women were heavily involved in maize processing, often for six to eight hours per day (Murdock and Provost 1973; Lancaster 1986; Hard 1990; Adams 1996; Hard et al. 1996). They used a broad range of grinding technologies, including various metates and open troughs, with one- and two-hand manos (Adams 1996). The specific technologies of such items as graters, grinders, pounders, mortars, and pestles suggest that women were grinding substances other than maize, such as various seeds and pigments. Females in mural tabloids carried burden baskets for wild plant collecting trips, prepared foods, transported fuel and water, and produced pottery (Brown 1970; Murdock and Provost 1973; Hibben 1975).

Mural portrayals of male hunters and over ten thousand recovered faunal remains that were brought in from surrounding grasslands and mountains indicate that men at Pottery Mound engaged in hunting. Because hunting is usually logistically organized, long-distance mobility of some portion of the male population is expected (Binford 1980). Targeting large mammals would require travel to the Manzano Mountains 32 km to the east and to the broken high mesa country approximately 16 km west of the site. These locations were excellent for mule deer, white-tailed deer, Merriam's elk, and bighorn sheep. Pronghorn antelope were locally available, ranging very close to the site in the Puerco Valley. Antelope drives were historically documented in this same prime location.

Ethnobiological research identified over fifty species of birds in the avifauna assemblage, many of which were depicted in murals (Emslie and Hargrave 1978). Pottery Mound was located on a "fly-way" for migrating birds, implying spring and fall bird hunting. The hundreds of birds and watercourses along the eastern edge of the site, in turn, attracted large numbers of mammals, providing additional hunting opportunities.

Surrounding montane zones with piñon pine (*Pinus edulis*) provided locales for piñon-gathering trips in the fall by more mobile members of the community. Piñon nuts are a high-return storable resource and facilitated "over-wintering" at a time when other wild resources were not available. Less mobile individuals, including older males and children, could perform tasks closer to home. This reduction in residential mobility with year-round site occupation enabled all members of the group to make labor contributions in this highly specialized prehistoric farming community.

The impressive richness of the site (including the elaborate murals and thousands of decorative potsherds from far-ranging locales) caused researchers to speculate that Pottery Mound was a center of trade and ceremonial significance in the American Southwest (W. H. Wills, personal communication, 1998). Only about half of the site has been excavated, so the extent of our knowledge of Pottery Mound will be improved by further excavations (Hibben 1996).

METHODS

The functional approach in this study uses an engineering beam model (Timoshinko and Gere 1972). The model can be appropriately applied to a structure that is long relative to its width, making it particularly suitable for structural analysis of the femoral diaphysis (Lovejoy et al. 1976; Ruff 1992; Ruff and Hayes 1983a). When the diaphysis is modeled as a beam, structural properties of the diaphysis can be calculated at biologically relevant section locations lying perpendicular to the long axis of the bone (Nagurka and Hayes 1980). Biomechanical analyses of cross sections indicate the strength or rigidity of the beam in resistance to the forces generated during repetitive physical activities (Timoshenko and Gere 1972). Biomechanical models predict that the changes in physical activities associated with subsistence shifts will be reflected in behaviorally conditioned skeletal morphology (Larsen 1997; Ruff 1992, 2000a, 2000b; Ruff and Larsen 1990).

Specific guidelines were followed during selection of individual specimens. Femora ($n = 199$) from 139 individuals were selected for this study. Femoral diaphyses with preserved 50% sections from the distal end of the shaft made it possible to generate cross-sectional property values. Reasonably complete long bones with fused epiphyses were required for estimation of length; thus only adults were included. Addressing biological questions related to the division of labor required the availability of associated pelvic remains for reliable sex assessment. Standard osteological techniques were employed to evaluate the sex of each individual (see Buikstra and Ubelaker 1994). Preference was always given to the reliable primary sexual characteristics seen in the pelvis.

The loading history of each femoral diaphysis was elucidated through an analysis of structural geometry at biologically relevant locations. It has been determined that the bone structure at the midpoint of femoral length most accurately reflects *in vivo* locomotor behavior (Ruff and Hayes 1983a, 1983b). Noninvasive computed tomography produced a digitized image of the internal and external contours of bone at the 50% location (Ruff and Leo 1986). The software program SLICE then quantified the amount and distribution of cortical bone in each cross section from the scanned image (Eschman 1992). These data provide the baseline information needed for statistical evaluation of relative mobility patterns (Ruff 1992).

The quantified geometric property values were employed in a cross-sectional shape index. The shape index, known as I_x/I_y, most accurately reflects *in vivo* locomotor activity and serves as a proxy for relative mobility (Ruff and Hayes 1983a, 1983b). The index is derived by dividing I_x (the amount and distribution of cortical bone in the medial/lateral femoral plane) by I_y (the amount and distribution of cortical bone in the anterior/posterior [A/P] femoral plane). Larger resultant values indicate greater bending strength in the A/P plane that translates to relatively higher levels of mobility.

RESULTS

Using subsistence as the independent variable, the I_x/I_y values were examined separately for each sex using one-way analysis of variance (ANOVA) with $\alpha = 0.05$. Any ANOVA comparison yielding a significant p-value was further examined using Fisher's Protected Least Significant Difference (PLSD) post-hoc pair-wise comparisons.

When examining males, the ANOVA indicates a significant difference ($p = 0.0217$) across the subsistence categories. Fisher's PLSD indicates that the I_x/I_y values for the agricultural males are significantly different from the values seen for both the forager ($p = 0.0320$) and transitional males ($p = 0.0229$; see Figure 7.3 and Table 7.1). Although the transitional males have the highest I_x/I_y values,

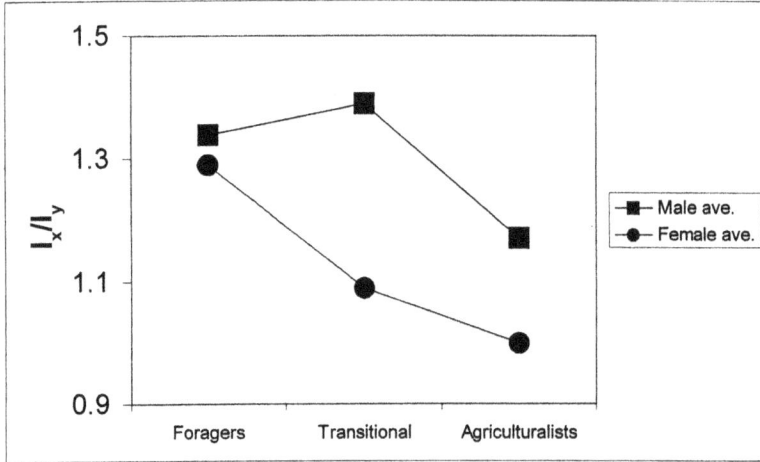

Figure 7.3. Femoral shape index (I_x/I_y) values reflecting decreasing mobility levels as dependence on maize increases. Note relative differences between the sexes in each subsistence category.

there is no significant difference between them and the forager males (see Figure 7.3 and Table 7.1). The femoral shape index values associated with the late agriculturalists reflect decreased mobility levels with maize intensification (see Figure 7.3).

When examining females, the ANOVA indicates a significant difference ($p <$ 0.0001) across the subsistence categories. Fisher's PLSD indicates that the I_x/I_y values for the forager females are significantly different from the values seen for both the transitional ($p = 0.0213$) and agricultural females ($p < 0.0001$; see Figure 7.3 and Table 7.1). Additionally, there is no significant difference between the transitional and late agricultural females (see Figure 7.3 and Table 7.1).

Males as a group always have larger values for the shape index than females as a group. Although femoral values for both sexes become increasingly smaller through time, they do so at very different rates (see Figure 7.3). During the foraging phase values for males and females are similar, with those for males being only slightly larger (see Figure 7.3 and Table 7.1). Femoral strength between

Table 7.1. Ix/Iy Values by Sex and Subsistence

Variable I_x/I_y	Foragers Males	Females	Transitional Males	Females	Agricultural Males	Females
Mean	1.34	1.29	1.39	1.09	1.17	1.00
S.D.	.29	.15	.37	.25	.22	.14
S.E.	.07	.05	.12	.102	.04	.03
C.V.	21.79	11.54	26.28	23.0	18.77	14.26

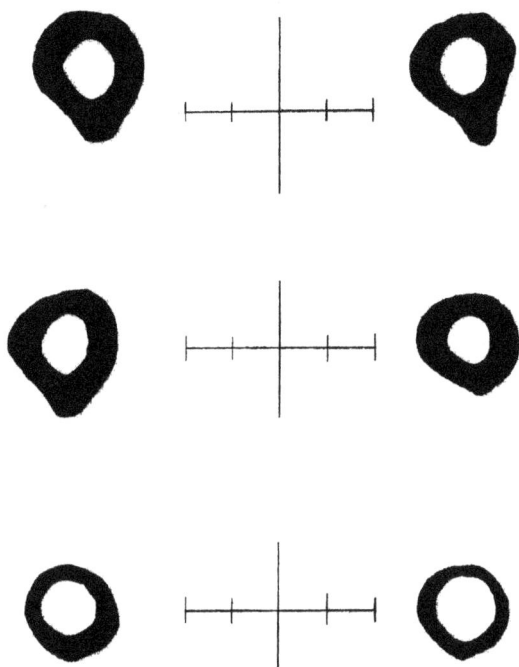

Figure 7.4. Right femoral cross sections at 50% of bone length, showing size and shape changes from foraging (top) to early agriculture (center) to late agriculture (bottom). Note structural differences between males (on the left) and females (on the right). Cross sections are oriented with anterior at the top, posterior at the bottom, medial to the right, and lateral to the left. Scale is in centimeters.

the sexes diverges markedly in the early agricultural phase, where male averages are in fact a little higher but not significantly higher than those seen in forager males. In females the averages for I_x/I_y sharply decline with the advent of agriculture (see Figure 7.3 and Table 7.1). With full-blown agriculture, index values for the sexes reconverge, driven by rapidly decreasing male values for I_x/I_y (see Figure 7.3). Agricultural female values continue to decline from transitional levels. These trends are qualitatively illustrated by the amount of cortical bone seen in the typical transverse cross section representative of each subgroup (see Figure 7.4).

RELEVANT RESEARCH

The long-term trend in decreasing femoral dimensions with increasingly more sedentary lifeways is documented from early *Homo* to anatomically modern human populations (Lovejoy and Trinkaus 1980; Trinkaus and Ruff 1989; Ruff et al. 1993; Ruff et al. 1994; Bridges 1995b; Ruff et al. 1999; Trinkaus et al. 1994; Holt and Churchill 2000; see the references cited in Larsen 1997). This long-term trend is an appropriate context in which to consider the variability in long bone structure seen in more recent archaeological samples.

The reduction in locomotor behavior with increasing dependence on maize is biologically manifested as lower I_x/I_y values (that is, more circular femoral diaphyses). Such short-term adaptation to subsistence change is consistently reported by biomechanical researchers, although bone strength may vary by regional setting due to local ecology, terrain, and variability in subsistence practices. Reduction in femoral dimensions with farming is reported from the Georgia coast (Ruff and Larsen 1990), the American Southwest (Ruff and Hayes 1983a, 1983b; Brock 1985; Brock and Ruff 1988; Ogilvie 2000), the Northern and Southern Great Plains (Ruff 1994), the Great Basin (Larsen and Kelly 1995; Ruff 2000b), the Tennessee River Valley (Bridges 1989, 1995a), and La Florida (Larsen et al. 1992; Ruff 1997).

DISCUSSION AND CONCLUSIONS

The biological evidence from the American Southwest demonstrates the trend in decreasing femoral dimensions with increased reliance on maize. It is of particular interest that application of these data to populations in economic transition proves useful for clarifying the processes by which early agricultural systems developed. The Trans-Pecos and Lower Pecos foragers were the most mobile of all the individuals examined. The males and females showed similar patterns in lower limb loading, suggesting a residentially mobile strategy. Based on archaeological evidence, ethnohistoric documentation, and local ecology in the southwest Texas region, a subsistence strategy employing mixed sex resource procurement and processing would be expected (Binford 2001).

The committed late agriculturalists from the sedentary setting of Pottery Mound, not surprisingly, were the least mobile. The late agricultural males, still more mobile than their female counterparts, likely engaged in small-scale hunting forays and agricultural pursuits at some distance from the main community (Ember 1983). The females displayed the lowest mobility levels in both sexes and all subsistence categories. Ethnographic projections suggest that they performed the sedentary domestic chores associated with the maintenance of agricultural production (Ember 1983; Rafferty 1985).

The occupants of the Tucson Basin during the late centuries BC were truly a population in economic transition. As a group, they demonstrated levels of locomotor behavior intermediate to those of the mobile foragers and sedentary late agriculturalists. Both the artifactual evidence and femoral index values suggest that Late Archaic groups in the Tucson Basin were still relying on wild resources to a great extent and were not yet fully committed to settled agriculture.

At the time of this economic transition the sexual division of labor was noticeably impacted. During the incorporation of early maize into the economy, male femoral structure reflected high levels of mobility equivalent to those of the mobile male Texas foragers. The maintenance of behaviors necessitating high mobility by males is compatible with a strategy of logistical resource targeting. Lithic assemblages analyzed from hunting camps in distant mountain ranges surrounding the Tucson Basin also support implications for long-range hunting forays by males (Vierra, personal communication, 2001). These Arizona males provide the exception to the expected pattern of decreasing mobility with agriculture.

Interestingly, the Tucson Basin females showed a marked decline in mobility at this time. The femoral data demonstrate that their locomotor behavior was not significantly different from that of the sedentary female agriculturalists. Comparisons with the female foragers, however, reveal dramatic behavioral differences. The reduction in I_x/I_y values in early agricultural females is compatible with a reduction in residential mobility. The gathering and preparation of wild plant resources has consistently been assigned to the domain of women's work across the ethnographic record of foragers. A reduction in residential mobility in females would provide more extended periods in proximity to residences, presumably underwriting the costs of the initial incorporation of early domesticates into the subsistence economy. A division of labor incorporating high logistical mobility in males and reduced residential mobility in females facilitated the addition of a new, perhaps more predictable, variant into the behavioral repertoire of late Archaic hunter-gatherer populations that would eventually lead to more sedentary lifeways. The female work effort, long overlooked in the anthropological literature, appears to have been the impetus for the most important economic, social, and biological transition in human history, the shift from food-collecting to food-producing systems.

ACKNOWLEDGMENTS

I want to thank the affiliated Native American tribes for their permission to conduct this research. Thanks are extended to the University of New Mexico Department of Radiology (Albuquerque), Maxwell Museum Laboratory of Human Osteology (Albuquerque), Desert Archaeology (Tucson), Arizona State Museum (Tucson), University Hospital (Tucson), Texas Archaeological Research Laboratory (Austin), Austin Radiological Association, and Witte Museum (San Antonio). Additional support was provided by the Arizona Archaeological and Historical Society (Tucson) and the Department of Anthropology, Office of Graduate Studies, and Student Research Allocations Committee at the University of New Mexico. Chris Ruff graciously permitted use of Figure 7.1. Production of the overview map was courtesy of Ron Stauber. All other figures were produced by Charles E. Hilton.

REFERENCES

Adams, J. L.

1996   Refocusing the Role of Food Grinding Tools as Correlates for Subsistence Strategies of Gatherers and Early Agriculturalists in the American Southwest. Paper presented at the 61st Society for American Archaeology Meetings, New Orleans, 1996.

Bailie, B.

n.d.   Life Zones and Crop Zones of New Mexico. *North American Fauna* 35: 32–38. In press.

Bayham, L. V., and D. H. Morris

1990   Thermal Maxima and Episodic Occupation of the Picacho Reservoir Dune Field. In *Perspectives on Southwestern Prehistory*, edited by P. E. Minnis and C. L. Redman, pp. 26–37. Westview Press, Boulder.

Bayham, L. V., D. H. Morris, and M. S. Shackley

1986   *Prehistoric Hunter-Gatherers of South Central Arizona: The Picacho Reservoir Archaic Project*. Anthropological Field Studies 13. Office of Cultural Resource Management, Department of Anthropology. Arizona State University, Tempe.

Binford, L. R.

1978   *Nunamiut Ethnoarchaeology*. Academic Press, New York.

1980   Willow Smoke and Dogs' Tails: Hunter-Gatherer Settlement Systems in Archaeological Site Formation. *American Antiquity* 45: 4–20.

1982   Archaeology of Place. *Journal of Anthropological Archaeology* 1: 5–31.

2001   *Constructing Frames of Reference: An Analytical Method for Archaeological Use of Hunter-Gatherer and Environmental Data Sets*. University of California Press, Berkeley.

Bridges, P. S.

1989   Changes in Activities with the Shift to Agriculture in the Southeastern United States. *Current Anthropology* 30(3): 385–394.

1995a   Biomechanical Changes in Long Bone Diaphyses with the Intensification of Agriculture in the Lower Illinois Valley. *American Journal of Physical Anthropology*, Supplement 20: 68.

1995b   Skeletal Biology and Behavior in Ancient Humans. *Evolutionary Biology* 4: 112–120.

Brock, S. L.

1985   Biomechanical Adaptations of the Lower Limb Bones through Time in the Prehistoric Southwest. Ph.D. dissertation. University of New Mexico, Albuquerque.

Brock, S. L., and C. B. Ruff

1988   Diachronic Patterns of Change in Structural Properties of the Femur in the Prehistoric American Southwest. *American Journal of Physical Anthropology* 75: 113–127.

Brody, J. J.

1964   Design Analysis of the Rio Grande Glaze Pottery of Pottery Mound. M.S. thesis. University of New Mexico, Albuquerque.

1970   The Kiva Murals of Pottery Mound. *Verhandlungen des XXXVIII Internationalen Amerikanistenkongresses* 2:101–110. Paper presented in Munich, Germany.

Brown, J. A., and R. K. Vierra

1983   What Happened in the Middle Archaic?: Introduction to an Ecological Approach to Coster Site Archaeology. In *Archaic Hunters and Gatherers in the American Midwest*, edited by J. L. Phillips and J. A. Brown, pp. 165–195. Academic Press, New York.

Brown, J. K.

1970   A Note on the Division of Labor by Sex. *American Anthropologist* 72: 1073–1078.

Brown, K. M.

1991   Prehistoric Economics at Baker Cave: A Plan for Research. In *Papers on Lower Pecos Prehistory*, edited by S. A. Turpin, pp. 87–140. University of Texas Press, Austin.

Bryant, V. M., Jr.

1974   Prehistoric Diet in Southwest Texas. *American Antiquity* 39: 407–420.

1986a   Pollen: Nature's Tiny Capsules of Information. In *Ancient Texans*, edited by G. Zappler, pp. 50–57. Texas Monthly Press, Austin.

1986b   Prehistoric Diet: A Case for Coprolite Analysis. In *Ancient Texans*, edited by G. Zappler, pp. 132–135. Texas Monthly Press, Austin.

Buikstra, J. E.

1992   Diet and Disease in Prehistory. In *Disease and Demography in the Americas*, edited by J. Verano and D. Ubelaker, pp. 87–101. Smithsonian Institution Press, Washington, D.C.

Buikstra, J. E., and D. Ubelaker (editors)

1994   *Standards for Data Collection from Human Skeletal Remains*. Research Series 44. Arkansas Archaeological Survey, Fayetteville.

Carmichael, D. L.

1990   Patterns of Residential Mobility and Sedentism in the Jornada Mogollon Area. In *Perspectives On Southwestern Prehistory*, edited by P. E. Minnis and C. L. Redman, pp. 122–134. Westview Press, Boulder.

Carter, D. R., T. E. Orr, and D. P. Fyhrie

1989   Relationships between Loading History and Femoral Cancellous Bone Architecture. *Journal of Biomechanics* 22: 231–244.

Chamay, A., and P. Tschantz

1972   Mechanical Influences in Bone Remodeling: Experimental Research on Wolff's Law. *Journal of Biomechanics* 5: 173–180.

Collins, M. B., and J. H. Labadie
1999   Excavation, Rock Art Recordation, Surface Feature Documentation, and Survey at Amistad National Recreation Area. *Texas Archaeology* 43(4): 3–7.

Cordell, L. S
1984   *Prehistory of the Southwest*. Academic Press, New York.
1997   *Archaeology of the Southwest*. Academic Press, San Diego.
n.d.   Pottery Mound: Values and Current Status. Unpublished report.

Crotty, H.
1979   The Kiva Murals and the Question of Mesoamerican Influence. Paper presented at the symposium on New Directions in Native American Art, Albuquerque.

Currey, J. D.
1990   Physical Characteristics Affecting the Tensile Failure Properties of Compact Bone. *Journal of Biomechanics* 23: 837–844.

Danielson, D. R., and K. J. Reinhard
1998   Human Dental Microwear Caused by Calcium Oxalate Phytoliths in the Prehistoric Diet of the Lower Pecos Region, Texas. *American Journal of Physical Anthropology* 107: 207–307.

Dering, J. P.
1979   Pollen and Plant Macrofossil Vegetation Recovered from Hinds Cave, Val Verde County, Texas. M.A. thesis. Texas A&M University, College Station.
1999   Earth Oven Plant Processing in Archaic Period Economies: An Example from a Semi-arid Savannah in South-central North America. *American Antiquity* 64(4): 659–671.

Dibble, D. S., and E. R. Prewitt
1967   *Survey and Test Excavations at Amistad Reservoir, 1964–1965*. Texas Archaeological Survey Report 3. University of Texas, Austin.

Draper, P.
1985   Two Views of Sex Differences in Socialization. In *Male-Female Differences: A Biocultural Perspective*, edited by R. L. Hall, pp. 5–25. Praeger Press, New York.

Draper, P., and E. Cashdan
1988   Technological Change and Child Behavior among the !Kung. *Ethnology* 27: 339–365.

Eddy, F. W.
1958   A Sequence of Cultural and Alluvial Deposits in the Cienega Creek Basin, Southeastern Arizona. M.S. thesis. University of Arizona, Tucson.

Eddy, F. W., and M. E. Cooley
1983   *Cultural and Environmental History of Cienega Valley, Southeastern Arizona*. Anthropological Papers 43. University of Arizona Press, Tucson.

Ember, C. R.
1983   The Relative Decline in Women's Contribution to Agricultural Intensification. *American Anthropologist* 85: 285–304.

Ember, M., and C. R. Ember

1976   The Conditions Favoring Matrilocal versus Patrilocal Residence. *American Anthropologist* 73: 571–594.

Emslie, S. D.

1981   Prehistoric Agricultural Ecosystems: Avifauna from Pottery Mound. *American Antiquity* 4: 853–861.

Emslie, S. D., and L. L. Hargrave

1978   An Ethnobiological Study of the Avifauna of Pottery Mound, New Mexico. Paper presented at the 43rd Annual Society for American Archaeology Meeting, Tucson.

Eschman, P.

1992   *SLCOMM*. Eschman Archaeological Services, Albuquerque.

Fish, S. K., and P. R. Fish

1991   Comparative Aspects of Paradigms for the Neolithic Transition in the Levant and the American Southwest. In *Perspective on the Past, Theoretical Biases in Mediterranean Hunter-Gatherer Research*, edited by G. A. Clark, pp. 396–410. University of Pennsylvania Press, Philadelphia.

Fish, S. K., P. R. Fish, and J. H. Matson

1990   Sedentism and Settlement Mobility in the Tucson Basin Prior to AD 1000. In *Perspectives on Southwestern Prehistory*, edited by P. Minnis and C. L. Redman, pp. 76–163. Westview Press, Boulder.

1992   Sedentism and Agriculture in the Northern Tucson Basin. In *The Marana Community in the Hohokam World*, edited by S. K. Fish, P. R. Fish, and J. H. Matson, pp. 11–19. Anthropological Papers 56. University of Arizona, Tucson.

Gaulin, S., and H. Hoffman

1988   Functional Significance of Sex Differences in Spatial Ability. In *Human Reproductive Behavior: A Darwinian Perspective*, edited by L. Betzig, P. Turke, and M. Borgerhoff-Mulder, pp. 129–152. Cambridge University Press, Cambridge.

Hard, R. J.

1990   Agricultural Dependence in the Mountain Mogollon. In *Perspectives on Southwestern Prehistory*, edited by P. E. Minnis and C. L. Redman, pp. 135–149. Westview Press, Boulder.

Hard, R. J., R. P. Mauldin, and G. R. Raymond

1996   Mano Size, Stable Carbon Isotope Ratios, and Macrobotanical Remains as Multiple Lines of Evidence of Maize Dependence in the American Southwest. *Journal of Archaeological Method and Theory* 3: 253–318.

Hartnady, P., and J. C. Rose

1991   Abnormal Tooth Loss Patterns among Archaic Period Inhabitants of the Lower Pecos Region, Texas. In *Advances in Dental Anthropology*, edited by M. A. Kelly and C. S. Larson, pp. 267–278. Wiley-Liss, New York.

Hemmings, E. T., M. D. Robinson, and R. N. Rogers
1968   Field Report on Pantano Site (AZEE: 2:50). M.S. thesis. University of Arizona, Tucson.

Hester, T. R.
1983   Late Paleoindian Occupation of Baker Cave, Southwestern Texas. *Bulletin of the Texas Archaeological Society* 53: 101–119.

Hibben, F. C.
1955   Excavations at Pottery Mound, New Mexico. *American Antiquity* 21: 179–180.
1960   Prehispanic Paintings at Pottery Mound. *Archaeology* 13: 267–271.
1961   The Dating of Pottery Mound. In *Illustrated World History*, edited by G. Rainerd. George Rainbird, London.
1966   A Possible Pyramidal Structure and Other Mexican Influences at Pottery Mound, New Mexico. *American Antiquity* 31: 522–529.
1967   Mexican Features of Mural Painting at Pottery Mound. *Archaeology* 20: 84–87.
1975   *Kiva Art of the Anasazi at Pottery Mound*. KC Publications, Las Vegas.
1996   The Prehistoric Site of Pottery Mound. Lecture presented in the Maxwell Museum Lecture Series. University of New Mexico, Albuquerque.
n.d.   Excavation Procedures Used in the Salvage Project at Pottery Mound. Unpublished field notes.

Hilton, C. E., and R. D. Greaves
1995   Mobility Patterns in Modern Human Foragers. Paper presented at the American Association of Physical Anthropology Meeting, Oakland. Abstract published in the *American Journal of Physical Anthropology*, Supplement 20: 111.

Holt, B., and S. Churchill
2000   Behavioral Changes in European Upper Paleolithic Foragers: Evidence from Biomechanical Analysis of the Appendicular Skeleton. *American Journal of Physical Anthropology*, Supplement 30: 182.

Huckell, B. B.
1977   Excavations at the Hasqin Site: A Multi-Component Preceramic Site near Ganado, Arizona. Report submitted to the Arizona Department of Transportation.
1983   Additional Chronometric Data on Cienega Valley, Arizona. In *The Cultural and Environmental History of Cienega Valley, Arizona*, edited by F. W. Eddy and M. E. Cooley, pp. 57–58. Anthropological Papers No. 43. University of Arizona, Tucson.
1984a   *The Archaic Occupation of the Rosemont Area, Northern Santa Rita Mountains, Southeastern Arizona*. Arizona State Museum Archaeological Series 147(1). University of Arizona, Tucson.
1984b   The Paleoindian and Archaic Occupation of the Tucson Basin: An Overview. *Kiva* 49(3–4): 133–145.
1988   Late Archaeology of the Tucson Basin: A Status Report. In *Recent Research on Tucson Basin Prehistory: Proceedings of the Second Tucson Basin Conference*, edited

by W. H. Doelle, and P. R. Fish, pp. 57–80. Anthropological Papers No. 10. Institute for American Research, Tucson.

1990 Late Preceramic Farmer-Foragers in Southeastern Arizona: A Cultural and Ecological Consideration of the Spread of Agriculture in the Arid Southwestern United States. Ph.D. dissertation. University of Arizona, Tucson.

1995 *Of Marshes and Maize: Preceramic Agricultural Settlements in the Cienega Valley, Southeastern Arizona.* University of Arizona Press, Tucson.

Huckell, B. B., and E. W. Haury

1988 *Excavations at Two Late Archaic Sites in the Cienega Valley, Southeastern Arizona.* University of Arizona, Tucson.

Huckell, L. B.

1995 Farming and Foraging in the Cienega Valley. In *Of Marshes and Maize: Preceramic Agricultural Settlements in the Cienega Valley, Southeastern Arizona*, edited by B. B. Huckell, pp. 74–97. University of Arizona Press, Tucson.

1996 Paleoethnobotany of Late Preceramic/Early Ceramic Sites along the Santa Cruz River, Tucson, Arizona. Paper presented at the 61st Society for American Archaeology Meeting, New Orleans.

Huebner, J. A.

1991 Cactus for Dinner, Again!: An Isotopic Analysis of Late Archaic Diet in the Lower Pecos Region of Texas. In *Papers on Lower Pecos Prehistory*, edited by S. A. Turpin, pp. 175–190. University of Texas Press, Austin.

1995a The Isotopic Composition and Ecology of Archaic Human Diet in the Eastern Chihuahuan Desert. Ph.D. dissertation. University of Texas, Austin.

1995b Stable Isotope Analysis of Bone and Soft Tissue from Four Mummies from the Eastern Chihuahuan Desert of Texas. Paper presented at the 2nd World Congress on Mummy Studies, Cartagena, Colombia.

Hurtado, A. M., and K. R. Hill

1990 Seasonality in a Foraging Society: Variation in Diet, Work Effort, Fertility, and the Sexual Division of Labor among the Hiwi of Venezuela. *Journal of Anthropological Research* 46(3): 293–345.

Hurtado, A. M., K. R. Hill, H. Kaplan, and I. Hurtado

1992 Trade-offs between Female Food Acquisition and Child Care among Hiwi and Ache Foragers. *Human Nature* 3(3): 185–216.

Kelly, R. L.

1983 Hunter-Gatherer Mobility Strategies. *Journal of Anthropological Research* 39: 277–306.

1992 Mobility/Sedentism: Concepts, Archaeological Measures, and Effects. *Annual Review of Anthropology* 21: 43–66.

1995 *The Foraging Spectrum: Diversity in Hunter-Gatherer Lifeways*. Smithsonian Institution Press, Washington, D.C.

Lancaster, J.

1986 Groundstone. In *Short Term Sedentism in the American Southwest: The Mimbres*

Valley Salado, edited by B. A. Nelson and S. LeBlanc, pp. 177–190, Appendix D. University of New Mexico Press, Albuquerque.

Larsen, C. S.

1997   *Bioarchaeology: Interpreting Behavior from the Human Skeleton.* Cambridge University Press, Cambridge.

Larsen, C. S., and R. L. Kelly

1995   *Bioarchaeology of the Stillwater Marsh: Prehistoric Human Adaptation in the Western Great Basin.* American Museum of Natural History, New York.

Larsen, C. S., C. B. Ruff, M. J. Schoeninger, and D. L. Hutchinson

1992   Population Decline and Extinction in La Florida. In *Disease and Demography in the Americas,* edited by J. W. Verano and D. H. Ubelaker, pp. 25–39. Smithsonian Institution Press, Washington, D.C.

Leviness, W. T.

1959   Pottery Mound Murals. *New Mexico Magazine* 37(3): 22–52.

Lovejoy, C. O., A. H. Burstein, and K. G. Heiple

1976   The Biomechanical Analysis of Bone Strength: A Method and Its Application to Platycnemia. *American Journal of Physical Anthropology* 44: 489–506.

Lovejoy, C. O., and E. Trinkaus

1980   Strength and Robusticity of the Neandertal Tibia. *American Journal of Physical Anthropology* 53: 465–470.

Lundelius, E. L., Jr.

1974   The Last 15,000 Years of Faunal Change in North America. *Museum Journal* 15: 141–160.

Mabry, J. B.

1998   *Archaeological Investigations of Early Village Life in the Middle Santa Cruz Valley: Analysis and Synthesis.* Anthropological Papers 10. Institute for American Research, Tucson.

Mabry, J. B., and J. J. Clark

1994   Early Village Life on the Santa Cruz River. *Archaeology in Tucson* 8(1): 1–5.

Mabry, J. B., D. L. Swartz, H. Wocherl, J. J. Clark, G. H. Archer, and M. W. Linderman

1995   *Archaeological Investigations of Early Village Sites in the Middle Santa Cruz Valley: Descriptions of the Santa Cruz Bend, Square Hearth, Stone Pipe, and Canal Sites.* Anthropological Papers 13. Center for Desert Archaeology, Tucson.

Marmaduke, W. S.

1978   Prehistoric Culture in Trans-Pecos, Texas: An Ecological Explanation. Ph.D. dissertation. University of Texas, Austin.

Matson, R. G.

1991   *The Origins of Southwestern Agriculture.* University of Arizona Press, Tucson.

Minnis, P. E.

1985   Domesticating Plants and People in the Greater American Southwest. In *Prehistoric Food Production in North America,* edited by R. I. Ford, pp. 309–340. Anthropological papers 75. University of Michigan, Ann Arbor.

1992   Earliest Plant Cultivation in the Desert Borderlands of North America. In *The Origins of Agriculture: An International Perspective*, edited by C. W. Watson and P. J. Watson, pp. 121–141. Smithsonian Institution Press, Washington, D.C.

Minturn, P. D.

1998   Osteology of the Santa Cruz Bend Site: AZAA:12:746 (ASM). In *Archaeological Investigations of Early Village Sites in the Middle Santa Cruz Valley: Analysis and Synthesis*, edited by J. B. Mabry, pp. 739–755. Anthropological Papers 19. Center for Desert Archaeology, Tucson.

Minturn, P. D., and L. Lincoln-Babb

1995   Bioarchaeology of the Donaldson Site and Los Ojitos. In *Of Marshes and Maize: Preceramic Agricultural Settlements in the Cienega Valley, Southeastern Arizona*, edited by B. B. Huckell, pp. 106–115. University of Arizona Press, Tucson.

Murdock, G. P., and C. Provost

1973   Factors in the Division of Labor by Cultural Analysis. *Ethnology* 12: 203–225.

Nagurka, M. L., and W. C. Hayes

1980   An Interactive Graphics Package for Calculating Cross-sectional Properties of Complex Shapes. *Journal of Biomechanics* 13: 59–64.

Ogilvie, M. D.

2000   A Biological Reconstruction of Mobility Patterns at the Foraging to Farming Transition in the American Southwest. Ph.D. dissertation. University of New Mexico, Albuquerque.

Rafferty, G.

1985 The Archaeological Record on Sedentariness, Recognition, Development, and Implications. In *Advances in Archaeological Method and Theory* 8: 113–147.

Reinhard, K. J., B. W. Olive, and D. G. Steele

1989   Bioarchaeology Synthesis: Study Unit 3, Southwestern Division Archaeological Overview, U.S. Army Corps of Engineers. In *From the Gulf to the Rio Grande: Human Adaptation in Central, South, and Lower Pecos, Texas*, edited by T. R. Hester, S. L. Black, D. G. Steele, B. W. Olive, A. A. Fox, K. J. Reinhard, and L. C. Bement, pp. 129–140. Arkansas Archaeological Survey, Fayetteville.

Riskind, D. H.

1970   Pollen Analysis of Human Coprolites of Parida Cave. In *Archaeological Investigations at Parida Cave, Val Verde County, Texas,* edited by R. K. Alexander, pp. 89–101. Papers of the Archaeological Salvage Project 19, Texas Archaeological Society, Austin.

Ruff, C. B.

1987   Sexual Dimorphism in Human Lower Limb Bone Structure: Relationship to Subsistence Strategy and Sexual Division of Labor. *Journal of Human Evolution* 16: 391–416.

1992   Biomechanical Analyses of Archaeological Human Skeletal Samples. In *Skeletal Biology of Past Peoples: Research Methods*, edited by S. R. Saunders and M. A. Katzenberg, pp. 37–58. Wylie-Liss, New York.

1994   Biomechanical Analysis of Northern and Southern Plains Femora: Behavioral Implications. In *Skeletal Biology in the Great Plains: Migration, Warfare, Health, and Subsistence*, edited by D. W. Owsley and R. L. Jantz, pp. 235–245. Smithsonian Press, Washington, D.C.

1997   Structural Analysis of Long Bones from La Florida: Interpreting Behavior. *American Journal of Physical Anthropology*, Supplement 24: 201.

2000a   Biomechanical Analyses of Archaeological Human Skeletal Material. In *Biological Anthropology of the Human Skeleton*, edited by M. A. Katzenberg and S. H. Saunders, pp. 71–102. Alan R. Liss, New York.

2000b   Skeletal Structure and Behavioral Patterns of Prehistoric Great Basin Populations. In *Understanding Prehistoric Lifeways in the Great Basin Wetlands: Bioarchaeological Reconstruction and Interpretation*, edited by B. E. Hemphill and C. S. Larsen, pp. 290–320. University of Utah Press, Salt Lake City.

Ruff, C. B., and W. C. Hayes

1983a   Cross-sectional Geometry of Pecos Pueblo Femora and Tibiae—A Biomechanical Investigation I: Method and General Patterns of Variation. *American Journal of Physical Anthropology* 60: 359–381.

1983b   Cross-sectional Geometry of Pecos Pueblo Femora and Tibiae—A Biomechanical Investigation II: Sex, Age, and Side Differences. *American Journal of Physical Anthropology* 60: 383–400.

Ruff, C. B., and C. S. Larsen

1990   Postcranial Biomechanical Adaptations to Subsistence Strategy Changes on the Georgia Coast. In *The Archaeology of Mission Santa Catalina de Guale: Bicultural Interpretation of a Population in Transition*, edited by C. S. Larsen, pp. 94–120. American Museum of Natural History, New York.

Ruff, C. B., C. S. Larsen, and W. C. Hayes

1984   Structural Changes in the Femur with the Transition to Agriculture on the Georgia Coast. *American Journal of Physical Anthropology* 64: 125–136.

Ruff, C. B., and F. P. Leo

1986   Use of Computed Tomography in Skeletal Structure Research. *Yearbook of Physical Anthropology* 29: 181–196.

Ruff, C. B., H. M. McHenry, and J. F. Thackeray

1999   The "Robust" Australopithecine Hip: Cross-sectional Morphology of the SK 82 and 97 Proximal Femora. *American Journal of Physical Anthropology* 109: 509–521.

Ruff, C. B., A. Walker, and E. Trinkaus

1994   Postcranial Robusticity in *Homo*, III: Ontogeny. *American Journal of Physical Anthropology* 95: 35–54.

Ruff, C. B., A. Walker, E. Trinkaus, and C. S. Larsen

1993   Postcranial Robusticity in *Homo*, I: Temporal Trends and Mechanical Interpretation. *American Journal of Physical Anthropology* 91: 21–53.

Schorsch, R. L. G.

1962 The Physical Anthropology of Pottery Mound: A Pueblo IV Site in West-Central New Mexico. M.A. thesis. University of New Mexico, Albuquerque.

Shafer, H. J.

1986a The Lower Pecos Environment: Evolution of the Present Landscape. In *Ancient Texans*, edited by G. Zappler, pp. 34–49. Texas Monthly Press, Austin.

1986b Lower Pecos Lifeways: Housing and Daily Rounds. In *Ancient Texans*, edited by G. Zappler, pp. 94–131. Texas Monthly Press, Austin.

Smith, B. D.

1992 *Eastern North American Horticulture: Rivers of Change*. Smithsonian Institution Press, Washington, D.C.

Speth, J. D.

1990 The Study of Hunter-Gatherers in the American Southwest: New Insights from Ethnology. In *Perspectives on Southwestern Prehistory*, edited by P. E. Minnis and C. L. Redman, pp. 15–25. Westview Press, Boulder.

Steele, D. G., and B. W. Olive

1989 Bioarchaeology of Region 3 Study Area: Study Unit 3, Southwestern Division Archaeological Overview, U.S. Army Corps of Engineers. In *From the Gulf to the Rio Grande: Human Adaptation in Central, South, and Lower Pecos, Texas*, edited by T. R. Hester, S. L. Black, D. G. Steele, B. W. Olive, A. A. Fox, K. J. Reinhard, and L. C. Bement, pp. 93–114. Arkansas Archaeological Survey, Fayetteville.

Stock, J. A.

1983 The Prehistoric Diet of Hinds Cave (41VV 456), Val Verde, County, Texas: The Coprolite Evidence. M.A. thesis. Texas A&M University, College Station.

Story, D. A., and V. M. Bryant, Jr.

1966 *A Preliminary Study of the Paleoecology of the Amistad Reservoir Area*. National Science Foundation Research Report G2667. University of Texas, Austin.

Taylor, W. W.

1964 Tethered Nomadism and Water Territoriality: An Hypothesis. In *Proceedings of the XXXV International Congress of Americanists, Mexico*, part 2: 197–203.

Thiel, J. H.

1996 Faunal Exploitation of Early Villagers in the Sonoran Desert. Paper presented at 61st Society for American Archaeology Meetings, New Orleans.

Timoshenko, S. P., and J. M. Gere

1972 *Mechanics of Materials*. Van Nostrand Reinhold, New York.

Trinkaus, E., S. Churchill, and C. B. Ruff

1994 Postcranial Robusticity in *Homo*, II: Humeral Bilateral Asymmetry and Bone Plasticity. *American Journal of Physical Anthropology* 93: 1–34.

Trinkaus, E., and C. B. Ruff

1989 Diaphyseal Cross-sectional Morphology and Biomechanics of the Fond-de-Forêt 1 Femur and the Spy 2 Femur and Tibia. *Bulletin of the Society of Royal Belgian Anthropology and Prehistory* 100: 33–42.

Turpin, S. A.
1988   Seminole Sink: Excavations of a Vertical Shaft Tomb, Val Verde County, Texas. *Plains Anthropologist Memoir* 22(33, Part 2).
1992   More about Mortuary Practices in the Lower Pecos River Region of Southwest Texas. *Plains Anthropologist* 37: 7–17.
1995   The Lower Pecos River Region of Texas and Northern Mexico. *Bulletin of the Texas Archaeological Society* 66: 541–560.

Turpin, S. A., M. Hennberg, and D. W. Riskind
1986   Late Archaic Mortuary Practices in the Lower Pecos River Region, Texas. *Plains Anthropologist* 31: 295–315.

Van Devender, T. R., J. Betancourt, and M. Wimberly
1984   Biogeographical Implications of a Packrat Midden Sequence from the Sacramento Mountains, South Central New Mexico. *Journal of Quaternary Research* 22: 344–360.

Van Devender, T. R., and W. G. Spaulding
1979   Development of Vegetation and Climate in the Southwestern United States. *Science* 222: 701–710.

Vierra, B. J.
1990   Archaic Hunter-Gatherer Archaeology in Northwestern New Mexico. In *Perspectives on Southwestern Prehistory*, edited by P. E. Minnis and C. L. Redman, pp. 57–70. Westview Press, Boulder.
1998   *41MV120: A Stratified Late Archaic Site in Maverick County, Texas.* Archaeology Studies Report 7, Archaeological Survey Report 251. Center for Archaeological Research, University of Texas at San Antonio.

Vivian, P. B.
1961   Kachina: The Study of Animism and Anthropomorphism within the Ceremonial Wall Paintings of Pottery Mound and Jeddito. M.A. thesis. Iowa State University, Ames.

Voll, C.
1967   The Glaze Paint Ceramics of Pottery Mound. M.A. thesis. University of New Mexico, Albuquerque.

Ward, C. G.
1992   Shelby Brooks Cave: The Archaeology of a Dry Cave in the Texas Trans-Pecos. M.A. thesis. University of Texas, Austin.

Williams-Dean, G.
1978   Ethnobotany and Cultural Ecology of Prehistoric Man in South Texas. Ph.D. dissertation. Texas A&M University, College Station.

Wills, W. H.
1988a   Early Agriculture and Sedentism in the American Southwest: Evidence and Interpretations. *Journal of World Prehistory* 2: 445–488.

1988b *Early Prehistoric Agriculture in the American Southwest.* School of American Research, Santa Fe.

1992 Plant Cultivation and the Evolution of Risk Prone Economies in the Prehistoric American Southwest. In *Transitions to Agriculture in Prehistory*, edited by A. B. Gebauer and T. D. Price, pp. 153–176. Prehistory Press, Madison.

Wills, W. H., P. Crown, J. Dean, and C. Laudton

1994 Complex Adaptive Systems and Southwest Prehistory. In *Understanding Complexity in the Prehistoric Southwest*, edited by G. Gumerman and M. Gell-Mann, pp. 297–339. Addison-Wesley, New York.

Wills, W. H., and B. B. Huckell

1994 Economic Implications of Changing Land-Use Patterns in the Late Archaic. In *Themes in Southwest Prehistory*, edited by G. J. Gumerman, pp. 33–52. School of American Research Press, Santa Fe.

8

The Grass Is Greener on the Other Side

A Study of Pastoral Mobility on the Eurasian Steppe of Southeastern Kazakhstan

CLAUDIA CHANG

The main objective of this chapter is to address the issue of pastoral mobility in the temperate grassland environments of the northern Tian Shan Mountains of southeastern Kazakhstan. In particular, ethnographic observations on contemporary Kazakh pastoral mobility are used to provide a set of working strategies and methods for reconstructing prehistoric pastoral lifeways from survey and excavation data collected by the Kazakh-American Talgar Project. I also evaluate some of the assumptions put forth by Soviet and post-Soviet archaeologists on the origins and development of nomadic steppe cultures. These archaeologists have characterized the formative stages for the evolution of pastoral nomadism in this region of the Eurasian steppe as the Bronze Age (1700 BC to 900 BC) and Iron Age (ca. 800 BC to AD 500) (Akishev 1990; Alexeev 1991).

I examine archaeological notions of pastoral mobility, how mobility can be found in the archaeological record, and the relationship between mobility and population density of pastoral adaptations. Indeed, the ethnographic and ethnoarchaeological observations on pastoral adaptations in this region of Eurasia suggest that certain forms of pastoral mobility should be extant in the archaeological record. Most notably, patterns of vertical transhumance—where herders managing sheep and goats, cattle, and horses move between the lowland steppe areas (ca. 1,100 to 550 m in elevation) for fall through spring grazing lands and the upland alpine meadows of the Tian Shan foothill regions (ca. 1,800 to 2,600 m in elevation) for summer grazing lands (July through September)—should be apparent in the archaeological record. Yet the models of pastoral mobility put forth by the Soviet scholars have been skewed toward documenting the distribution of burial kurgans and graves, while overlooking sites or places used as pastoral loci (such as habitations, campsites, herding facilities, and shrines).

The Andronovo culture of the Bronze Age has been described as a steppe-based nomadic pastoral adaptation that brought innovation and change to the agricultural settlements of the desert-oases of Central Asia proper (Hiebert

1994). The Saka, an early Iron Age nomadic culture, have been characterized as horse-riding populations who practiced nomadic, semisedentary, and sedentary ways of life, depending upon their adaptations to the diverse conditions of the desert-oases and outlying steppe regions (Yablonsky 1995: 229). These prehistoric reconstructions of pastoral nomadism have been based upon materials excavated from burial mounds.

Little is known about the general settlement patterns or lifeways of the "reputed nomadic" populations of the Bronze and Iron Ages of Southeastern Kazakhstan. Furthermore, answers to questions concerning the nature of mobility in these pastoral nomadic cultures are inferred rather than documented. Two central cultural historical questions have guided the archaeology of this region: (1) How did the pastoral nomads of this area contribute to large-scale migrations of cultural groups that reputedly took place in both periods? and (2) Were the Iron Age pastoral nomads such as the Saka and Wusun the catalysts for change and innovation over the vast grasslands of Eurasia? The goal of my research has been to question and reinvestigate these models of pastoral nomadism.

The Study Area

The study area is 25 to 80 km east of Almaty, the largest city of the Republic of Kazakhstan, along the foothills and alluvial fan areas of the Zailiisky Alatau, a northern range of the Tian Shan Mountains. We specifically chose two distinct environmental zones for survey: (1) the Talgar fan, an alluvial fan or apron formed from the Talgar River that extends 10 to 15 km from the edge of the Talgar foothills; and (2) the Turgen/Asi upland valleys, a series of broad alpine river valleys nestled between the high glacier peaks of the Zailiisky Alatau.

The Talgar fan ranges in elevation from 550 m to 1,200 m and consists of grasslands, forested areas, cultivated cereal fields, orchards, and urban and rural housing development and infrastructure. The upper Turgen valley (known as Oi Jailau) and the Asi River valley range in elevation from 2,200 m to 2,600 m and are vegetated with meadows, steppe grasslands, and conifer forests on the northern exposures of the mountain slopes. These upland valleys are situated in a national forest and thus have been set aside for recreational and pastoral land use. Herd management has been collectivized since the 1930s; Kazakh herders keep mixed herds of sheep and goats, cattle, and horses in both environmental zones. Usually the contemporary herders practice short-distance vertical transhumance, moving between the upland pastures of Asi and Turgen (areas of summer pasture) and the alluvial fans (areas of winter pasture) of the Talgar, Issyk, Turgen, and Chilik Rivers.

Contemporary Kazakh Pastoral Mobility

When the topic of mobility is addressed by Western archaeology, examples are usually drawn from contemporary hunter-gatherers or horticulturalists that continue to practice mobility (Kelly 1992; Hard and Merrill 1992). The models used for examining mobility generally rely upon Binford's (1980, 1982) distinctions between residential mobility and logistical mobility. For a clear illustration of these distinctions and their pertinence to hunter-gatherer populations, I quote Kelly (1992: 44):

> *Collectors* move residentially to key locations (e.g. water sources) and use long logistical forays to bring resources to camp. *Foragers* "map onto" a region's resource locations. In general, foragers do not store food; they make frequent residential moves and short logistical forays. Collectors store food; they make infrequent residential moves but long logistical forays.

Does the distinction between "foraging" and "collecting" apply to pastoral adaptations? Kazakh herders, like all mobile herders, move their "food" or at least their livestock to them. In the "raw sense" of food-getting logistics, pastoral nomads are more like collectors than they are like foragers. Yet pastoral mobility is contingent upon moving the animals to adequate grazing territories and to available water as well as avoiding conflicts with other herders who also compete for the best grazing and water resources. Thus most animal husbandry systems require a kind of "mapping on" strategy, by which the herders lay claim to territories by moving from their camps (usually in a fixed place) on short logistical forays. Of course, the major difference between hunter-gatherers and pastoralists is that pastoralists always move with their food source but must maximize the general health and condition of their herds and flocks by establishing some means for claiming the best grazing lands and water sources for themselves. Transhumant herders in northern Greece lay claim to grazing areas by attempting "to pack their flocks" in a given territory marked by the location of their animal folds (Chang and Tourtellotte 1993). Herders in southern Greece, when confronted with an invader attempting to usurp grazing lands in the village communal lands, might resort to violence or fold burnings (Koster 1977).

My ethnographic observations of contemporary Kazakh herders (who maintain sheep and goats, cattle, and horses on the same landscapes as did their ancient predecessors) suggest that they employ a mobility strategy similar to Binford's "collectors." Contemporary Kazakh herders practice short-distance vertical transhumance (wintering in lowland areas and summering in upland areas) (Akishev 1990). Pastoral transhumance as I have observed it on the rural landscapes of northern Greece and southeastern Kazakhstan usually involves two fixed residential places as well as a series of places to which herders travel on logistical forays, in search of grazing land and water. If herders travel far from their encampments, they settle at night with their herds at a corral or place that can be protected from wolves.

The single most important factor in choosing where to move a flock or herd is population pressure. The herder must consider carrying capacity—the number of animals that can be supported on a given area of grazing lands (Barth 1961). Too many animals on any given grazing territory result in degradation of the pastures and ultimately reduce the carrying capacity of the land. Most herders realize this, especially if they milk their animals, because the quality of milk declines as the forage quality declines (Koster 1977). The Kazakh keep cattle and horses that they milk in order to produce cheese or *kumiss* (mare's milk), so individual herders note the impact of poor forage on the quality and quantity of milk production.

Kazakh herders practice what Kelly (1992: 45) describes as *territorial* or *long-term mobility*. Building on Binford's (1982, 1983) definition, Kelly describes territorial or long-term mobility as cyclical movements of a group utilizing a set of territories over a long period, such as a decade. We have observed Kazakh herders who use the upland plateaus of Turgen and Asi, returning every summer to the same *jailau* (summer pasture) in June through August. The location of the upland grazing territories might occasionally shift, although such changes have consequences. Herders who attempt to stake out a new grazing territory (closer to the dirt track, for example) may discover that they are invading someone else's territory and must therefore compete with others for available pasture. In the summer of 2002, when rainfall was plentiful in the upland Asi Valley (ca. 2,200 m in elevation), several households shared a broad valley along the Asi River approximately 1 km long that had been previously occupied by a single household and its herd. The son of one of the "newcomers," a herd owner of over six hundred sheep and goats, cattle, and horses, informed us that his father intended to return to his previous territory the following year.

Although cyclical or long-term mobility is practiced, the more typical pattern of mobility involves territories that are fixed between two known points: the summer yurt and the winter residence in a small town, village, or collective. Importantly, however, summer pasture territories involve more flexible patterns of use-rights, because upland grazing areas are considered to be "open" territories or common pasture lands.

Usually the main facilities—the felt yurt, a corral for holding the animals during the night, and milking facilities—-mark the location of a herding household's grazing territory. The system of herding requires the separation of the mixed herd into different grazing units by species. For example, twenty horses may be tended by one household member, forty cattle by another, and five hundred sheep by yet another. If the grazing territory is exceptionally rugged or inhabited by wolves, experienced herders are put in charge of the flock. The composition of the Kazakh herding household may fluctuate in the summer pasture area. In 1997 we met an older man who was spending his summer with his two daughters-in-law and their children and their herd of a thousand animals, while his sons cultivated their agricultural holdings in the lowlands. The following year the sons and father returned to the same territory with their herd, but without their wives. These ethnographic observations suggest the variations with regard to household labor, mobility patterns, and camp locations.

Researching Mobility in the Archaeological Record

The current models of pastoral mobility adopted by the Soviet archaeologists are based upon two systems of pastoral movement on the Eurasian steppe. The first is a long-distance system of horizontal movement across the steppes, where summer pastures are located in the south and winter pastures are located in the north; distances can range from 100 km to 1,000 km. The second is a short-distance system of vertical movement from the mountains and foothills to the lowland valleys, where summer pastures are located in the uplands and winter pastures are located in the lowlands; distances between pastures can range from 50 km to 100 km (Akishev 1990).

Contemporary pastoral transhumance between the upland valleys of Turgen and Asi (ca. 2,200 to 2,600 m in elevation) and the lowland steppe of Talgar, Turgen, Issyk, and Chilik (ca. 1,100 to 550 m in elevation) suggests that the current models of pastoral mobility for the Bronze and Iron Age are inadequate and lack sufficient empirical evidence. Since 1994 the Kazakh-American Talgar Project has conducted surface surveys and excavations with the expectation that the economies and land-use strategies of the ancient nomads can be reconstructed from archaeological data. Our methods, although standard for Western archaeology, differ from the long tradition of Russian and Soviet period archaeological research on the Eurasian steppe, which has been based upon evidence drawn from ancient texts and from the archaeology of mortuary complexes. Our survey and excavations are "works in progress" that have their own limitations. In the following sections I discuss how we designed our research, the theoretical framework and methods we employed, and the preliminary results of this research.

Ideally, an archaeological project designed to research pastoralism—both as an economy and as a land-use strategy—would require fine-grained chronological sequences and full spatial coverage of the study area through the use of sampling designs. Until the mid-1990s (when we introduced the practice of pedestrian surface survey) artifact scatters of ceramics, animal bones, and grinding stones found on the surface of plowed fields were overlooked and ignored. Even the preliminary reconnaissance surveys yielded information on artifact scatters that fall into the general chronological categories of the Bronze Age, Iron Age, or medieval period, based on ceramic typologies of the surface materials. The local archaeologists have been able to place the burial mounds into chronological periods on the basis of dimensions (height and diameter of burial mounds) and surface features (stone circles, slab-lined cists, stone or soil matrix of the mound). Some archaeological features such as the foundations of sod-houses known as *zimovki* (winter dwellings) have also been identified, as well as stone-coursed architecture from the Bronze and Iron Ages.

A definitive phase designation within broad chronological periods has yet to be constructed for the Bronze Age or Iron Age ceramic sherds of the Semirechye region. We can only make rough estimates that place artifact scatters or isolated ceramic finds into the following chronology: (1) Bronze Age (ca. 1700–900 BC); (2) Iron Age (ca. 700 BC–AD 500); (3) Turkic Period (ca. AD 600–900); (4) Medieval Islamic Period (ca. AD 800–1250); (5) Mongol Period (AD 1250–1500); and (6) historic Kazakh Period (ca. AD 1700 to present). Obviously the lack of a more precise chronological framework within these broad labels limits our ability to define subphases and subsequently to sort out palimpsests.

Archaeologists working in the western hemisphere have noted the methodological and theoretical problems with using surface survey data to infer ancient settlement patterns (Dewar 1991; Plog 1973; Rouse 1972). In particular, Robert Dewar (1991) has commented extensively on the fact that survey data have been misused in settlement-pattern analysis. He points out that settlement-pattern analysis often treats archaeological components within a single phase or period as contemporaneous, although sites within a given period or phase may not be contemporaneous or may even represent overlapping occupational periods. While archaeologists are fully aware that survey data represent remnants of past settlement-systems (Dewar 1991: 604), they still use such data to derive population estimates and the spatial distribution of settlements across given landscapes. In the case of our data, archaeological sites that conceivably span a thousand years but are placed within a single period (for example, the Iron Age) can hardly be considered suitable for detailed settlement-pattern analysis.

SURVEY METHODOLOGY AND RESULTS

The Kazakh-American Talgar Project conducted pedestrian surveys from 1997 through 1999 on the Talgar alluvial fan, a broad delta formed by the north-flowing Talgar River (approximately 150 sq km). We walked over 287 transects in plowed fields and along stream cuts. Our goal was to cover as broad an area of the fan as possible, using the Talgar River as a natural boundary and dividing the fan into eastern and western sections. We did so with a random sampling strategy.

From 1994 through 2002 the Kazakh-American Talgar Project also excavated four Iron Age sites on the Talgar alluvial fan and one Iron Age and one multicomponent Iron Age and Bronze Age site in the upland Turgen and Asi valleys. Most of the Iron Age sites show multiple occupation levels based upon *in situ* radiocarbon-dated contexts. These excavated sites usually have four to eight stratigraphic levels, indicating repeated occupations. Tuzusai (a small village hamlet) has evidence of at least six different horizons and four occupation levels. At Tseganka 8 six to eight occupational and building sequences have been documented for architectural features such as pit houses. The Taldy Bulak 2 site has six different occupational levels designated for different activity areas and features. The meaning of these sequences of site reoccupation is not entirely clear. They could represent (1) shifting locations of hamlets and small residential camps spanning a given occupational period; (2) repeated seasonal occupation by groups of mobile pastoralists or mixed herding-farming groups; or (3) abandonment and then reoccupation by sedentary groups.

What is particularly significant about the excavations at Tuzusai, Tseganka 8, and Taldy Bulak 2 (all Iron Age sites from the Talgar alluvial fan) is the overlapping radiocarbon sequence of Phases I–VI, spanning from 775 BC to AD 75 (Chang et al. 2002). All three sites appear to have overlapping periods of occupation in Phase V, spanning from 400 to 40 BC. These preliminary data suggest that the demographic expansion of the Iron Age, as represented by burial sites and settlement sites, might have taken place during Phase V. For the regional cultural history of Semirechye these two periods of occupation are of special interest, since the splendid Golden Warrior tomb, located in Issyk (about 20 km to the east of the Talgar fan) dates from 400 to 200 BC. If indeed Phase V does represent the peak of Iron Age settlement and demographic expansion, this suggests that it is also the formative period when the height of Saka wealth and status differentiation took place.

The single most important factor for testing these assumptions about demographic expansion and the evolution of hierarchy (as apparent from the archaeological remains of an extensive mortuary complex) is to develop a tight chronological framework with phase designations that can address the issues of (1) frequency of population relocation and (2) length of phases (Dewar 1991: 605). If our assumption that the Iron Age populations of the Talgar and Turgen/Asi area were mobile or at least semisedentary is correct, however, the

Table 8.1. Chronology of Tuzusai 1 (Excavated 1992–1996), Tseganka 8 (Excavated 1998–2000), and Taldy Bulak 2 (2001)

Stratigraphic sequence	Radiometric date	Calibrated result (2 Sigma, 95%)
VIII		
Tuzusai, Occupation 5, Unit V-11, Fire Pit 1	140±70 BP	Cal AD 1650 to 1950
VII		
Tuzusai, Burial 1, Animal bone collagen from sheep scapula	650±50 BP*	Cal AD 1275 to 1410
VI		
Tuzusai, Pit 24 fill	2020±40 BP* (Oxford)	Cal BC 100 to AD 75
Tuzusai, Pit 17 fill	2070±40 BP* (Oxford)	Cal BC 180 to AD 25
VIa		
Tseganka 8, Unit V-10, on subsoil	2190±80 BP*	Cal BC 400 to 40
V		
Tuzusai, Pit 30 B	2170±30 BP	Cal BC 335 to 290 and BC 230 to 115
Tseganka 8, Pit 13 bottom	2130±80 BP	Cal BC 350 to 300 and Cal BC 220 to 50
Taldy Bulak 2, Fire Pit 2, Horizon 3	2280±40 BP*	Cal BC 400 to 350 and Cal BC 310 to 210
Tuzusai, Pit 29	2230±30 BP	Cal BC 380 to 190
Tuzusai, Unit V-13, ash deposit	2170±60 BP	Cal BC 380 to 40
Tseganka 8, Pithouse 2, Floor 2	2130±40 BP*	Cal BC 385 to 100
IV		
Tuzusai, Pit 22	2310±50 BP* (Oxford)	Cal BC 415 to 345 and BC 310 to 210
Tuzusai, Pit 8—'92	2320±40 BP* (Groningen)	Cal BC 410 to 260 and Cal BC 230 to 115
III		
Tseganka 8, Pithouse 3, Floors 3a/b	2390±70 BP	Cal BC 775 to 370
Taldy Bulak 2, Unit D-8, Ash Pit, Horizon 4	2400±70 BP	Cal BC 780 to 370
II		
Tseganka 8, Pithouse 3, Floor 4	2190±40 BP	Cal BC 350 to 310 and Cal C 210 to 40
Tseganka 8, Storage Pit '98	2300±80 BP*	Cal BC 740 to 710 and BC 535 to 80

Note: The radiometric dating and calibrations were done by Beta Analytic, Inc. (Stuiver 1998). *AMS (Accelerated Mass Spectrometry) dates—obtained at Beta Analytic, Inc., unless specified.

problem of inferring settlement patterns from survey and excavation data will be even more serious. The paleoethnobotanical and faunal materials collected from Taldy Bulak 2, Tuzusai, and Tseganka 8 clearly demonstrate the presence of a mixed economy based upon cereal cultivation of wheat, millet, and barley and a herding system reliant upon sheep and goats, cattle, and horses (with very small percentages of camels) (Benecke 1999–2000; Rosen et al. 2000). Almost all of the Talgar fan Iron Age settlements currently excavated show multiple occupation levels, thus indicating that a single location was used, abandoned, and then reused over a period of six hundred or seven hundred years. Does this represent long-term pastoral mobility, seasonal mobility, or repeated sequences of use, abandonment, and reoccupation? Such questions can be posed but not answered by our data.

Some sites such as Tseganka 8 appear to be occupied in a confined locale, where repeated floors and occupation levels show dense concentrations of artifacts. Other sites such as Taldy Bulak 2 and Tuzusai are distributed over a large territory (up to 1 km^2). Does this suggest that Taldy Bulak 2 and Tuzusai were temporary encampments that were reoccupied year after year, in the same manner that the contemporary Kazakh yurts and encampments are reoccupied? We have noted that the Kazakh encampments might shift 10 to 100 m from season to season in the upland plateaus of Asi and Turgen. Low-density sites spread over large areas could represent seasonal temporary encampments with many different activity areas, while tightly packed pithouse sites with repeated occupations in a single confined locale represent hamlets occupied on a permanent, year-round basis. In Pithouse 3 at Tseganka 8 we noted the thick packing of one floor level over the next and recorded over six to eight different flooring layers, often with little or no fill levels between some of the floor levels (3a and 3b). Could these repeated floor layers represent episodes of continuous occupations or periodic remodeling of dwellings that were occupied on a permanent or semipermanent basis?

Until we have excavated more sites and noted the variation in features, occupation episodes, and their palimpsest nature, we cannot determine the settlement patterns for these agro-pastoral populations. Yet the wide variety of features, artifact distributions, and densities at the three excavated Iron Age sites overlapping in time suggests a wide range of variation in the types of settlements and episodes of use in any give phase or subphase. Settlement packing (of both people and animals) must have been a problem during the height of the Iron Age occupation of the Talgar alluvial fan (Phases IV through VI). The majority of burial mounds in Talgar, Issyk, and Bes Shatyr also date to this period (ca. 400 BC–100 AD), often labeled the Saka-Wusun period by Soviet archaeologists (Moshkova 1992). Clearly the relationship between high numbers of mortuary

sites (burial mounds) and settlement sites during these phases should be indicative of the height of demographic expansion on the alluvial fans of Semirechye.
It would also make sense that the increased population density found in these
phases corresponds to increasing patterns of social stratification (noted from the
inventories found at the burial mounds) and greater reliance upon agrarian food
production (Akishev and Kushaev 1963; Chang and Tourtellotte 1998).

THE PROBLEM OF SITE OR PLACE VISIBILITY
IN THE ARCHAEOLOGICAL RECORD

Local archaeologists have conducted their own survey reconnaissance and
have documented and inventoried known archaeological sites, plotting these
locations on a 1:500,000 scale *Arheologiskaya Karta Kazakhstana* (Ageeva et
al. 1960). Western-style archaeological surveys, however, have only been introduced since the mid-1990s by the Kazakh-American Talgar Project (Chang et al.
1999). Surface surveys were made by pedestrian walking in upland areas where
site visibility (especially stone outlines of houses, burial mounds, and graves) is
high and on the alluvial fan in plowed fields and along river and erosion cuts.
Isolated finds of ceramics, grinding stones, and other artifacts were recorded,
as well as artifact scatters and architectural features. Each cluster of artifacts or
architectural features within a 100 m² area was recorded as a locus. The results of
our survey have been reported elsewhere (Chang and Tourtellotte 2000). Here
I provide data on our preliminary results from these surveys.

Table 8.2. Survey Results from the Talgar Alluvial Fan (1997–1999) and the Turgen-
Asi (1997–2002) Surveys

Locus category	Talgar alluvial fan	Turgen-Asi uplands
Iron Age sites	59	7
Probable Iron Age sites (sherd scatters)	29	
Iron Age kurgans	182	60
Loci with 1 sherd	100	
Loci with 2 sherds	16	
Loci with 3 sherds	25	
Loci with grinding stone	16	
Bronze Age sites		6
Bronze Age kurgans		29
Medieval sites	2	
Kurgans of unknown period		85
Totals	427	187

An area of approximately 150 km^2 in the Talgar alluvial fan yielded a total of 427 loci, a density of about 2.8 loci per km^2. About 270 of these loci (63%) were identified as probable Iron Age sites. The remaining were Bronze Age, medieval, or of indeterminate period. In an area of approximately 46 km^2 in the Turgen/Asi upland valleys, approximately 187 loci were recorded, a density of about 4.1 loci per km^2. About 35% of these loci are Iron Age, and about 19% are Bronze Age. Only six Bronze Age settlements and seven Iron Age settlements were found; the remaining loci are kurgans (burial mounds).

Site visibility was a far greater problem on the Talgar alluvial fan than in the upland valleys of Turgen and Asi. We occasionally found Iron Age artifacts in the profile cuts of streambeds but not on the plowed surfaces, so it is clear that sites on the Talgar fan were often deeply buried under the wind-blown and river-deposited loess soils. The loci that yielded the highest number of surface artifacts (sherds, grinding stones, and bones) were often the most disturbed and destroyed by modern-day agricultural activities. We also discovered that the relationship between surface finds and subsurface remains was skewed, as noted by Jack Nance and Bruce Ball (1986) when developing test pit sampling strategies for the discovery of buried sites. Nance and Ball (1986) conclude that sites with more surface artifacts probably represent a higher density of buried artifacts. These sites are more likely to be recovered by using test pit sampling than by surface surveying. In our surface surveys, it may indeed be the case that both low-density sites (surface and subsurface) and high-density sites are present, but deeply buried sites will be invisible and therefore not found on surface surveys (Shott 1995; Wandsnider and Camilli 1992). Many of our surface surveys on the Talgar fan were done on plowed fields, already indicating the skewed nature of these artifact scatters.

Michael Shott (1995: 478) notes: "But surface documents [artifact scatters] are not merely limited and slightly skewed samples of underlying records [buried artifact deposits]. They also can contain a strong random element, such that successive episodes of cultivation do not necessarily expose similar numbers, distributions, or kinds of artifacts."

Our ethnographic observations of Kazakh corrals and campsites in the upland valleys of Asi and Turgen indicate that the pastoral nomadic camps, especially summer camps, have a low artifact density and would probably not be visible after abandonment. A prime pastoral location used successively over many years, however, should yield a higher artifact density over time. But these repeated occupations might show up as low-density surface remains over a large area (as at Taldy Bulak 2) rather than as dense concentrations of artifacts within a confined area (as at Tseganka 8). The most distinguishing element of the pastoral site is the corral (marked by an architectural feature and deposits of animal dung), a feature that might not preserve well in the archaeological record. Ironically, the Bronze and Iron Age mortuary sites in the uplands are more visible, because the surface topography and geomorphology of the uplands contribute to high site visibility. We may have been unable to find many settlement sites in the uplands because the artifact scatters associated with such buried sites are invisible in these upland grasslands. Bronze Age burials include cist-lined graves and ossuaries placed inside rectangular stone wall structures, while graves from the Iron Age through the Turkic and medieval periods are marked by burial mounds. The Iron Age to medieval burial mounds could conceivably be used as indications of population density, particularly because it was the custom to place only one or two burials in each mound.

In contrast, the alluvial fan areas north of the Tian Shan Mountains have been exposed to processes of rapid soil deposition by wind or water. The small village and hamlet sites of the Iron Age are deeply buried, usually under 0.5 to 1.0 m of loess. Deep plowing and bulldozing of prime agricultural land since the 1960s have exposed many archaeological sites in the Talgar alluvial fan. We continue to be puzzled, however, by the lack of Bronze Age materials found there. Is this due to the geomorphology of the alluvial fan (Bronze Age sites could be covered by 1 m or more of loess), or does it represent the lack of Bronze Age settlement on the lowland steppe areas?

Insights into Pastoral Mobility on the Eurasian Steppe

From this preliminary research, it appears that the Bronze Age populations of Semirechye were pastoral nomads, while the Iron Age populations were semisedentary agro-pastoralists. First, the locations of the Bronze Age settlements and mortuary complexes in the upland plateaus of Asi and Kurgen but not in the fertile lowlands suggest that the populations of that period were practicing some kind of vertical transhumance. The contemporary climate (rainfall and temperature) statistics show that Asi and Turgen are located in areas without enough frost-free days to sustain the cultivation of crops. The contemporary Kazakh herders who pasture their animals during July through September in these upland valleys certainly do not practice any form of cultivation. Whether present-day climatic conditions are a good indicator of the existing climate during the Bronze Age clearly needs further investigation.

The number of lowland settlements and burial mounds and their overall density on the Talgar alluvial fan indicate a demographic expansion of populations in the thousand-year period of the Iron Age on the fertile alluvial fan. The paleoethnobotanical data at Tuzusai, Tseganka 8, and Taldy Bulak 2 all indicate the presence of cultivated species of millet, wheat, barley, and possibly rice

(Arlene Rosen, personal communication). Yet these data may be a product of the limitations of our survey methods and results. We have yet to locate Bronze Age deposits on the Talgar alluvial fan. It is probable that agro-pastoralists or nomadic pastoralists of the Andronovo period settled or utilized these alluvial fans, which would have provided good grassland environments for pastures or for foraging and for incipient agriculture.

Our current paleoethnobotanical and zooarchaeological research on the excavated Iron age sites of Tuzusai, Tseganka 8, and Taldy Bulak 2 indicates a mixed cereal economy of millet, wheat, barley, and possibly rice and a herding component of sheep and goats, cattle, horses, and camels. The upland Bronze Age sites of Asi 1 and Asi 2 (occupied in the early Andronovo period, ca. 1600 to 1400 BC) show no evidence of cultivated plants but do contain evidence for the herding of sheep, goats, and cattle. The Iron Age site of Kizil Bulak 3 in the Turgen Valley is a seasonal site with over 12 different periods of occupation that shows evidence of sheep, goats, cattle, and horses and wild species of plants. Some grinding stones found at Kizil Bulak 3 also indicate processing of gathered plants or agricultural grains brought from the lowland agricultural areas.

From such data we can infer that the upland valleys of Turgen and Asi were best suited for seasonal transhumance, most likely during the summer months. That does not necessarily mean, though, that the Bronze Age populations of Semirechye did not practice agriculture in other areas. It suggests that we have found the place where agriculture would have been most productive in both the Bronze and Iron Ages: on the alluvial fans and in the lowlands. The Iron Age steppe sites represent a mixed cereal and animal husbandry economy and a settlement-pattern that indicates a year-round or seasonal occupation during the summer when agrarian activities took place. The Iron Age sites found in the uplands were not agricultural sites; the paleoethnobotanical data show only evidence of wild plants (Rosen, personal communication). Such sites represent the use of the uplands for pastoral activities during the summer months.

The Bronze Age excavations at Asi 2 suggest the existence of a nomadic pastoral economy. Whether the Bronze Age nomadic pastoral populations practiced agriculture is still unclear. If Bronze Age settlements could be identified on the Talgar alluvial fan, we might be able to test this. Still, beyond these generalizations, neither the excavations nor the surveys of the Talgar alluvial fan or the Turgen/Asi upland valleys are sufficient to produce an accurate reconstruction of mobility patterns during the Bronze Age or Iron Age in Semirechye.

The following questions should guide future research in the Semirechye region of southeastern Kazakhstan.

(a) What types of species were predominantly herded by the early Bronze Age populations?

(b) Were these species herded over sets of territories in a system of short-distance vertical transhumance (radius of 50 km or less) or in a system of long-distance horizontal transhumance (radius of 50 to 500 km or more)?

(c) What was the system of spatial organization used by agro-pastoralists in exploiting the Talgar alluvial fan?

(d) Did the agro-pastoralists spend the whole year at a given village or hamlet, or did they circulate seasonally or yearly over a set of territories?

(e) Do the repeated occupations at the Iron Age settlement sites of Talgar represent abandonment and reuse over a long period or within short-term phases?

As more surveys and excavations are conducted on the Talgar alluvial fan and in the upland Turgen and Asi Valleys, the issue of pastoral mobility needs to be considered. A larger, regional coverage of both areas may allow us to posit the existence of a system of either long-distance horizontal or short-distance vertical transhumance. We must look for broad patterns of spatial organization and indicators of seasonality at sites found in both environmental zones. For example, if short-distance vertical transhumance was practiced during the Iron Age of Semirechye, then there should be evidence for winter occupation on the Talgar alluvial fan and summer occupation in the upland Turgen and Asi Valleys. In the same vein, if semisedentary agro-pastoralism was practiced in the Iron Age, permanent year-round villages and hamlets should be found on the Talgar alluvial fan, while temporary campsites or house structures should be found in the upland Turgen and Asi Valleys. Such inferences must be drawn from a comparison of the excavated materials found at Iron Age sites in both the upland and lowland zones in conjunction with surface surveys. Indicators such as shared ceramic styles, seasonal usage, and tool assemblages of sites in both zones might then allow comparative analysis of settlement patterns and overcome some of the problems created by site visibility and palimpsests.

Perhaps the greatest single stumbling block for the archaeology of this region is the assumption that pastoral nomadism was the sole economic base of the Bronze Age and Iron Age populations. The concept of pastoral mobility became a means by which the Soviet period archaeologists could avoid conducting settlement pattern-analysis. I agree wholeheartedly with Dewar (1991) and others who question settlement-pattern analysis when length of occupations at given sites is not considered adequately. But a necessary first step of any study on pastoral nomadism in prehistory is detailed analysis of the distribution of sites across a physical space, especially space that can be demarcated into different environmental zones. We know from the many studies of pastoral nomadism, semisedentary pastoralism, and sedentary agro-pastoral groups that there is tremendous variation in the spatial organization of places in a given landscape utilized by people who spend some portion of their lives herding and husbanding animals (see Chang 1992). Thus it seems mandatory that archaeologists consider the spatial distribution of archaeological loci across these landscapes, even if the sites themselves represent at best remnants of the past settlement-system.

ACKNOWLEDGMENTS

The Kazakh-American Talgar Project has been generously funded by the Wenner-Gren Foundation for Anthropological Research and the National Geographic Society from 1994 to 1996. Field research from 1994 through 2002 was funded by the National Science Foundation Archaeology Program Grant SBR-9603361, "The Social Evolution of Steppe Communities in Southeastern Kazakhstan and the Rise of Civilization." K. M. Baipakov, F. P. Grigoriev, and A. N. Mariashev, all of the Republic of Kazakhstan, were of invaluable assistance during the fieldwork. P. A. Tourtellotte directed the surveys described in this chapter and provided the maps and photographs. I wish to thank Pei-Lin Yu, Frédréic Sellet, and Russell Greaves for inviting me to participate in the symposium where this work was first presented. I welcomed the comments of the two anonymous reviewers who read this manuscript, who urged me to consider the problem of mobility within both the ethnoarchaeological and archaeological contexts of nomadic pastoral societies.

REFERENCES

Ageeva, E. I., K. A. Akishev, G. A. Kushaev, A. G. Maksimova, and T. N. Senigova (compilers)
1960   *Arheologicheskaya Karta Kazakhstana*. Alma-Ata: Izdatel'stvo Akademenni Nauk Kazakhskoi SSR.
Akishev, K. A.
1990   Les nomades à cheval du Kazakhstan Antique. In *Nomades et sédentaires en Asie Centrale: Apports de archéologie et de l'ethnologie*, edited by H.-P. Francfort, pp. 15–18. Éditions du Centre National de la Recherche Scientifique, Paris.
Akishev, K. A., and G. A. Kushaev
1963   *Drevnyaya Kul'tura Sakov n Usunei Dolini Reka Ili* (Ancient Saka and Wusun Culture of the Ili River Valley). Alma-Ata: Nauka.
Alexeev, V. P.
1991   A Brief Cultural History. Lectures given at Harvard University, transcribed by Geraldine Reinhardt-Waller. Chapter 7, Bronze Age in Eurasia, lectures 7 and 8, delivered July 17 and 29, 1991, and chapter 8, Iron Age in Eurasia, delivered August 5, 1991.
Barth, F.
1961   *Nomads of South Persia*. Little, Brown, Boston.

Benecke, N.

1999–2000   Unpublished Reports on the Faunal Remains from Iron Age Sites in the
Talgar Alluvial Fan. On file in the Department of Anthropology and Sociology,
Sweet Briar College.

Binford, L. R.

1980   Willow Smoke and Dogs' Tails: Hunter-Gatherer Settlement Systems and Archaeological Site Formation. *American Antiquity* 45(1): 4–20.

1982   The Archaeology of Place. *Journal of Anthropological Archaeology* 1: 5–31.

1983   *In Pursuit of the Past*. Thames and Hudson, London.

Chang, C.

1992   Archaeological Landscapes: The Ethnoarchaeology of Pastoral Land Use in the
Grevena Province of Greece. In *Space, Time, and Archaeological Landscapes*, ed-
ited by J. Rossignol and L. Wandsnider, pp. 65–89. Plenum Press, New York and
London.

Chang, C., and P. A. Tourtellotte

1993   The Ethnoarchaeological Survey of Pastoral Transhumance Sites. *Journal of Field
Archaeology* (3): 249–264.

1998   The Role of Agro-pastoralism in the Evolution of Steppe Culture in the Semirechye Area of Southern Kazakhstan during the Saka/Wusun Period (600 BCE–
400 CE). In *Bronze Age and Early Iron Age Peoples of Central Asia*, edited by Victor
H. Mair, pp. 264–279. Institute of Man Press and the University of Pennsylvania
Press, Washington, D.C., and Philadelphia.

2000   The Kazakh-American Talgar Project Archaeological Field Surveys in the Talgar
and Turgen-Asi Areas of Southeastern Kazakhstan: 1997–1999. In *Kurgans, Ritual Sites, and Settlements: Eurasian Bronze and Iron Age*, edited by J. Davis-Kimball, E. M. Murphy, L. Koryakova, and L. T. Yablonsky, pp. 83–88. BAR (British
Archaeological Reports) International Series 890.

Chang, C., P. A. Tourtellotte, K. M. Baipakov, and F. P. Grigoriev

1999   The Kazakh-American Talgar Archaeological Surveys in 1997 and 1998 in the
Talgar Region. *Izvestia, obshestvennye nauk* (219): 168–184.

2002   The Evolution of Steppe Communities from the Bronze Age through Medieval
Periods in Southeastern Kazakhstan (Zhetysu). A. Kh. Margulan Institute of Archaeology, Sweet Briar/Almaty.

Dewar, R. E.

1991   Incorporating Variation in Occupation Span into Settlement-Pattern Analysis.
American Antiquity 56(4): 604–620.

Hard, R. J., and W. L. Merrill

1992   Mobile Agriculturalists and the Emergence of Sedentism: Perspectives from
Northern Mexico. *American Anthropologist* n.s. 94(3): 601–620.

Hiebert, F. T.

1994   *Origins of the Bronze Age Oasis Civilization in Central Asia*. American School of

Prehistoric Research Bulletin 42. Peabody Museum of Archaeology and Ethnology, Harvard University, Cambridge, Mass.

Kelly, R. L.

1992   Mobility/Sedentism: Concepts, Archaeological Measures, and Effects. *Annual Review of Anthropology* 21: 43–66.

Koster, H. A.

1977   The Ecology of Pastoralism in Relation to Changing Patterns of Land Use in the Northeast Peloponnese. Ph.D. dissertation. University of Pennsylvania.

Moshkova, M. G.

1992   *Stepnaya polosa Asiatskoi chasti SSSR v skifo-sarmatskoi vremi* (Steppe Region of the Asiatic Part of the SSSR in the Scythe-Sarmanti Time). Nauka Izdatel'stvo, Moscow.

Nance, J. D., and B. F. Ball

1986   No Surprises?: The Reliability and Validity of Test Pit Sampling. *American Antiquity* 51(3): 457–483.

Plog, F. T.

1973   Diachronic Anthropology. In *Research and Theory in Current Archaeology*, edited by C. L. Redman, pp. 181–198. John Wiley, New York.

Rosen, A. M., C. Chang, and F. P. Grigoriev

2000   Paleoenvironments and Economy of Iron Age Saka-Wusun Agro-pastoralists in Southeastern Kazakhstan. *Antiquity* 74: 611–623.

Rouse, I.

1972   Settlement Patterns in Archaeology. In *Man, Settlement, and Urbanism*, edited by P. J. Ucko, R. Tringham, and G. W. Dimbleby, pp. 95–107. Duckworth, London.

Shott, M. J.

1995   Reliability of Archaeological Records on Cultivated Surfaces: A Michigan Case Study. *Journal of Field Archaeology* 22(4): 475–490.

Stuiver, M., et al.

1998   INTCAL98 Radiocarbon Age Calibration. *Radiocarbon* 40(3): 1041–1083.

Wandsnider, L., and E. L. Camilli

1992   The Character of Surface Archaeological Deposits and Its Influence on Survey Accuracy. *Journal of Field Archaeology* 19(2): 169–188.

Yablonsky, L. T.

1995   The Material Culture of the Saka and Historical Reconstruction, Chapter 14. In *Nomads of the Eurasian Steppes in the Early Iron Age*, edited by J. Davis-Kimball, V. A. Bashilov, and L. T. Yablonsky, pp. 201–240. Zinat Press, Berkeley.

9

From Atlatl to Bow and Arrow

Implicating Projectile Technology in Changing Systems of Hunter-Gatherer Mobility

PEI-LIN YU

In regional culture histories, the transition from large, broad-based projectile points to smaller, lightweight forms is often cited as a shift in the launching method from spearthrower (or atlatl) to bow and arrow (cf. Aikens and Higuchi 1982: 109; Cabrera Valdez 1984: 279; Grayson 1994: 250). Much research on the projectile transition has focused on intrinsic attributes of points in order to separate them into discrete types (Bettinger and Eerkens 1999: 231; Beck 1998: 21). In addition to formal variation in stone points, the reduction method may differ. Analysis of one sample from northeastern North America showed that dart points are typically reduced from cores and arrow points from flakes (Nassaney and Pyle 1999: 251–252).

Projectile point types are often used as chronological markers or "guide fossils" (Huckell 1996: 326), although variation in form may also result from mechanically conditioned behaviors, such as breakage, repair, and resharpening (Huckell 1996: 327). In a global survey, Pierre Cattelain (1997: 232) found that projectile point form alone is not correlated with hafting contexts or means of launching. Charlotte Beck's (1998) analysis of examples from Gatecliff Shelter indicates that neck width is acted upon by selective forces and is useful in distinguishing darts from arrow points. Statistical tests for many archaeological sequences show that small, lightweight points replace or augment large, heavy, broad-based points (Shott 1997).

The projectile transition occurred at different times, and at different rates, throughout the world. The transition to bow and arrow never occurred in Australia. Atlatls and bows and arrows were used in tandem in the recent past in the Arctic, the North American Southeast, and parts of Mesoamerica. Recent efforts to explain the variation in scope and timing of the projectile transition have focused on distinguishing between *in situ* development and diffusion, especially in North America (cf. Bettinger and Eerkens 1999; Nassaney and Pyle 1999), then proceeding to test models for different modes of transmission. Robert L. Bettinger and Jelmer Eerkens (1999: 240) explicitly define the projectile transition as a cultural evolutionary stage indicative of a culture's versatility and innovativeness.

Technology, however, unlike cultural markers such as style, articulates closely with adaptive strategies of resource use and mobility and therefore is expected to co-vary with selective forces (Beck 1998: 23; Knecht 1997: 2). If selective forces are less intense in a given case, technological behaviors may be adopted and incorporated as part of the preexisting system. In this case the material record would show the new and old technologies coexisting, at least for a time. More intense selective forces and *in situ* development of a new technology may lead to total replacement of the old system and corresponding replacement in the archaeological record.

This chapter builds on the substantial body of work on projectile point classification and development of regional chronologies by exploring the larger adaptive context of the projectile transition. In order to do this, it is necessary to compare archaeological sequences that share similar projectile transitions but are geographically distinct. Defining and explaining changes in a varied set of adaptive contexts should shed new light on the significance of the projectile point transition for the larger issue of cultural evolution.

AREAS OF STUDY

The archaeological records of northeast Asia, southern Europe, and the North American Great Basin all indicate that large projectile points were either replaced by or augmented by smaller, lightweight points.

The decrease in sites containing only large points is a defining characteristic of the projectile transition, regardless of time or place (Figure 9.1).

The following cultural sequences were selected because they include well-dated projectile point transition sites or levels within multicomponent sites: (1) Late Paleolithic/Initial Jomon periods of central Japan and Hokkaido; (2) Late Solutrean/Early Magdalenian periods of Cantabrian Spain; and (3) Late Archaic/Early Prehistoric periods of the North American Great Basin. Table 9.1 shows the sites discussed in this chapter.

ESTABLISHING ARCHAEOLOGICAL RELATIONSHIPS

Basic information on lithic tool types and counts, faunal species present, and methods of excavation is routinely recorded in study area archaeological site reports. These data are useful in determining the nature of the tool kit, the animals targeted, and the depositional context of artifacts. Study area sites or site levels that bracket the projectile transition were selected.

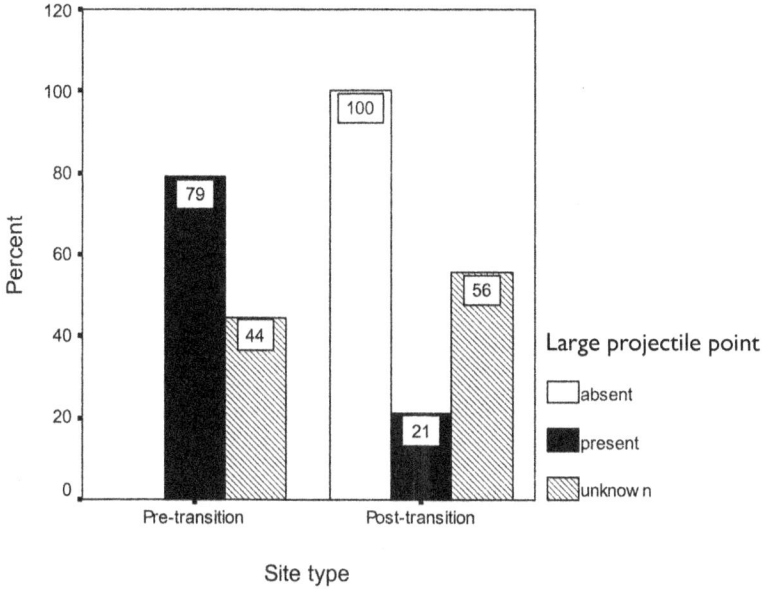

Figure 9.1. Bar graph of pre- and post-transitional projectile points for the study areas.

Lithic tool density (*n* tools per cubic meter of excavated sediment) in the study areas tends to stay within a range of about 0–100 tools/m³ (Figure 9.2). Post-transition sites show a Y-shaped distribution after ca. 17,000 BP, in which they become densely deposited or continue to be sparsely deposited.

Detailed information on sediment accumulation was not available for all sites, but light lithic tool deposition is apparently an important characteristic of

Table 9.1. Projectile Point Transition Sites in the Analysis

Coastal Spain	Japan	North American Great Basin
La Riera	Suzuki	Corn Creek Dunes
El Rascano	Nogawa	Danger Cave
El Juyo	Uenodaira	Hogup Shelter
Ekain	Shirataki-Hattori dai	Sudden Shelter
Altamira	Tachikawa	Cowboy Cave
Ambrosio	Kidosaku	
El Castillo	Horokazawa	
El Pendo	Midorigaoka	
Chufin	Fukui	
Cueto de la Mina	Natsushima	
	Kamikuroiwa	
	Kosegasawa	
	Yubetsu-Ichikawa	

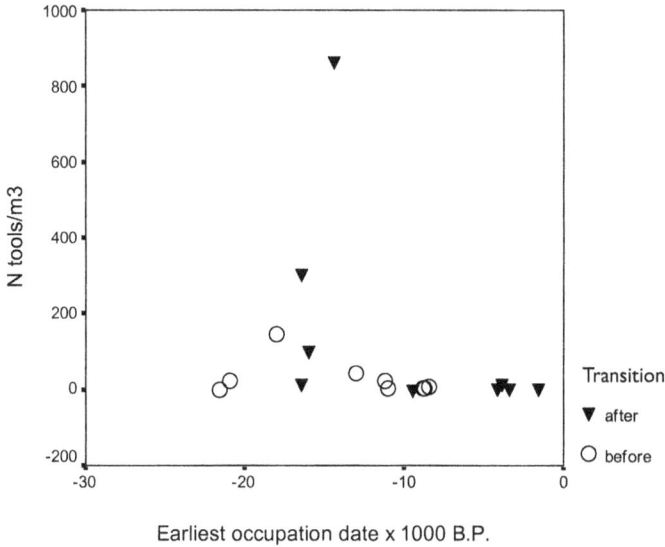

Figure 9.2. Transition by lithic tool density and ^{14}C date.

pre-transition sites in all regions, regardless of age. The consistency of lithic tool density in pre-transition sites is remarkable, with a mean of 27.1 tools/m^3. Post-transition sites show a bimodal pattern of light and heavy deposition, with heavy values averaging 425.18 tools/m^3. Post-transition sites also exhibit regional clustering, with densest lithics in the Spanish Magdalenian cave sites, intermediate values in Japanese sites, and lowest values in the Great Basin.

Lithic tool density may reflect the frequency or intensity of occupations or site formation processes such as differential sediment accumulation. Hunter-gatherer landscape use is responsive to the timing and location of resources (Binford 1983). If lithic tool density reflects human behavior rather than site formation processes, it is expected that tool density will correlate with archaeological traces of subsistence.

In all three study areas the presence or absence of animal species is the most commonly reported class of subsistence data. Lithic tool density shows a Y-shaped distribution when plotted against the proportion of large to all terrestrial mammal species (Figure 9.3). When the total assemblage of terrestrial mammal species is composed of about 38% large animals, the Spanish cave sites of Ekain, El Juyo, and El Rascaño exhibit a sudden increase in lithics density. Japanese Paleolithic sites are not represented due to poor bone preservation of volcanic soils. Great Basin sites show the lowest values for both lithic tool density and presence of large mammal species.

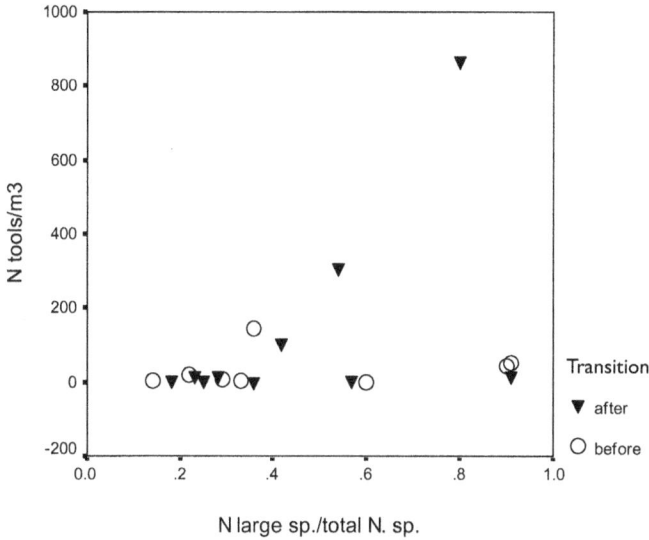

Figure 9.3. Transition by lithic tool density and large mammal species frequency.

In study sample sites the relationship between lithic tool class density and the frequency of large mammal species demonstrates that varying tool densities reflect real differences in site use rather than site formation processes. If this is true, then pre-transition sites in all three study areas share low tool densities and therefore low discard rates despite variation in chronology, environment, and subsistence. Post-transition sites show a Y-shaped distribution in which Spanish cave sites diverge toward a new pattern of high tool-discard rates. This pattern suggests more frequent or longer occupations. Great Basin sites consistently show low tool-discard rates and low representation of large game, and Japanese sites are intermediate.

Assemblage richness, or the number of tool classes, is divided by the number of tools to provide a measure for lithic tool class diversity. In pre-transition sites lithic tool diversity stays within a range of 0.01 to 0.03 regardless of geographic area or chronology (Figure 9.4), with one exception: the pre-transition level of Cowboy Cave in Utah. This site contained very low numbers of tools but a moderate number of tool classes, which increased the value of the ratio between the two. Post-transition sites show a bimodal pattern in tool diversity as with tool density, in which diversity either rises quickly or stabilizes at the low, pre-transition range.

In sum, pre-transition sites or site levels show surprisingly consistent sparse lithic tool density and low tool class diversity. Post-transition sites show a lithic tool density threshold at a ratio of 0.38 large mammals and at site dates of ca.

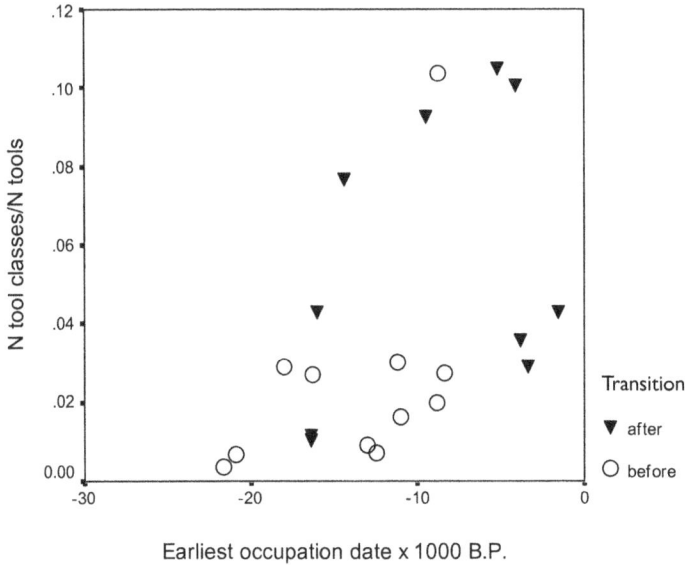

Figure 9.4. Transition by *n* tool classes and ^{14}C date.

17,000 BP. Two trajectories are visible, one showing a drastic increase in density and the other maintaining stable pre-transition levels. Post-transition site tool class diversity shows a similar pattern.

The study sample indicates that archaeological correlates of the pre-transition stage are consistent across a broad range of chronologies and environments. Low tool discard rates and low tool class diversity are important defining characteristics of pre-transition behavior that suggest short-term or infrequent use of sites and a conservative tool kit.

In contrast, post-transition sites show significant divergence between "old-fashioned" low values for artifact discard and tool class diversity and new, very high values. The high values may indicate long-term or more frequent site use and a more diverse tool kit.

USING PROJECTILE WEAPONRY

These archaeological patterns suggest significant changes in site use and technological strategies during the projectile transition. Projectile points may be used for many different purposes, but design constraints are generally imposed by their function as hunting tools (Greaves 1997: 313). Therefore ethnographic information on hunting practices is a useful framework for understanding changes in site use and technology.

Many authors work from the assumption that large lanceolate and small triangular points respectively represent two specific hunting tools, the atlatl and the bow and arrow (Grayson 1994: 253; Straus 1983: 90; Sugihara 1973 in Aikens and Higuchi 1982: 71). The best available comparative information on these two forms of launching comes from experimental studies of projectile ballistics and from ethnographic hunting accounts.

Experimental Data: Range, Accuracy, and Momentum

Experiments in projectile technology highlight the contrasting abilities of atlatls and bows and arrows. Cattelain (1997: 218) summarizes several experiments in central and northern Australia to arrive at a potential throwing distance for spearthrowers of up to 180 m. Australian hunters' preferred throwing distances with a spearthrower, however, are between 10 and 20 m (summarized in Cundy 1989: 17).

Potential bow and arrow throwing distance is about 100 m using traditional bows, but preferred throwing distances with bow and arrow are summarized by Cattelain (1997: 227) at about 9–25 m, beyond which accuracy is compromised. Subtracting the actual from the potential throwing distance provides an estimate of the force differential between these weapons: the spearthrower potential minus actual distance is ca. 160 m; for bow and arrow the difference is ca. 75 m or less. Therefore, holding range and projectile weight constant, the spearthrower delivers over twice as much force upon impact, presuming equal weights for projectiles.

This estimate can be refined by looking at the simple momentum of different projectiles. Deceleration due to drag has been shown to be inversely proportional to the projectile's mass (Cundy 1989: 34), so that high velocity increases the kinetic energy of a projectile but also increases drag by an equivalent amount. Momentum from heavy mass overcomes the drag factor; therefore heavier projectiles will penetrate farther than lighter ones when all other factors are held constant (Cundy 1989: 34; Dietrich 1996: 41).

Academics and modern bowhunters agree that mass is crucial to the constraints and opportunities of different projectiles. Although light weight and higher velocity produce a flatter trajectory and better accuracy at long range, greater mass delivers the greatest penetration and the best wounding potential (Dietrich 1996: 54). Launching velocity of ancient projectile points cannot be known for certain, but point weights can be compared to approximate mass.

Great Basin dart shaft dimensions are known from archaeological specimens at Lake Winnemucca, Humboldt and Lovelock Caves, and Leonard Rockshelter (Spencer 1974: 41). An experimental untipped atlatl dart mainshaft based on those measurements had a length of 57.5 cm and weight of 67.5 gm (Spencer

1974: 41). No foreshaft was manufactured. The mean weight of Elko points from Danger Cave and Hogup Shelter (Aikens 1970: 46–47) is 4.4 gm. The total estimated weight of a Great Basin dart-mainshaft assembly is therefore around 72 gm.

Darts from Cattelain's ethnographic sample (1997: 229) fall into two size classes; between 30 and 63 cm in length and 55–200 gm in weight for Arctic examples and between 190 and 460 cm in length and 250–600 gm in weight for Australian examples. Arrows are divided similarly: in open temperate or forested tropical areas they are 43–110 cm long and weigh 15–40 gm. Arrows used in open tropical (usually savanna) areas are 110–210 cm long and weigh 35–88 gm (Cattelain 1997: 229).

Among smaller projectiles, arrows are 0.55 the weight of atlatl darts; among larger projectiles, arrows are 0.15 the weight of atlatl darts. The average of these values is a 0.35 ratio for arrow to atlatl dart weights. The relationship between the masses of the two projectiles can be approximated by the relationship between their weights. Given that momentum = mass × velocity, the momentum of darts is from two to seven times greater than arrows, with an average of about 3.5 times greater, if velocity is held constant.

This ratio clearly affects the size of intended prey. An atlatl dart delivers a more forceful blow and penetrates deeper. The point must be correspondingly robust to withstand high impact. Large animals have thicker hides, more meat mass, and denser bones, all of which may deflect or break the point. In order to maximize lethal effect, it is important to decrease the drag of the projectile head once it has struck so that it penetrates deeper (Cundy 1989: 34). Two characteristics can do this: a very sharp cutting edge and a broad head that accommodates the shaft as it follows (Cundy 1989: 34).

Little energy is needed once the skin is penetrated, however, and large animals can be killed effectively by small arrows (Cundy 1989: 36; Knight 1975: 253 in Cundy 1989). If atlatls and arrows are equally lethal, why did some hunters retain atlatls long after bows and arrows came into use?

Ethnographic Data: Contexts of Use

Ethnographic studies show that the spearthrower is used almost exclusively in open environments (Cattelain 1997: 219). At the time of European exploration, atlatls were commonly used in only two regions of the world: desert Australia and the Arctic (Cattelain 1997: 215). Atlatls are shock weapons (Churchill 1993: 18; Hutchings 1998: 13) that rapidly debilitate prey and reduce the need for tracking. Australian hunters have completely impaled large game such as kangaroos with atlatl spears (Tindale 1925). Large arrows in open tropical settings work the same way. The Pumé of Venezuela use long arrows whose weight adds shock to impact, and length inhibits movement after the animal is hit. This is particularly important in prey of high escape risk, such as arboreal, flying, or swimming animals (Oswalt 1976: 81). For example, coastal Eskimo hunters use the bow and arrow for terrestrial game and the atlatl for marine mammals and aquatic birds (Cattelain 1997: 215).

In contrast, bows are used in every environment and exhibit great variation in form and mechanics. Small arrows work as delivery devices for small, sharp points and are often armed with lethal substances in the case of large prey (for example, by desert hunters of South Africa) or arboreal prey (as in the case of the South American tropics). Bows and arrows offer a flexibility of use contexts that cannot be matched by the atlatl. Atlatl spears are bulkier than arrows: 1.3 to 3 times longer and 2 to 5 times heavier. Norman B. Tindale (1925: 98) noted that Australian hunters prefer bamboo for spear shafts if available, in order to increase the number of missiles that they can carry. A hunter can carry fewer spears than arrows, all other things being equal. In addition, the spearthrower itself is bulkier, and often heavier, than a typical bow (Cattelain 1997: 217).

Hunters who use atlatls or long arrows must carry their projectiles by hand, which limits the number that can be carried on trips to four to seven (Bartram 1997: 335; Greaves, personal communication, 1999). Up to twenty arrows might be carried in a quiver, however (Hitchcock and Bleed 1997: 348). More projectiles allow for more shots per trip and longer search times. Search time duration is directly related to potential encounters with resources on hunting trips (Greaves 1997: 310). Also, less time is lost in resharpening or other repair, which may not be necessary if enough arrows are carried.

Modern bowhunting and archaeological experiments indicate that smaller projectiles are flatter in trajectory and more accurate on target (Cattelain 1997: 230; Churchill 1993: 18; Dietrich 1996:54). Thus bows and arrows, while still lethal for larger game, allow for smaller target sizes if the range is held constant (Churchill 1993: 20).

Finally, the atlatl requires space for the throw and preferably space in front to step forward. Although Eskimo atlatls are small and may be launched from a seated position in a kayak (Nelson 1899: 151–152), the normal position among Australian hunters stalking terrestrial game is standing and walking (Cattelain 1997: 219). The bow and arrow are held close to the body and may be launched when standing still or seated, requiring less space (Figure 9.5).

Hunters attempting to narrow the distance between themselves and prey have two options: concealment and disguise (Dietrich 1996: 160). Disguise involves carrying extra gear, but concealment requires only the presence of minimal obstacles in the prey's line of sight. In a comparison between large and small bows and arrows, Lawrence E. Bartram (1997: 340) found that large arrows hamper

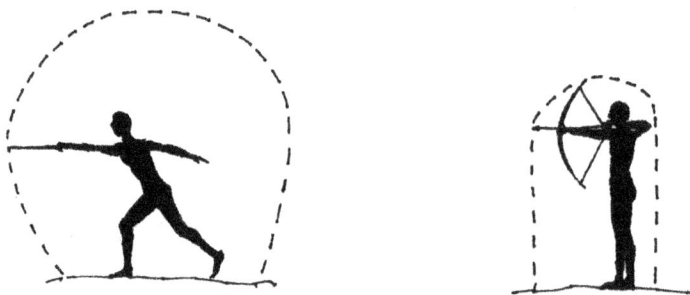

Figure 9.5. Space requirements for launching different projectiles.

the hunter's ability to stalk by crawling or shoot while hidden in vegetative cover. Small bows and arrows permit the hunter to fire on prey while in heavy vegetative cover or to crawl closer if cover is sparse.

To summarize, ethnographic accounts show that atlatls immobilize prey upon impact and thus reduce the risk of escape. This is particularly important with large, fast animals or animals that can escape into environments beyond the hunter's reach, such as air and water. At some point, these benefits became outweighed by the arrow's light weight, flatter trajectory, and versatility in launching. Smaller missiles such as bow and arrows offer the hunter longer search times and more shots per trip. Smaller prey can be targeted successfully, and stalking is possible in dense or ground-level cover. The hunter might spend less time in repair or maintenance when equipped with many projectiles.

Extended trip times, higher number of shots, shrinking prey size, and exploitation of previously unused habitats suggest the hunter's need to maximize the number of successful shots per trip. This in turn indicates widening diet breadth and the need to decrease pursuit time averaged across a growing number of prey classes (Churchill 1993: 21). Pre-transition sites and post-transition sites that stay at pre-transition values are characterized by selective use of the landscape and less need to hunt in heavily vegetated areas. For example, Great Basin hunters may not have needed to shoot while standing in dense vegetation but found the bow and arrow useful for targeting the small game characteristic of the desert.

CLIMATIC CONDITIONERS OF THE TRANSITION

The transition to bow and arrow occurred at different times and at different rates in the three study areas, but both Japanese and Spanish transitions predate the Great Basin. Why? Use contexts of the bow and arrow suggest that the answer lies in the hunter's choice to include heavily vegetated terrain in the normal subsistence round. An early projectile transition predicates a high percentage of heavily forested habitat in the general environment.

Temperature and Growing Seasons

As mentioned above, small projectile points appear early in Cantabrian Spain and Japan (ca. 15,000–14,000 BP and ca. 14,000–12,000 BP, respectively). The transition is distinct and is used in both regions as a chronological marker. In the Great Basin the transition took place from ca. 5000 to 1700 BP and was so gradual that the dates are still subject to debate.

Effective temperature (ET: average of temperatures taken at the beginning and end of the growing season) in the three study regions has varying effects on the transition (values taken from Binford's database; also see Binford 2001: 258). The "bow and arrow" line can be traced along effective temperature values, showing the latest transition in the Great Basin at a value of about 13.5° C in the left of the figure (Figure 9.6). Japan's effective temperatures fall on both sides of the Great Basin, and Spain has the highest effective temperature values on the right of the figure. The bow and arrow line suggests that the transition is delayed at ET values of about 13° C.

Precipitation and Vegetative Cover

Vegetative growth is regulated by the interaction between temperature and rainfall. Figure 9.7 shows how rainfall conditions the net above-ground productivity, or grams of plant material added per m² per year, in each study area (values are from Binford's database). Overall, higher rainfall results in higher vegetative growth rates. Southern Japan and Spain have higher values, northern Japan is intermediate, and the Great Basin shows very low rainfall and vegetative growth.

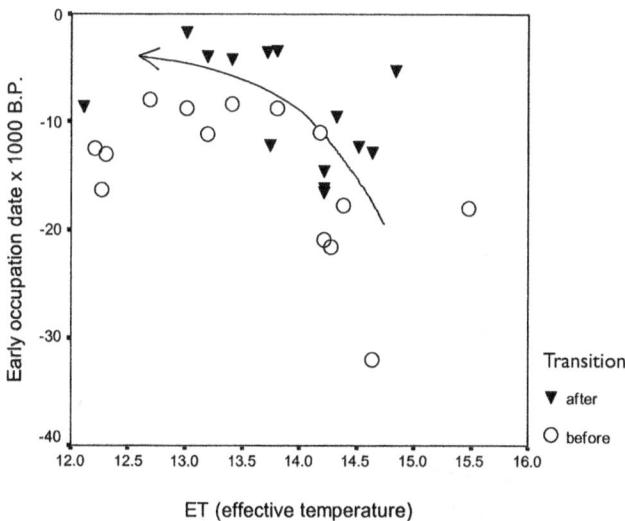

Figure 9.6. Transition by effective temperature and ^{14}C date.

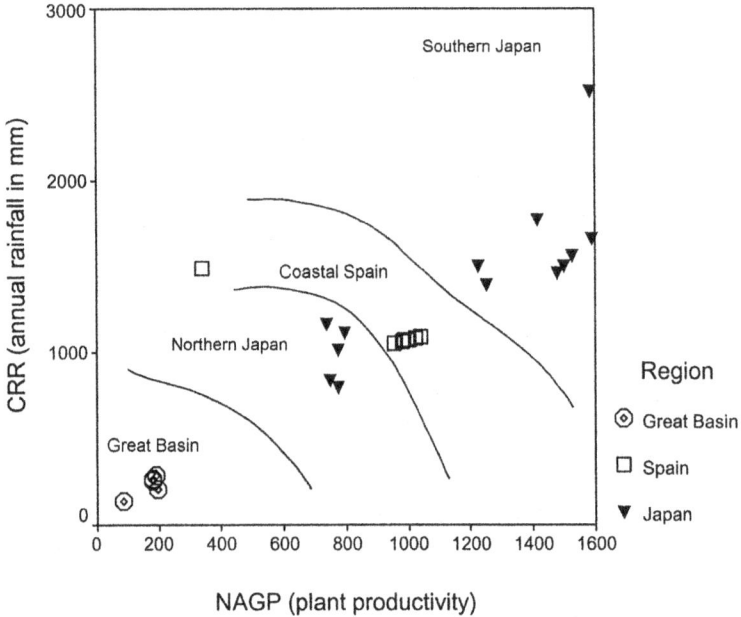

Figure 9.7. Study regions by annual rainfall and net above-ground productivity.

Earliest transitions occur in Spanish and southern Japanese sites, which both are characterized by high annual rainfall, high net above-ground productivity, and high effective temperatures. The Great Basin sites, with lowest annual rainfall, low net above-ground productivity, and moderate effective temperatures, consistently show the latest transition. Northern Japan and Hokkaido, with coldest temperatures in the sample and low to moderate rainfall, are intermediate to Spain and the Great Basin. These results indicate that high aridity and cold temperatures delay the transition but that aridity is a more significant deterrent.

The Great Basin's position along the bow and arrow line strongly indicates a different selective context from Cantabrian Spain and Japan. Recent research indicates that the transition in the Great Basin may have been a historical event of adoption rather than an adaptive process of innovation (Blitz 1988). If this is the case, then selective forces were not intense enough to encourage projectile innovation; high aridity and low plant biomass are associated with low population density and high mobility among hunter-gatherers (Binford 1983). The blurred Great Basin transition across several thousand years and significant geographic variation indicates that the selective forces favoring bow and arrow technology were in play, but at a lower degree of intensity.

Paleoclimatic Triggers

If paleoenvironments affected the chronology of the projectile transition, climatic changes in the study areas may have provided discernible trigger events.

In Japan the projectile point transition occurred at ca. 14,000 BP in the south and moved north over time (Aikens and Higuchi 1982: 87). The terminal glacial climate of the late Paleolithic and early Jomon was several degrees cooler than at present (Pearson 1977: 1239). Pollen records indicate that Hokkaido was open parkland with spruce, beech, and elm and that the main body of Japan was thickly forested with oak, elm, and beech. In the early Holocene these forests began to shift to broadleaf evergreen forests, and climatic warming initiated marine transgression of the Japanese coastline (Pearson 1977: 1240). This transgression probably reached its maximum during the early Jomon, ca. 9000 BP (Pearson 1977: 1240), swamping the coast, moving into topographically varied areas, and offering a greater diversity of coastal swamps, bays, and estuaries than before (Aikens et al. 1986: 16).

The Spanish Solutrean period overlaps with a glacial maximum (ca. 20,000–17,000 BP) and shifts to the Magdalenian period during interstadial warming (Straus 1991: 89). Warming glacial climates in the open coastal plains along the coast of Cantabrian Spain allowed small thickets of trees to develop in wind-sheltered areas (Straus and Clark 1986: 136). Marine transgression had begun by the early Magdalenian, but little coastline was lost (Straus and Clark 1986: 136).

The projectile point transition in the Great Basin occurred much later than in the other two areas, in the late Holocene. During this time the desert underwent cooling temperatures and increasing moisture from an aridity maximum at about 7000–6000 BP. Pollen data indicate that juniper (which follows moisture) became more frequent and that piñon pine moved into the Great Basin from the south (Grayson 1994: 221–222). Overall, cooling temperatures and moisture in the eastern Great Basin caused grass to replace sage and scrub in open areas and allowed the treeline to move down in elevation (Grayson 1994: 221–222).

In all three study areas an environmental rebound from climatic extremes of either Pleistocene cold or mid-Holocene aridity accompanies the projectile point transition. In Japan and Spain, which were more densely vegetated to begin with, the transition occurred earliest. In the desert of the Great Basin post-Pleistocene warming did not constitute a rebound and in fact led to the prolonged drought of the middle Holocene. Only in the late Holocene did the Great Basin return to early Holocene vegetative productivity (Grayson 1994).

Through analysis of the three study areas this chapter has restated relationships that are well known to human ecologists: climate conditions landscape characteristics such as resource distribution, and human hunters adjust their mobility to resource availability in time and space. Technology reflects these adjustments.

Although projectile technology is observed to change at different times and at different rates, these differences may be quantitative rather than qualitative, similar processes playing out in different theaters (cf. Binford 2001: 33).

Another important factor conditions hunter mobility and has the potential to affect the timing of technological shifts: the proximity of other hunters. This brings the analysis full circle. The archaeological values of lithic tool density and assemblage richness, and ethnographically known use contexts of hunting gear, can be linked with environmental characteristics, using human demographics.

INFERRING POPULATION GROWTH AND REORGANIZATION OF HUNTING TERRITORIES

One class of post-transitional sites in the study areas shows evidence of redundant, frequent site use (high lithic tool density tied to discard) and generalized subsistence (high assemblage richness tied to diverse processing needs). The bow and arrow have been demonstrated to suit longer trip times, more shots per trip, and targeting of small prey in previously unfavorable forested terrain. All other things being equal, these adaptations are consistent with decreases in territory for a foraging group (Binford 1983: 211; Kelly 1995: 151).

If hunter-gatherers use mobility as a security option to be abandoned only under duress (Binford 1983: 210), then the factor most likely to decrease residential mobility is the proximity of neighbors. This should also hold true for hunting mobility. Projected hunter-gatherer population densities were derived for the study sites based on ethnographically known hunter-gatherer densities in similar environments (values from Binford 2001). These values show that high population density would have been reached first in Spain, followed by northern Japan, southern Japan, and finally the Great Basin, which patterns well with known dates of the projectile point transition (Figure 9.8).

In arid environments where territories are large and mobility high, packing thresholds are not expected to be reached very quickly. But changing population distributions in highly productive environments may have fueled early projectile transitions. In areas where preferred hunting niches filled and hunting territory overlap increased, hunters could maximize hunting returns in shrinking territories by including more stalking habitats and diverse prey sizes. This favored a lightweight projectile of flexible use context and high accuracy. Features of the landscape, including hunting territories and associated residential sites, began to be used more often or for longer periods. Diversification of tool classes suggests a shift from procurement to processing; various scrapers and awls for hide working constitute most of the assemblage richness in Spain and ground stone and crescent chipped stone forms for vegetable processing in Great Basin. Such

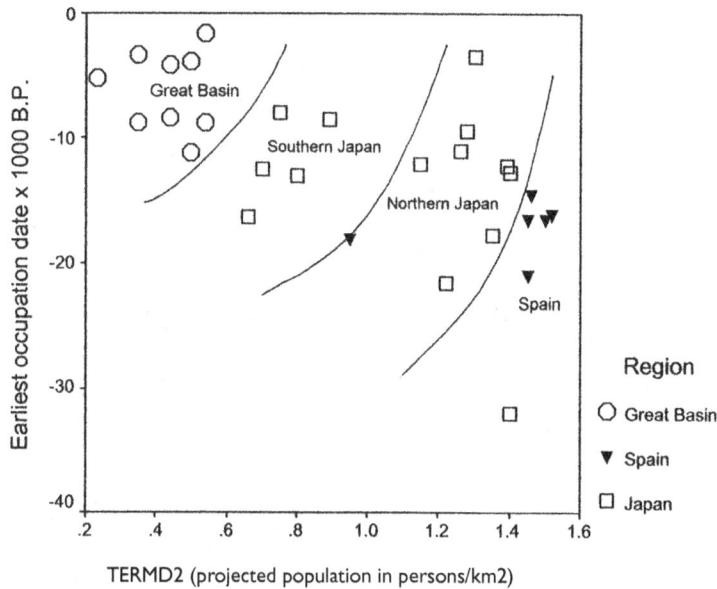

Figure 9.8. Regional variation in projected forager population density by ^{14}C date.

tactical shifts do not necessarily signify an absolute decrease in mobility, but they do reflect strategic adjustments in scope and scheduling as a response to gradual packing of the landscape.

Changing Modes of Warfare

Changes in hunting strategies and technology certainly affected the conduct of warfare (Nassaney and Pyle 1999: 258). The results presented in this chapter suggest that arrows allowed the development of ambush or guerrilla tactics. This kind of warfare, which relies on stealth and flexibility, is less formalized and more lethal than melée-style warfare of atlatl users such as those in desert Australia (Gason 1879). Arrows as weapons may also have permitted longer war forays and added the potential to launch more projectiles per foray.

CONCLUSION

Changes in tactics or behavior reflect adaptive responses to certain conditions, which the archaeologist may discern through the application of relevant bodies of information to archaeological data. The projectile transition represents a tactical adjustment to changing resource availability: the addition of a new hunting niche. The timing and the rate of this process are predictably associated with the intensity of selective forces acting upon human populations; factors such as population density, mobility, and resource availability are all conditioned by local environments.

In conclusion, defining the projectile transition (and other changes in type and frequency of materials associated with mobility, technology, and subsistence) in strictly evolutionary terms may not be warranted. Making a strategic adjustment to changing conditions, or choosing not to do so, does not in itself constitute an evolutionary threshold. The true Rubicon, the ability to adopt or develop new strategies of mobility and resource use and to exploit those strategies situationally, was crossed long before the projectile transition took place.

REFERENCES

Aikens, C. M.

1970   *Hogup Cave*. University of Utah Anthropological Papers 93. University of Utah Press, Salt Lake City.

Aikens, C. M., K. M. Ames, and D. Sanger

1986   Affluent Collectors at the Edges of Eurasia and North America: Some Comparisons and Observations on the Evolution of Society among Northern-Temperate Coastal Hunter- Gatherers. In *Prehistoric Hunter-Gatherers in Japan*, edited by T. Akazawa and C. M. Aikens, pp. 3–26. University of Tokyo Press, Tokyo.

Aikens, C. M., and T. Higuchi

1982   *Prehistory of Japan*. Academic Press, New York.

Altuna, J., and J. M. Merino

1980   *El yacimiento prehistórico de la Cueva de Ekain, Deba, Guipúzcoa* (The Prehistoric Site of Ekain Cave, Deba, Guipúzcoa). Sociedad de Estudios Vascos, Guipúzcoa, Spain.

Altuna, J., and L. G. Straus

1976   The Solutrean of Altamira: The Artifactual and Faunal Evidence. *Zephyrus* 26–27: 175–182.

Barandiaran, I., L. G. Freeman, J. Gonzalez-Echegeray, and R. G. Klein

1985   *Excavaciones en la Cueva del Juyo* (Excavations in El Juyo Cave). Ministerio de Bellas Artes y Archivos, Madrid.

Bartram, L. E.

1997   A Comparison of Kua (Botswana) and Hadza Bow and Arrow Hunting. In *Projectile Technology*, edited by H. Knecht, pp. 321–344. Plenum Press, New York.

Beck, C.

1998   Projectile Point Types as Valid Chronological Units. In *Unit Issues in Archaeology: Measuring Time, Space, and Material*, edited by A. Ramenofsky and A. Steffen, pp. 21–40. University of Utah Press, Salt Lake City.

Bettinger, R. L., and J. Eerkens

1999   Point Typologies, Cultural Transmission, and the Spread of Bow and Arrow Technology in the Prehistoric Great Basin. *American Antiquity* 64(2): 231–242.

Binford, L. R.

1983   *In Pursuit of the Past.* Thames and Hudson, New York.

2001   *Constructing Frames of Reference.* University of California Press, Berkeley.

Blitz, J. H.

1988   Adoption of the Bow and Arrow in Prehistoric North America. *North American Archaeologist* 9: 123–145.

Cabrera Valdez, V.

1975   El yacimiento Solutrense de Cueva Chufín (Riclones, Santander). *XIV Congreso Nacional de Arqueología*: 157–164.

1984   El yacimiento de la Cueva de "El Castillo" (Puente Viesgo, Santander). Biblioteca Praehistorica Hispana, Madrid.

Cattelain, P.

1997   Hunting during the Upper Paleolithic: Bow, Spearthrower, or Both? In *Projectile Technology*, edited by H. Knecht, pp. 213–240. Plenum Press, New York.

Churchill, S.

1993   Weapon Technology, Prey Size Selection, and Hunting Methods in Modern Hunter-Gatherers: Implications for Hunting in the Paleolithic. In *Hunting and Animal Exploitation in the Later Paleolithic and Mesolithic of Eurasia*, edited by G. Peterkin, H. Bricker, and P. Mellars, pp. 11–24. Archeological Papers of the American Anthropological Association 4. Washington, D.C.

Conde de la Vega del Sella

1916   *Paleolítico de Cueto de la Mina (Asturias).* Comisión de Investigaciones Paleon-tológicas y Prehistóricas, Memoria No. 13. Museo Nacional de Ciencias Natura-les, Madrid.

Corchón, M.

1971   Materiales solutrenses de la Cueva Santanderina de El Pendo. *Zephyrus* 21(11): 7–21.

Cundy, B. J.

1989   *Formal Variation in Australian Spear and Spearthrower Technology.* BAR Series 546. BAR, Oxford.

Dietrich, D.

1996   *Bowhunting Big Game.* North American Bowhunting Association, Chicago.

Echegaray, J. G., and I. Barandiaran-Maestu

1981   *El Paleolítico Superior de la Cueva del Rascaño (Santander).* Ministerio de Cultura, Santander.

Esaka, T., and S. Nishida

1967   Aichi-ken Kamikuroiwa iwakagi. In *Nihon no doketsu iseki*, edited by Nihon Kok-ogaku kyokai koketsu iskei chosa tokubetsu Iinkai, pp. 224–236. Heibonsha, To-kyo.

Gason, S.

1879   The Manners and Customs of the Dieyerie Tribe of Australian Aborigines. In *The Native Tribes of South Australia*, edited by J. D. Woods, pp. 253–307. E. S. Wigg and Son, Adelaide.

Grayson, D. K.

1994   *The Desert's Past*. Smithsonian Institution Press, Washington, D.C.

Greaves, Russell D.

1997   Hunting and Multifunctional Use of Bows and Arrows: Ethnoarchaeology of Technological Organization among the Pumé of Venezuela. In *Projectile Technology*, edited by H. Knecht, pp. 287–320. Plenum Press, New York.

Hayashi, K.

1968   The Fukui Microblade Technology and Its Relationships in Northeast Asia and North America. *Arctic Anthropology* 5(1): 128–190.

Hitchcock, R., and P. Bleed

1997   Each According to Need and Fashion: Spear and Arrow Use among San Hunters of the Kalahari. In *Projectile Technology*, edited by H. Knecht, pp. 345–370. Plenum Press, New York.

Huckell, B. B.

1996   The Archaic Prehistory of the North American Southwest. *Journal of World Prehistory* 10(3): 305–373.

Hutchings, Karl W.

1998   The Identification of Paleoindian Fluted Point Delivery Technology through the Analysis of Lithic Fracture Velocity. Paper presented at the 63rd Annual Society for American Archaeology Meetings, Seattle.

Jennings, J. D.

1980   *Cowboy Cave*. University of Utah Anthropological Papers 104. University of Utah Press, Salt Lake City.

1957   *Danger Cave*. University of Utah Press, Salt Lake City.

Jennings, J. D., A. R. Schroedl, and R. N. Holmer

1980   *Sudden Shelter*. University of Utah Anthropological Papers 103. University of Utah Press, Salt Lake City.

Kato, S., and H. Oi

1961   The Midorigaoka Site, Kunneppu, Tokoro County, Hokkaido. *Minzokugaku Kenkyu* (Tokyo) 26(1): 24–30. In Japanese with English summary.

Kellar, J. H.

1955   *The Atlatl in North America*. Prehistory Research Series 3. Indiana Historical Society, Indianapolis.

Kelly, R. L.

1995   *The Foraging Spectrum*. Smithsonian Institution Press, Washington, D.C.

Knecht, Heidi

1997   Introduction. In *Projectile Technology*, edited by H. Knecht, pp. 1–3. Plenum Press, New York.

Knight, B.

1975   The Dynamics of Stab Wounds. *Forensic Science* 6: 249–255.

Nakamura, K.

1960   *Kosegasawa doketsu.* Nagaoka shiritsu kagaku hakubutsukan Kenkyu chosa Ho-koku, Vol. 3. Translated in Aikens and Higuchi 1982.

Nassaney, M. S., and K. Pyle

1999   The Adoption of the Bow and Arrow in Eastern North America: A View from Central Arkansas. *American Antiquity* 64(2): 243–263.

Nelson, E. W.

1899   *The Eskimo about the Bering Strait.* Eighteenth Annual Report of the Bureau of American Ethnology to the Secretary of the Smithsonian Institution, 1896–1897. Smithsonian Institution, Washington, D.C.

Oswalt, W. H.

1976   *An Anthropological Analysis of Food-Getting Technology.* John Wiley and Sons, New York.

Pearson, R.

1977   Paleoenvironment and Human Settlement in Japan and Korea. *Science* 197/4310: 1239–1245.

Ripoll-Lopez, S.

1988   *La Cueva de Ambrosio, Almería, Spain.* BAR International Series S462(i). BAR, Oxford.

Serizawa, C., and Y. Kamaki

1967   Fukui Cave, Nagasaki Prefecture. In *Nihon no doketsu iseki*, edited by Select Committee for the Investigations of Cave Sites. Select Committee for the Investigations of Cave Sites, Tokyo. In Japanese with English summary.

Shirataki Research Group

1967   *Shirataki iseki no kenkyu.* University of Hokkaido, Sapporo. In Japanese with English summary.

Shott, M.

1993   Spears, Darts, and Arrows: Late Woodland Hunting Techniques in the Ohio Valley. *American Antiquity* 58: 425–443.

1997   Stones and Shafts Redux: The Metric Discrimination of Chipped Stone Dart and Arrow Points. *American Antiquity* 62: 86–101.

Spencer, L.

1974   Replicative Experiments in the Manufacture and Use of a Great Basin Atlatl. In *Great Basin Atlatl Studies*, edited by T. R. Hester, M. P. Mildner, and L. Spencer, pp. 37–60. Ballena Press, Ramona, Calif.

Straus, L. G.

1983   From Mousterian to Magdalenian: Cultural Evolution Viewed from Vasco-Cantabrian Spain and Pyrenean France. In *The Mousterian Legacy*, edited by E. Trinkaus, pp. 73–111. BAR International Series 164. BAR, Oxford.

1991   Epipaleolithic and Mesolithic Adaptations in Cantabrian Spain and Pyrenean France. *Journal of World Prehistory* 5(1): 83–104.

Straus, L. G., and G. A. Clark

1986   *La Riera Cave.* Anthropological Research Papers No. 36. Arizona State University, Tempe.

Sugihara, S.

1973   *Point Tool Culture of Uenodaira, Nagano Prefecture, Japan.* Meiji University, Tokyo. In Japanese with English summary.

Sugihara, S., and C. Serizawa

1957   Shell Mounds of the Earliest Jomon Culture at Natsushima, Kanagawa Prefecture, Japan. *Memoirs of the Tokyo Archaeological Society* 3(2): 1–33. In Japanese with English summary.

Sugihara, S., and M. Tozawa

1975   *Microlithic Culture of Shirataki-Hattoridai.* Meiji University, Tokyo. In Japanese with English summary.

Suzuki, M.

1974   Chronology of Prehistoric Human Activity in Kanto, Japan, Part II: Time-Space Analysis of Obsidian Transport. *Journal of the Faculty of Science* (University of Tokyo) section 5: anthropology 4, no. 4: 396–469.

Thomas, D. H.

1978   Arrowheads and Atlatl Darts: How the Stones Got the Shaft. *American Antiquity* 43: 461–472.

Tindale, N. B.

1925   Natives of Groote Eylandt. *Records of the South Australian Museum* 3: 61–134.

Williams, P. A., and R. I. Orlins

1963   *The Corn Creek Dunes Site: A Dated Surface Site in Southern Nevada.* Nevada State Museum Anthropological Papers No. 10. Nevada State Museum, Carson City.

Yoshizaki, M., S. Segawa, and M. Ishikawa

1960   Hakodate: Hokkaido shiritsu hakodate hakubutsukan (no English title or page numbers available). *Kiyo* 6. In Japanese with English summary.

Yubetsu-Ichikawa Research Group

1982   *Yubetsu-Ichikawa Iseki* (The Yubetsu-Ichikawa Site). Yubetsu-cho kyoiku Iinkai, Yubetsu, Japan. In Japanese, translated in Aikens and Higuchi 1982.

10

Two Steps Forward, One Step Back

The Inference of Mobility Patterns from Stone Tools

FRÉDÉRIC SELLET

Among prehistoric hunter-gatherers, Paleoindians are widely regarded as the champions of mobility (for example, Amick 1996b: 419; Kelly and Todd 1988: 234). This argument has merit, but there is also much more to the story. Recent studies have focused on extracting more detailed information on how often Paleoindian groups moved, on the range and type of mobility, and, more importantly, on the impact of mobility strategies on their technology (Amick 1999; Anderson 1996; Bamforth 1985, 2002; Bamforth and Becker 2000; Beck and Jones 1990; Boldurian 1991; Deller and Ellis 1992; Hofman 1992, 1999; Ingbar 1992, 1994; Jones et al. 2003; Seeman 1994; Shott 1986; Storck 1997). Given that Paleoindian groups are said to epitomize mobile lifeways, and in view of the universality of mobility as a survival scheme, it can be argued that a careful study of Paleoindian mobility strategies will enhance our understanding of the archaeological record of other small-scale prehistoric societies. With this goal in sight, this chapter provides a reflection on the relationships between mobility and archaeological variability in cultural material and illustrates the discussion with an example from a Paleoindian camp site.

Mobility has often been recognized as one of the critical factors affecting archaeological variability (for example, see Binford 1979, 1980; Gamble 1991:1; Kelly 1992; Odell 1996b, 2001:62; Parry and Kelly 1987:300). Recognizing its effects archaeologically has not proven an easy task, however. Binford (1983: 112–113), for instance, notes: "The problem is that archaeologists have been approaching their research on Palaeolithic sites from a modern sedentary view of the world. Since the hunting and gathering peoples which we are trying to study probably did not share that view, we must try to bring our perspective closer to reality."

Yet reality itself is ambiguous; middle-range observations based on modern hunter-gatherer groups have revealed the intricacies of mobility strategies (for example, Politis in this volume). The many types of mobility—including seasonal, residential, and logistical—make it hard to evaluate hunter-gatherers' moves in a coherent and consistent manner (Kelly 1995: 111). It has been shown, for example, that the frequency and range of hunter-gatherer movements are influenced by relationships between individual foraging and environment (Binford 1979, 2001; Greaves 1997; Kelly 1983, 1992, 1995). At the end, however, the greater understanding of mobility strategies gained from ethnographic observations has not always had the expected impact on archaeological studies.

This difficulty is particularly obvious in lithic analyses. Stone tools are the most conspicuous and sometimes the only trace of ancient nomadic ways of life. They are the main body of data from which inferences about past mobility strategies are drawn. Nevertheless, in the case of stone tools, ethnographic analogies are limited if not altogether lacking. We cannot rely on modern hunter-gatherers to learn about how stone tool production, use, and discard relate to overall mobility strategies. Reflecting on the archaeological manifestations of mobility, Kelly (1992: 56) shares Binford's frustration and remarks: "At present, then, many interpretations of stone tool assemblages as indicators of mobility are subjective, intuitive, and sometimes contradictory." In light of this commentary, as well as the pessimism expressed by others (see Torrence 1994: 126), it seems legitimate to question our ability to extract information relevant to mobility from stone tools. The following discussion addresses this issue and explores basic correlations between prehistoric mobility strategies and lithic assemblages. Are we really always one step behind in our understanding of past behavior?

THE PROBLEM: CAN WE CONNECT MOBILE ROCKS TO MOBILE PEOPLE?

When it comes to explaining the outcome of technological strategies, there is a consensus that a wide variety of factors account for stone tool variability (Nelson 1991: 59). Mobility is only one of them. Robin Torrence (1994: 127), for instance, condemns the dominance of settlement-subsistence system thinking in North American archaeology and the role of mobility as an explanatory device in technological variability:

> The two strategies, mobility and technology, are equally important to survival, but they solve different problems related to food getting. One provides access to resources and the other ensures that the prey is not lost once encountered. In addition, these two strategies work in conjunction with other kinds of behavior such as information processing and social exchange, but each has different potentials for achieving desired ends as well as different costs. The overemphasis of mobility as the primary or even sole way to achieve goals is dangerous because it obscures the complexity of how cultural groups put together packages composed of various mixes of strategies.

Torrence's skepticism does not contradict the idea that a nomadic lifestyle would affect the organization of lithic technology (see Torrence 1983, 1989 for instance). Rather, it highlights our inability to recognize the part played by mobility in shaping the organization of lithic systems. Central to the argument is the problem of equifinality (Torrence 1994: 129; see also Kelly 1992: 55; Meltzer 1989; Odell 1994, 1996a: 76; Sassaman 1994: 99).

Equifinality implies that different causes can produce the same effect. For instance, a number of variables impact stone tool manufacture use and discard, including tool function, hafting technique, raw material availability, subsistence strategies, time stress, and social factors. Some of these variables call for similar technological responses and thus could result in similar outcomes. At the end it is difficult to recognize the role of each in determining archaeological variability. With lithic studies, the problem is obviously exacerbated by the lack of middle-range data on stone tool manufacture and use.

Lithic studies have approached the problem of equifinality, implicitly or explicitly, from a variety of viewpoints; but for the sake of this discussion we will only consider a broad dichotomy. The first group of studies includes attempts at documenting the idiosyncrasies of specific settings in order to make site comparisons meaningful. By contrast, the second group of studies is more inclusive and avoids singularities entirely by enlarging the scale of analysis to a series of sites. In effect, both methods attempt to identify regularities in the archaeological record. They are equally valid, within their own limitations.

Regional analyses, for example, are governed by the second of the two principles outlined above; they require a large sample of sites. The argument behind regional analyses of stone tools is that widening the breadth of the study will make idiosyncrasies disappear, overshadowed by patterned behaviors. In many ways this is analogous to looking at a photograph at different scales. Under high magnification only the grain (the color pixels) is visible. It is impossible to reconstruct the whole picture. But the grain dissolves if the picture is seen at some distance, and only then does the subject become recognizable.

The principal difficulty in a large-scale study lies in the need to keep site function constant and to control the chronological framework of the site's occupations. These requisites impose many restrictions on the nature of the data that can be integrated into the analysis. In the case of stone tools, a chronological control is generally possible only for projectile points (at least in North America, where they are often the only temporal markers). This is especially true for Paleoindian archaeology.

Projectile technology represents only a single aspect of the overall lithic system, however, and a very specific one. The technical requirements of point manufacture and replacement differ from those of other stone tools. For example, points were hafted in an elaborate way and hence might have stayed in the technological system longer. Similarly, their manufacture was a complex process that required proper training, great skills, and adequate tools, raw materials, or time. A bifacial Paleoindian point was anything but quickly made. In addition, projectiles exemplify only the extractive function of technology; other stone tools were used to process food, work wood or hide, and so forth. Accordingly, observations made from projectile points alone might not be indicative of general subsistence strategies (see Bamforth 2002: 85).

Another critical aspect of large-scale intersite comparison is the need to control for site function. Unfortunately, this is not always possible with large-scale analyses, which often rely on surface collections. These represent a biased artifact collection, with no contextual information. With few exceptions, surface collections cannot provide an accurate account of the formation history of a site (is the site a single occupation or a series of palimpsests?) and, perhaps more importantly, of the activities taking place at the site.

The second approach to controlling for equifinality employs a much smaller analytical scale, because it involves site studies. It is based on the recognition that each archaeological setting was unique and required a special blend of strategies for solving the particular problems faced by the social group (for example, Binford 1979; Nelson 1991: 84; Torrence 1994: 129). That is the approach taken in this chapter, which attempts to identify the role played by each part of an adaptive system. The premise of this argument is that the organization of a given technology should unravel the systemic connections between external stimuli and human responses. The basic assumptions are outlined by E. E. Ingbar and J. L. Hofman (1999: 102): "The concept of 'technological organization' is that there is a regular relationship between tools, activities, and activity planning. Activities and activity planning in turn are determined by mobility, labor forces, fixed and mobile resources targets, procurement time frames, as well as situational variables such as season and terrain."

In the attempt to contextualize technological decisions, the anticipation of needs becomes a legitimate target of archaeological inquiry (Sellet 2004). Because of the constraints associated with a nomadic way of life, hunter-gatherer groups had to anticipate future needs and transport multipurpose tools (see Politis in this volume). Bifaces, for example, have been interpreted as an extremely flexible and transportable technology that was designed to perform a wide array of tasks (Kelly 1988; Kelly and Todd 1988). For mobile groups, the appropriate scheduling of weaponry manufacture, maintenance, and replacement equates to a form of risk minimization (Kuhn 1992: 185). In a sense, the scheduling of lithic activities controls time and energy expenditures and permits a greater investment in other tasks more important to the survival of the group, such as food extraction and processing (Jochim 1981).

Anticipating the replacement of stone tools was critical because of the wasteful nature of lithic activities and the ensuing need for raw material. A good analogy can be found in the making of a Paleoindian projectile point, which required high-quality, fine-grained raw materials and produced hundreds of waste flakes that might have been too broken or too small to have been used in any task. Stone tools also had a limited life; maintenance or resharpening reduced their size and usefulness. Planning therefore had to balance the advantages of having readily available tools with the aggregate cost of making them in advance, maintaining them, and transporting them.

Ultimately, planning the refurbishment of the tool kit is only one element in the array of adaptive decisions that hunter-gatherers had to make. It is imperative to consider that the decisions impacting technology were highly dependent upon the spatial and timing requirements of all other subsistence-linked activities (Parry and Kelly 1987: 300). The specific constraints put on the technology by limited access to resources in space and time and the resulting instability are even more important in the case of highly mobile people (Wobst 1974: 153). Considering the requisites of a mobile way of life and the implications for strategies of tool production, use, and discard, we can reasonably expect that different tools would play different roles in the overall technology and would not be affected in the same way by scheduling. Therein lies one of the most obvious weaknesses of technological studies at the site level: the difficulty of identifying how lithic activities are segmented in space and time—or, in other words, the difficulty of recognizing the results of scheduling (a dynamic process) on the stone tools (static objects).

The following study takes data from a single site and tries to place the inferred technological strategies into a broader context of human organization. It contrasts the transported and expedient Paleoindian tool kits at the Hell Gap site and maps out how three major tool types—gravers, endscrapers, and sidescrapers—were produced, transported, and discarded.

Method of Analysis

The transport of tools is often directly inferred from the occurrence of exotic materials at a given locality (for example, Goodyear 1989: 7; Dincauze 1993: 286). Nevertheless, while correlating occurrence of exotic materials with high mobility is intuitive, it is not completely accurate (see Bamforth 2002: 85). Recent investigations, approaching the issue from the point of view of technological organization, have demonstrated that strategies of tool manufacture, use, and replacement produce a complex mix of raw material types (Ingbar 1992, 1994; Amick 1999). There seems to be more to a reconstruction of mobility strategies than a straightforward equation of human movement with lithic raw material movement (see also Brantigham 2003; Thacker in this volume).

Consequently, this research attempts to investigate strategies of tool production, transport, and discard independent of raw material type(s). To do so, the lithic collection was first sorted according to raw material types; then each group was further subdivided, based on differences in color or inclusions (Larson and Kornfeld 2000; Sellet 1999, 2004). The resulting analytical units, or nodules, should ideally represent the reduction of a single chunk of raw material. The nodule method is comparable to refitting but is a less secure inference, because it does not rely on physically conjoining various pieces. In a sense, it is a form of "virtual refitting." Once the preliminary sorting is achieved, two types of nodules can then be identified: nodules that contain a single artifact—single item nodules (SIN)—and nodules with more than one artifact—multiple item nodules (MIN).

Single item nodules are more likely to have been reduced elsewhere, transported, and discarded at the site, because there is no associated by-product of their manufacture *in situ*. Multiple item nodules, in contrast, presumably represent on-site manufacture. The entire range of nodules at a given site provides a glimpse of the segmentation of lithic activities in space and time. They map out immediate and anticipated needs.

The nodule method is not without limitations. It is dependent upon the amount of variability both within and between raw materials. In this case, visual differences between nodules were sufficiently clear to validate the approach. Furthermore, the nodule method is especially valuable here because raw material proportions and tool kit structure do not change significantly through time (at least for the early Paleoindian levels). Raw material proportion and tool kit structure therefore indicate little about changes in site function, subsistence strategies, and ultimately mobility strategies. The finer-grained analysis provided by nodules is necessary to extract information about these potential behavioral dynamics.

The bulk of the raw material is of local origin. More than 95% of the raw materials originated from the Hartville uplift (see Table 10.1) and could have been picked up in the direct vicinity of the site or at the famous Spanish Diggings quarries nearby (Saul 1969). Only a negligible fraction of the tool assemblage was made on exotic stones. In this regard, the Hell Gap site contrasts sharply with most Paleoindian localities. This raises an interesting question: did Paleoindian groups at the Hell Gap site behave expediently and have a limited mobil-

Table 10.1. Proportion of Local and Exotic Raw Materials, in Number of Nodules

	Level 1		Level 2		Level 2e		Level 6	
Exotic materials	7	4.3%	3	3.5%	3	2.4%	2	1.2%
Local materials	155	95.6%	85	96.5%	120	97.6%	168	98.8%

ity range? This would be a logical conclusion in view of the local/exotic model. The argument, however, does not take into account the scheduling of tool production and replacement, as discussed below.

SITE BACKGROUND

The Hell Gap site is in fact a series of localities nestled in a small valley that opens onto the plains of southeastern Wyoming. The focus of this analysis is the main locality, locality one. It was excavated in the 1960s by a team composed of H. Irwin, C. Irwin-Williams, G. Agogino, and C. V. Haynes. The Hell Gap excavation yielded a wealth of data, which subsequently served to establish the entire traditional Paleoindian chrono-cultural sequence (Irwin-Williams et al. 1973).

Originally the site was interpreted as a series of discretely stacked campsites, each characterized by a different type of projectile point. A recent reevaluation of the stratigraphy challenged this reconstruction, however, and revealed a complex taphonomic history (Sellet 1999, 2001) in which Paleoindian points are interstratified. The present study encompasses the four oldest archaeological components of locality one, which correlate with the early Paleoindian occupations (Goshen, Folsom, Agate Basin, and Hell Gap in age).

Level 1 is the deepest archaeological unit recovered at the site and one of the richest. It contains Goshen and Folsom points. A less dense zone just above it, level 2, yielded Agate Basin and Folsom points. At the same depth but east of level 2 is level 2e. It is a rich archaeological horizon that yielded Agate Basin points only. Finally, the remaining levels, 3 to 6, vary in artifact density and thickness. They were lumped together for the purpose of this study, because levels 3, 4, and 5 contain a very small number of artifacts. Level 3–6 produced several Hell Gap points (level 6), as well as a few Goshen and Folsom diagnostics (levels 3–5).

RESULTS OF THE ANALYSIS

The respective proportions of the three unifacial tool types (endscrapers, sidescrapers, and gravers) in single or multiple item nodules for the four levels considered here are shown in Table 10.2. Although the basic structure of the Paleoindian tool kit remains the same through time (sidescrapers are always the

Table 10.2. Hell Gap Locality One, Number of Tools According to Nodule Type

	Endscrapers		Sidescrapers		Gravers		Others		Total	
	MIN	SIN	MIN	SIN	MIN	SIN	MIN	SIN	MIN	SIN
Level 1	5	7	12	15	2	0	7	3	26	25
Level 2	0	3	3	8	2	0	1	1	6	12
Level 2E	1	5	20	10	1	0	11	1	33	16
Level 3–6	2	5	2	16	0	1	2	1	6	23
Total	8	20	37	49	5	1	21	6	71	76

Note: MIN: multiple item nodule; SIN: single item nodule.

most abundant tool type, followed by endscrapers and then gravers), the details of assemblage composition are not identical.

The sorting of the collection into nodules permits the identification of transported tools. The numbers in Table 10.2 indicate that some of the tools—the gravers—were regularly made on the spot and rarely carried. At the opposite of the spectrum, endscrapers were seldom manufactured expediently. Endscrapers, unlike gravers, belong to the transported tool kit. Finally, sidescrapers had a more flexible role: they were both transported and expediently manufactured.

At Hell Gap tool types clearly performed different functions within the tool kit and responded differently to scheduling strategies. Transport costs, manufacturing costs, overall utility, design, and function of the tools all explain the pattern. These factors formed a complex web of interactions that served as the basis for operationalizing tool production and replacement strategies. Hafting, for instance, could explain why endscrapers remained in the system longer than gravers did.

Another reason may lie in the specific requirements of tool production: the manufacture of endscrapers necessitated bigger and thicker flake blanks that had the ability to withstand heavy pressure without failing. Gravers, in contrast, could have been (and generally were) made on thinner flakes. Those flake blanks, unlike the ones required for endscrapers, were readily found among the by-products of biface manufacture.

These results bring into question the rationale behind a correlation of local or exotic raw material types with transport of tools and then with human mobility. That a piece of raw material was transported does not mean that the tool made from it was too. One could imagine, for instance, the case of a graver produced expediently on a flake detached from a biface. The graver could be made from exotic material yet never have been transported. In the same vein, the fact that a tool was made on local material does not imply that it was expedient, as shown by the Hell Gap assemblage.

One of the main criticisms of the local/exotic model is the difficulty of identifying direct or indirect procurement in lithic assemblages (Close 2000: 51). Some have argued that exotic raw materials could have been exchanged through trade and that distance to raw material source would therefore not be indicative of a group's mobility (for example, Ellis 1989; see also Meltzer 1989 for a more comprehensive discussion of the issue). While this is certainly true, the present example points to an additional level of danger behind the local/exotic raw material rhetoric. It reinforces the need to reconstruct the structure and composition of prehistoric tool kits before extracting information on either human or tool movement.

The nodule partitioning in Table 10.2 provides further evidence regarding Paleoindian mobility and, more precisely, Paleoindian land-use strategies. When all tools are considered, regardless of type, two distinct patterns can be observed. In some of the levels (level 2 and level 3–6) the majority of tools were manufactured off site. By contrast, most tools that were recovered in level 1 and especially in level 2E were made *in situ*. This suggests that levels 2 and 3–6 probably represent short-term occupations, while the other levels indicate a more durable stay at the site.

These data seem to indicate that Paleoindian groups exploited the valley in different ways: logistically, for short hunting or gathering forays, and also as a residential base-camp location. Even though it is not possible to reconstruct settlement systems from a single locality or to extend the present results to all Paleoindian adaptations, the contrasting strategies of exploitation of the Hell Gap area at last provide clues to the complexity of Paleoindian mobility patterns.

Considering the versatile role of sidescrapers in the Paleoindian tool kit, it is possible that they could serve as a relative measure of length of occupation, everything else being equal. This argument runs counter to earlier conclusions regarding tool proportions as indicators of cultural differences (Irwin and Wormington 1970). Under this formula, a larger quantity of sidescrapers among the discarded tools would point toward a longer stay in the Hell Gap valley.

PALEOINDIAN MOBILITY OR THE NEED FOR SPEED

Over the years most ideas about Paleoindian mobility have arisen from the basic tenet that Paleoindians colonized the American continents quickly. Because most current and past models of colonization provide little time for the first settlers to move from Alaska to Tierra del Fuego, Paleoindians have to be able to travel rapidly, regardless of the environments that they would encounter. Clovis points, traditionally considered the markers of that first human wave, are found over most of the North American continent and are tightly dated to about 11,500 to 10,900 years BP. For Clovis groups to jump from one biotope to another in so little time was only possible in a context of high residential mobility. In Paleoindian archaeology, mobility has therefore been an explanatory device for colonization.

Exotic materials were recovered at numerous Paleoindian localities, which in turn fueled the perception that the pattern was universal. In a recent article, D. B. Bamforth (2002: 87) remarks:

> The inference that Paleoindian groups did uniformly move over large territories is founded in the recognition of long-distance transport of large amounts of stone at sites like Shoop and Blackwater Draw that were discovered early in the history of Paleoindian research. The links that have been forged since that time between Paleoindian technology and "high mobility" depend substantially on taking this mobility as a given.

In view of these historical considerations, it is not surprising that Paleoindians are believed to be highly mobile. But is this accurate? In other words, can ideas about the first settlers entering an empty continent be extended to the entire Paleoindian record? And perhaps even more importantly: how can we improve upon that traditional depiction of Paleoindian mobility?

An isolated site (even one as rich as the Hell Gap site) will not solve the puzzle of Paleoindian mobility. Nevertheless, among the range of archaeological localities, stratified ones like Hell Gap are extremely valuable. They provide a big slice of time and simultaneously offer the advantage of keeping a variety of factors that might otherwise affect mobility constant, such as access to lithic raw material sources, access to food resources (assuming that season of occupation is identical), access to other resources such as water and firewood, and finally the topography of the terrain.

The analysis of Paleoindian technological organization at the Hell Gap site demonstrates that Paleoindian mobility might not have been uniform across space and time. Unfortunately, stratified Paleoindian campsites such as Hell Gap are rare. Our perception of mobility strategies has been almost entirely shaped by studies of kill sites or ephemeral occupations where mostly projectile points (often made on exotic materials) are recovered (Kelly and Todd 1988: 237; Meltzer 1984: 4, 1989: 38; Goodyear 1989: 7). These highly specialized localities provide a narrow window on the organization of technology, which is accompanied by a similarly biased perspective on general subsistence strategies. This characterization, in turn, has strongly affected our interpretation of Paleoindian settlement systems.

A good example of how this depiction has shaped ideas on mobility can be found in an overview of Paleoindian adaptations on the plains by W. R. Wedel (1964: 198): "These early hunters were free-ranging, moving wherever large game animals were available and perhaps not regularly returning to reoccupy earlier camp locations." Wedel's statement is now more than half a century old. His model has been since challenged (Meltzer 1993; Kornfeld 1988; Amick 1996a; Bamforth 1985, 2002), but Plains Paleoindians are still often described as moving from camp to camp in the migratory hunting of bison (for example, Jodry and Stanford 1992: 158, Kelly and Todd 1988: 236). A high residential mobility is said to be necessary to maintain contact with the mobile bison herds.

Bison was also the staple of the diet of modern Plains Indian groups. Unfortunately, ethnographic data about the subsistence and mobility patterns of these recent populations cannot be used as a direct analogy for Paleoindian adaptations, mainly due to the use of horses by historic groups. Ethnographic reports do, however, provide useful information on the flexibility and variability of their mobility strategies across seasons.

Most Plains Indians were sedentary for much of the cold season. Winter camps among the Blackfeet, for instance, lasted from early November to April (Ewers 1955: 124). The seasonal sedentary lifestyle was rendered possible by the storage of large quantities of dried meat, secured during the fall hunt (Kelly 1983: 298). The processed meat would then be carried by horse to the winter camp.

Another answer to the problem of winter shortage is illustrated by Martin Garretson (1938: 182), who mentions groups in eastern Wyoming that would camp for three months near a buffalo jump. In this case, rather than moving the food to the people (by storing quantities of processed meat for future consumption), people were moved to the source of food.

Not all historic groups relied on horses for transport. The Assiniboines had fewer horses than other Plains tribes and used mostly dogs to carry burdens. In this context, the historic description of Assiniboine subsistence strategies by Edwin Thompson Denig (1961) potentially provides a more accurate analogy for pre-horse adaptations, including Paleoindian ones. Describing the paucity of horses, Denig (1961: 96) remarks: "The inability to carry destroys the desire to lay up provisions and militates against any economy in the article of meat. They are compelled to follow the buffalo at all times when over one day's travel from their camp."

Facing food shortage during cold months as a result of the impossibility of securing large quantities of processed food, the Assiniboine adaptive response was high residential mobility, not unlike Wedel's (1964) portrayal of Paleoindian subsistence.

The Paleoindian archaeological record contains little indication of the intense marrow processing that would point to a reliance on storage strategies (Todd 1987; Todd et al. 1992: 162). But patterns of limited carcass exploitation are not universal; new evidence from the Jim Pitts site in South Dakota (Sellet, unpublished data), from the Hell Gap site (Rapson and Niven n.d.), and from the Folsom level at the Agate Basin (Hill 2001) site in Wyoming suggests that in a few cases bones were extensively broken up. All three sites are cold-month occupations (Frison and Stanford 1982; Rapson and Niven n.d.). Although more work is needed before a secure conclusion can be reached, these few examples could reflect a distinct winter subsistence organization and therefore discrete mobility strategies.

D. S. Amick (1994) similarly postulated a Paleoindian subsistence organization that incorporated discrete seasonal mobility patterns. He pointed to the absence of large Folsom kill sites in the mountain-basin range of New Mexico and used variability in the raw material of projectile points to propose a seasonal east-west movement of small hunting bands that would be logistically organized in the mountain-basin region and residentially organized in the plains. Amick's study was geographically confined to the southern plains and carried the limitations of regional approaches summarized earlier; but it set the stage for a comprehensive model of assemblage variability that incorporates temporal fluctuations of resources. Whether Late Pleistocene climates included such seasonal fluctuations, however, is still debated (see Todd et al. 1990: 824).

In the end we have to consider the possibility that there may not have been a unique Paleoindian mobility strategy. The ethnographic data cited above suggest that contemporaneous groups living in the same area and exploiting the same food resources devised multiple adaptive solutions, all involving distinct mobility patterns. The above-mentioned cases are also just a few examples of the total behavioral repertoire available to hunter-gatherers.

Our knowledge of Paleoindian mobility has long relied on inferences drawn from raw material economy (the presence of exotic raw materials) and from generalizations about subsistence organization (Paleoindians were big game hunters). It has also been colored by ideas about the first settlers that were unfairly extended to the entire Paleoindian record. As a result, Paleoindians were thought to epitomize high mobility. Yet long-distance transport of raw materials is not unique to Paleoindians (see Reeves 1990: 184). More importantly, as demonstrated in this study and elsewhere, it is not an accurate measure of mobility. The archaeological record provides little support for a uniform Paleoindian adaptive system. Were Paleoindians the champions of mobility? The most we can say is that some of them may have been.

Conclusions

It is widely agreed that Paleoindians are highly mobile, but beyond this general-ization there is no clear consensus on the specificity of Paleoindian mobility. The ambiguity stems for the most part from methodological shortcomings in the way in which inferences are drawn from stone tools. With regard to this issue, several points need to be stressed.

First, raw material economy does not directly represent mobility. Although one can hypothesize fairly securely that tools made from exotic materials were transported to a site, the antithesis is not true: tools made from local materials are not always manufactured on site. It follows that transported and expedient tool kits cannot be reconstructed from a simple local/exotic material dichotomy. Most arguments about Paleoindian mobility are built on inferred relationships between tool movement and human movement that are precarious at best. As a result, the local/exotic model fails to reveal the intricacies of the processes that affect the realm of lithics.

Second, assemblage composition does not directly reflect mobility either. In fact, tool types have distinct use-lives and respond differently to scheduling. Thus the functional role of each type of tool influences discard and replacement strategies and, ultimately, assemblage composition. A comprehensive study of assemblage variability should isolate the role of each tool in the tool kit and fac-tor it into a reconstruction of technological decisions.

It follows from these remarks that not all tool types contain an equal level of information. Projectile points, for instance, are unlikely to suggest anything but a panoramic view of mobility strategies: they are among the longest-lasting tools in the tool kit. They might relate to residential mobility—although the type of contingency is unclear—but will not tell much about logistical forays. Conversely, the Hell Gap site showed that the role of sidescrapers is more flex-ible. They provide a finer-grained resolution of mobility-related decisions. The respective proportions of sidescrapers manufactured on site versus imported ones hint at the duration of occupation and at the specificity of mobility strate-gies at a given locality.

All told, the problems raised by equifinality suggest that the relationship be-tween mobility strategies and technology cannot be abbreviated into simple cor-relations. Inferring mobility strategies from stone tools is not an easy task. Kelly and Torrence's doubts outlined at the beginning of this chapter are legitimate. While we can formulate broad conclusions about mobility strategies, it is more challenging to expose the interactions between mobility strategies and potential technological responses. The systemic approach advocated in this chapter should help us move away from empirical generalizations that hide the complexities at play.

REFERENCES

Amick, D. S.

1994   Technological Organization and the Structure of Inference in Lithic Analysis: An Examination of Folsom Hunting Behavior in the American Southwest. In *The Organization of North American Prehistoric Chipped Stone Tool Technologies*, edited by P. Carr, pp. 9–34. Archaeological Series No. 7. International Monographs in Prehistory, Ann Arbor.

1996a   Folsom Diet Breadth and Land Use in the American Southwest. Ph.D. dissertation. University of New Mexico, Albuquerque.

1996b   Regional Patterns of Folsom Mobility and Land Use in the American Southwest. *World Archaeology* 27(3): 411–426.

1999   Raw Material Variation in Folsom Stone Tool Assemblages and the Division of Labor in Hunter-Gatherer Societies. In *Folsom Lithic Technology*, edited by D. Amick, pp. 169–187. Archaeological Series No. 12. International Monographs in Prehistory, Ann Arbor.

Anderson, D. G.

1996   Modeling Paleoindian and Early Archaic Settlement in the Southeast: A Historical Perspective. In *The Paleoindian and Early Archaic Southeast*, edited by D. G. Anderson and K. E. Sassaman, pp. 29–57. University of Alabama Press, Tuscaloosa.

Bamforth, D. B.

1985   The Technological Organization of Paleoindian Small Group Bison Hunting on the Llano Estacado. *Plains Anthropologist* 30: 243–258.

2002   High Tech Foragers? Folsom and Later Paleoindian Technology on the Great Plains. *Journal of World Prehistory* 16(1): 55–98.

Bamforth, D. B., and M. S. Becker

2000   Core/Biface Ratios, Mobility, Refitting, and Artifact Use-Lives: A Paleoindian Example. *Plains Anthropologist* 45(173): 273–290.

Beck, C., and G. T. Jones

1990   Toolstone Selection and Lithic Technology in Early Great Basin Prehistory. *Journal of Field Archaeology* 17(3): 283–299.

Binford, L. R.

1979   Organization and Formation Processes: Looking at Curated Technologies. *Journal of Anthropological Research* 35: 255–272.

1980   Willow Smoke and Dogs' Tails: Hunter-Gatherer Settlement Systems and Archaeological Site Formation Processes. *American Antiquity* 45(1): 4–20.

1983   *In Pursuit of the Past.* Thames and Hudson, New York.

2001   *Constructing Frames of Reference.* University of California Press, Berkeley.

Boldurian, A.

1991   Folsom Mobility and Organization of Lithic Technology: A View from Blackwater Draw, New Mexico. *Plains Anthropologist* 36: 281–295.

Brantingham, P. J.

2003   A Neutral Model of Stone Raw Material Procurement. *American Antiquity* 68(3): 487–510.

Close, A. E.

2000   Reconstructing Movement in Prehistory. *Journal of Archaeological Method and Theory* 7(1): 49–77.

Deller, D. B., and C. J. Ellis

1992   *Thedford II: A Paleoindian Site in the Ausable River Watershed of Southwestern Ontario.* Museum of Anthropology Publications, Memoir 24. University of Michigan, Ann Arbor.

Denig, E. T.

1961   *Five Indian Tribes of the Upper Missouri.* University of Oklahoma Press, Norman.

Dincauze, D. F.

1993   Fluted Points in the Eastern Forests. In *From Kostenki to Clovis,* edited by O. Soffer and N. Praslov, pp. 279–292. Plenum Press, New York.

Ellis, C. J.

1989   The Explanation of Northeastern Paleoindian Lithic Procurement Patterns. In *Eastern Paleoindian Lithic Resource Use,* edited by C. Ellis and J. Lothrop, pp. 139–164. Westview Press, Boulder.

Ewers, J. C.

1955   *The Horse in Blackfoot Indian Culture.* Smithsonian Institution Press, Washington, D.C.

Frison, G. C., and D. J. Stanford

1982   *The Agate Basin Site: A Record of Paleoindian Occupations in the Northwestern Plains.* Academic Press, New York.

Gamble, C. S.

1991   An Introduction to the Living Spaces of Mobile People. In *Ethnoarchaeological Approaches to Mobile Campsites,* edited by C. S. Gamble and W. A. Boismier, pp. 1–24. Archaeological Series 1. International Monographs in Prehistory, Ann Arbor.

Garretson, M.

1938   *The American Bison.* New York Zoological Society, New York.

Goodyear, A. C.

1989   A Hypothesis for the Use of Cryptocrystalline Raw Material among Paleoindian Groups of North America. In *Eastern Paleoindian Lithic Resource Use,* edited by C. J. Ellis and J. C. Lothrop, pp. 1–10. Westview Press, Boulder.

Greaves, R. D.

1997   Hunting and Multifunctional Use of Bows and Arrows: Ethnoarchaeology of Technological Organization among Pumé Hunters of Venezuela. In *Projectile Point Technology,* edited by H. Knecht, pp. 287–320. Plenum Press, New York.

Hayden, B.

1982   Interaction Parameters and the Demise of Paleoindian Craftsmanship. *Plains An-thropologist* 27(96): 109–123.

Hill, M. G.

2001   Paleoindian Diet and Subsistence Behavior on the Northwestern Great Plains of North America. Ph.D. dissertation. Department of Anthropology, University of Wisconsin, Madison.

Hofman, J. L.

1992   Recognition and Interpretation of Folsom Technological Variability on the Southern High Plains. In *Ice Age Hunters of the Rockies*, edited by D. Stanford and J. Day, pp. 193–224. Denver Museum of Natural History, Denver.

1999   Folsom Fragments, Site Types and Prehistoric Behavior. In *Folsom Lithic Technology*, edited by D. Amick, pp. 122–143. Archaeological Series No. 12. International Monographs in Prehistory, Ann Arbor.

Ingbar, E. E.

1992   The Hanson Site and Folsom on the Northwestern Plains. In *Ice Age Hunters of the Rockies*, edited by D. Stanford and J. Day, pp. 168–192. Denver Museum of Natural History, Denver.

1994   Lithic Material Selection and Technological Organization. In *The Organization of North American Prehistoric Chipped Stone Tool Technologies*, edited by P. Carr, pp. 45–56. Archaeological Series 7. International Monographs in Prehistory, Ann Arbor.

Ingbar, E. E., and J. L. Hofman

1999   Folsom Fluting Fallacies. In *Folsom Lithic Technology*, edited by D. Amick, pp. 98–110. Archaeological Series No. 12. International Monographs in Prehistory, Ann Arbor.

Irwin, H. T., and H. M. Wormington

1970   Paleo-Indian Tool Types in the Great Plains. *American Antiquity* 35(1): 24–34.

Irwin-Williams, C., H. T. Irwin, G. A. Agogino, and C. V. Haynes

1973   Hell Gap: Paleo-Indian Occupation on the High Plains. *Plains Anthropologist* 18(59): 40–53.

Jochim, M. A.

1981   *Strategies for Survival: Cultural Behavior in an Ecological Context*. Academic Press, New York.

Jodry, M. A.

1999   Folsom Technological and Socioeconomic Strategies: Views from Stewart's Cattle Guard and the Upper Rio Grande Basin, Colorado. Ph.D. dissertation. American University, Washington, D.C.

Jodry, M. A., and D. J. Stanford

1992   Steward's Cattle Guard Site: An Analysis of Bison Remains in a Folsom Kill-Butchery Campsite. In *Ice Age Hunters of the Rockies*, edited by D. Stanford and J. Day, pp. 101–168. Denver Museum of Natural History, Denver.

Jones, G. T., C. Beck, E. Jones, and R. E. Hughes

2003 Lithic Source Use and Paleoarchaic Foraging Territories in the Great Basin. *American Antiquity* 68(1): 5–38.

Kelly, R. L.

1983 Hunter-Gatherer Mobility Strategies. *Journal of Anthropological Research* 39: 277–306.

1988 The Three Sides of a Biface. *American Antiquity* 53(4): 717–734.

1992 Mobility and Sedentism: Concepts, Archaeological Measures and Effects. *Annual Review of Anthropology* 21: 43–66.

1995 *The Foraging Spectrum: Diversity in Hunter-Gatherer Lifeways.* Smithsonian Institution Press, Washington, D.C.

Kelly, R. L., and L. C. Todd

1988 Coming into the Country: Early Paleoindian Hunting and Mobility. *American Antiquity* 53(2): 231–244.

Kornfeld, M.

1988 The Rocky Folsom Site: A Small Folsom Assemblage from the Northwestern Plains. *North American Archaeologist* 9(3): 197–222.

Kuhn, S.

1992 On Planning and Curated Technology in the Middle Paleolithic. *Journal of Anthropological Research* 48: 185–214.

Larson, M. L., and M. Kornfeld

2000 Chipped Stone Nodules: Theory, Method, and Examples. *Lithic Technology* 22(1): 4–18.

Meltzer, D. J.

1984 On Stone Procurement and Settlement Mobility in Eastern Fluted Point Groups. *North American Archaeologist* 6: 1–24.

1989 Was Stone Exchanged among Eastern North American Paleoindians? In *Eastern Paleoindian Lithic Resources*, edited by C. J. Ellis and J. Lothrop. pp. 12–39. Westview Press, Boulder.

1993 Is There a Clovis Adaptation? In *From Kostenki to Clovis*, edited by O. Soffer and N. Praslov, pp. 293–310. Plenum Press, New York.

Nelson, M.

1991 The Study of Technological Organization. *Archaeological Method and Theory* 3: 57–100.

Odell, G. H.

1994 Assessing Hunter-Gatherer Mobility in the Illinois Valley: Exploring Ambiguous Results. In *The Organization of North American Prehistoric Chipped Stone Tool Technologies*, edited by P. Carr, pp. 70–86. Archaeological Series No. 7. International Monographs in Prehistory, Ann Arbor, Mich.

1996a Economizing Behavior and the Concept of "Curation." In *Stone Tools, Theoretical Insights into Human Prehistory*, edited by G. Odell, pp. 51–79. Plenum Press, New York.

1996b *Stone Tools and Mobility in the Illinois Valley.* Archaeological Series 10. International Monographs in Prehistory, Ann Arbor.

2001 Stone Tool Research at the End of the Millennium: Classification, Function, and Behavior. *Journal of Archaeological Research* 9(1): 45–100.

Parry, W. J., and R. L. Kelly

1987 Expedient Core Technology and Sedentism. In *The Organization of Core Technology*, edited by J. K. Johnson and C. A. Morrow, pp. 285–304. Westview, Boulder.

Rapson, D., and L. Niven

n.d. Reevaluating Hell Gap: Bison Dentition Studies and Analysis of the Hell Gap Site Faunal Assemblage. Manuscript on file at the Department of Anthropology at University of Wyoming, to be included in the final monograph for the Hell Gap site.

Reeves, B. O. K.

1990 Communal Hunters of the Northern Plains. In *Hunters of the Recent Past*, edited by L. B. Davis and B. O. K. Reeves, pp. 168–194. Unwin Hyman, London.

Sassaman, K. E.

1994 Changing Strategies of Biface Production in the South Carolina Coastal Plain. In *The Organization of North American Prehistoric Chipped Stone Tool Technologies*, edited by P. Carr, pp. 99–117. Archaeological Series No. 7. International Monographs in Prehistory, Ann Arbor, Mich.

Saul, J. M.

1969 Study of the Spanish Diggings Aboriginal Flint Quarries of Southeastern Wyoming. In *National Geographic Research Reports* (1964 projects), pp. 183–199.

Seeman, M. F.

1994 Intercluster Lithic Patterning at Nobles Pond: A Case for "Disembedded" Procurement among Early Paleoindian Societies. *American Antiquity* 59(2): 273–287.

Sellet, F.

1999 A Dynamic View of Paleoindian Assemblages at the Hell Gap Site, Wyoming: Reconstructing Lithic Technological Systems. Ph.D. dissertation. Department of Anthropology, Southern Methodist University. University Microfilms, Ann Arbor.

2001 A Changing Perspective on Paleoindian Chronology and Typology: A View from the Northwestern Plains. *Arctic Anthropology* 38(2): 48–63.

2004 Beyond the Point: Projectile Manufacture and Behavioral Inference. *Journal of Archaeological Science* 31: 1553–1566.

Shott, M. J.

1986 Technological Organization and Settlement Mobility: An Ethnographic Examination. *Journal of Anthropological Research* 42: 15–51.

Storck, P.
1997   *The Fisher Site*. Museum of Anthropology Publications, Memoir 30. University of Michigan, Ann Arbor.

Todd, L. C.
1987   Analysis of Kill-Butchery Bonebeds and Interpretation of Paleoindian Hunting. In *The Evolution of Human Hunting*, edited by H. Nitecki and D. Nitecki, pp. 225–266. Plenum Press, New York.

Todd, L. C., J. L. Hofman, and C. B. Schultz
1990   Seasonality of the Scottsbluff and Lipscomb Bison Bonebeds: Implications for Modeling Paleoindian Subsistence. *American Antiquity* 55(4): 813–827.
1992   Faunal Analysis and Paleoindian Studies: A Reexamination of the Lipscomb Bison Bonebed. *Plains Anthropologist* 37(139): 137–165.

Torrence, R.
1983   Time, Budgeting and Hunter-Gatherer Technology. In *Hunter- Gatherer Economy in Prehistory: A European Perspective*, edited by G. Bailey. pp. 11–22. Cambridge University Press, Cambridge.
1989   Tools as Optimal Solutions. In *Time, Energy and Stone Tools*, edited by R. Torrence, pp. 1–6. Cambridge University Press, Cambridge.
1994   Strategies for Moving On in Lithic Studies. In *The Organization of North American Prehistoric Chipped Stone Tool Technologies*, edited by P. Carr, pp. 123–133. Archaeological Series 7. International Monographs in Prehistory, Ann Arbor.

Wedel, W. R.
1964   The Great Plains. In *Prehistoric Man in the New World*, edited by J. Jennings and E. Norbeck, pp. 193–220. University of Chicago Press, Chicago.

Wheat, J. B.
1972   *The Olsen-Chubbuck Site: A Paleoindian Bison Kill*. Society for American Archaeology Memoirs 26.

Wobst, H. M.
1974   Boundary Conditions for Paleolithic Systems: A Simulation Approach. *American Antiquity* 39(2): 147–178.

11

Local Raw Material Exploitation
and Prehistoric Hunter-Gatherer Mobility

PAUL T. THACKER

Stone artifacts are the most common and longest-lasting elements of the archaeological record, a quality that helps explain the focus on lithic artifact analysis in Paleolithic and Paleoindian archaeology. Many of the earliest archaeological reconstructions of Late Pleistocene hunter-gatherer movement interpreted linear distance to exotic raw material sources as a direct indicator of mobility, initiating a set of assumptions that remain widely employed today. Yet the distance between the geological source and an artifact's recovery location is only a measure of the movement of a specific stone fragment. Such decontextualized raw material data rarely correlate with individual or group mobility, because numerous other processes influence raw material use and transport within hunter-gatherer lithic economies.

This chapter examines the interpretive significance of local raw material exploitation by incorporating insights from evolutionary ecology and resource use models within a diachronic regional approach. Constructing dynamic models of changing prehistoric hunter-gatherer mobility using contextualized flaked stone assemblages is theoretically possible, but the task is much more difficult than is commonly assumed. Middle-range arguments developed from diet-breadth ecological models imply that current hypotheses of mobility in Late Upper Paleolithic and Epipaleolithic Portuguese Estremadura warrant reassessment.

Transporting Artifacts, Individual Movement, and Residential Mobility: Scales and Significance

Obstacles to reconstructing prehistoric mobility stem from the very nature of the endeavor: understanding characteristics of a dynamic human settlement system by using the archaeological record, essentially a static residue. Ethnographic studies and ethnoarchaeology are critical and productive sources for pattern recognition and modeling of hunter-gatherer variability, including lithic organization and mobility strategies. Unfortunately, many archaeologists apply these models to the archaeological record without full consideration of their structural implications.

The well-used (and probably abused) forager-collector continuum described by Binford (1980) is an example of an anthropological model with complex inferences for discerning settlement strategies in archaeology. Residential movement is a critical concept for archaeological interpretation in the forager/collector framework. Foragers move camp more frequently and utilize a daily foraging radius, while collectors move base camps less frequently but rely on logistical forays to acquire resources located beyond the daily radius. Several studies of ethnographic groups have found relationships between number of annual moves, resource acquisition, and ranging behavior (Binford 2001; Kelly 1995).

As Kelly pointed out (1992), archaeologists are often vague in their definitions of mobility. The annual range traveled by a specific band is rarely correlated with the total territory exploited during an individual's lifetime. In many cases, individuals within the same band have significantly different foraging radii. A group may relocate its campsite many times during a year but not move very far, and vice versa. The number of residential moves per year is rarely directly related to annual range, except when other environmental relationships are considered (Binford 2001).

In addition to the manifold problem of frequency of residential movement versus overall range, archaeologists face a mire of equifinality upon realizing that in some cases certain raw materials in the form of *specific tools* may move farther when people move base camp less often. For example, specialized tools may be associated with tasks conducted during special-purpose, long-distance logistical forays. This behavioral association results in the transport of certain raw materials (used for these specialized tools) far from their geological sources even when residential mobility (base camp movement) is low. In contrast, lithic materials may move greater distances *in cobble or core form* in a system exhibiting increasing residential mobility (Binford 1979; Binford and O'Connell 1984).

Much confusion within archaeological middle-range theory concerning raw material exploitation originates in misconceptualizations of the archaeological record as a snapshot of ethnographic ranging behavior. Using Binford's caveat that organizational structure is the critical component for interpreting the structure of the archaeological record, archaeologists are wise to assume that sites do not represent events within an ethnographic range. Rather, the settlement and subsistence of a hunter-gatherer group had organizational consequences for raw material exploitation, site functions, and assemblage variability. The identification and explanation of structured patterning in the organization of a lithic technology can, in certain cases, anchor hypotheses of prehistoric hunter-gatherer systemic mobility (Gamble and Boismier 1991; Kuhn 1989). This undertaking is very different from attempting to document historical group movements (Close 1996, 2000).

In order to avoid these analytical pitfalls, archaeological evidence of raw material exploitation must be contextualized. Contextualization begins by evaluating lithic raw material exploitation practices within the geographical distribution of available lithic resources, including local and nonlocal materials. Against this characterized background, inter- and intra-assemblage variability and technological organization must supplement indicators of activities/site functions on a regional level. Contextualizing lithic exploitation and technological organization is a necessary prerequisite for incorporating lithic raw material data into systemic mobility reconstructions or interpretations.

LITHIC RAW MATERIALS AS ECOLOGICAL RESOURCES

Within the interdisciplinary exchange between evolutionary ecology and anthropology, a number of studies have productively modeled human hunter-gatherers as resource managers (Williams and Hunn 1982; Smith and Winterhalder 1992). Rather than being real in any sense, such models are useful heuristically for developing possible explanations of archaeological patterning (Smith 1987) and identifying situations where human groups behave in notably counterintuitive ways.

Several concepts from evolutionary ecology have been adapted for use in lithic analysis, most visibly the attempts to discern the effects of time stress and buffering mechanisms on stone tool design and use-life (Torrence 1989; Edmonds 1987). The Paleoindian and Paleolithic archaeological community's emphasis (perhaps overemphasis) on indicators of risk management may be linked to the understandably seductive allure of concepts such as curation and design reliability. These concepts were not developed for analysis of stone tool technology. Thus valid critiques of them should not be surprising (Nash 1996; Odell 1996; Shott 1996); nor should the setbacks of such approaches indict the application of evolutionary ecological perspectives to hunter-gatherer prehistory. These archaeological applications have met with limited success precisely because of the problem of assemblage context. Isolated, single-site, or otherwise decontextualized lithic data are insufficient for robust modeling of prehistoric mobility strategies (Hill 1994; Odell 1994).

Patch-choice and diet-breadth models are two mainstays of food-resource exploitation models in anthropological evolutionary ecology (Bettinger 1991). These models are suitable foundations for building an ecological model of lithic resource exploitation, granted that appropriate modifications due to differences between organic food resources and inorganic lithic materials are necessary. Resource models in optimal foraging theory are geographical, describing behavior across a landscape of unevenly distributed resources of varying types

(Winterhalder 2001). Mapping lithic resource occurrences and characterizing distributions is a necessary first step for understanding prehistoric raw material exploitation practices (Church 1994).

Lithic resources are significantly different from organic counterparts on the landscape in those source areas:

(a) they are more predictably located;
(b) they are rarely exhausted and do not require rejuvenation;
(c) they are more predictable in terms of resource-return rates;
(d) they are not usually as time-dependent or time-variable (for example, seasonality).

Because of these crucial differences from biotic resources, patch-choice models have limited application for lithic studies. Diet-breadth models, however, hold more potential for understanding stone tool assemblage organization.

Several ethnographic food resource studies have demonstrated that the specific nature of an individual resource is an important determinant of strategies of exploitation and subsequent mobility decisions (Bettinger et al. 1997; Binford 2001). Specifically, diet-breadth models reconstruct possible food resources available for human consumption and compare these options with the resources actually exploited. The value in such an approach extends beyond confirming reasonable decisions by humans occupying a region; the approach frequently results in the discovery of counterintuitive relationships, such as the use of a lower-ranked resource when higher-ranked resources are plentiful (Bird 1997; Madsen and Schmitt 1998). Often human hunter-gatherers behave in a less than optimal manner, demonstrating the fundamentally social nature of information flow and the constructed elements of landscape and subsistence (Holt 1996; Mithen 1989, 1990).

Archaeological application of these ecological insights facilitates an explicitly geographic raw material selection model similar to those successfully applied to organic resources (Bonzani 1997; Grayson and Delpech 1998). Considering the diversity and availability of knappable raw materials is not new in archaeology (see Odell 1984; Reid 1997) and is an important consideration in the lithic sourcing literature (Church 1995; Shackley 1998).

This ecological analysis assumes that prehistoric groups knew the distribution of lithic resources when occupying or traversing a region. Such an assumption may not be warranted in all archaeological cases but is certainly reasonable in the vast majority of prehistoric situations (Torrence et al. 1996). Pioneering or migrating groups moving across an absolutely unknown territory were rare occurrences in prehistory; in those exceptional circumstances, most accepted evolutionary ecological resource models contain the same (if not more damaging) theoretical obstacles as this proposed raw material model (Clark 1994).

Generalizing handling costs is problematic for lithic resources as well as for food resources, and cost rankings and comparisons are necessarily left to the specific regional variability/technologies under investigation. In many cases, handling costs of stone may include quarrying/mining activities, differential decortification requirements, heat treatment, and the like. Fortunately most handling cost activities involving lithic materials generate recognizable signatures in the archaeological record, an advantage not shared by many organic resource procurement and processing activities.

The embedded nature of most lithic procurement observed ethnographically is a critical difference between lithic resource acquisition and many organic counterparts (Binford 1979). Rarely is stone the only or even the primary purpose of a hunter-gatherer's excursion (Gould and Saggers 1985; Haury 1994), so it is reasonable to consider the opportunity cost of procurement activities to be minimal relative to other foraging activities. That is, it is unnecessary to consider choosing to procure stone as calculated against obtaining food resources. Different stone sources (for example, an outcrop versus secondary occurrence in a gravel) or material types (fine quartzite versus quartz or chert), however, do have varying opportunity costs (Elston 1990). These differential costs, especially when both local and exotic stone are utilized in assemblages, underlie the archaeological interest in distance to raw material sources.

A prehistoric group's choice not to knap a locally available quartzite is of significance within this approach. As Kelly (1992: 55) argues, "To know what one resource offers means knowing what it offers relative to others." Ironically, this methodology, derived from evolutionary ecology, may discern elements of stone tool production and use that are *not* explainable except through a more complex theory of social action and/or reproduction (Clark 1999). In order to interpret the full relevance of raw material procurement choices, it is important to return again to context: namely, inter- and intra-assemblage variability in stone use.

TECHNOLOGICAL ORGANIZATION AND RAW MATERIAL SELECTION

The majority of lithic assemblages contain numerous raw material types, unevenly distributed across tool classes, reduction classes, core classes, and other categories. Rather than compressing variability into raw material trends within the assemblage as a whole, lithic analysis must focus attention on the distribution of raw materials across different assemblage classes and within varying reduction sequences (Hayden et al. 1996; Ingbar 1992). Often local raw materials were used for different purposes than exotic lithic resources (that is, locals may constitute the majority of certain tool classes); or, more obviously, prehistoric knappers may have employed a different reduction strategy based on the type of raw material (Andrefsky 1998; Montet-White and Holen 1991). While separating assemblages into raw materials for analysis of technological organization is commonly done, middle-range theory to explain resulting patterns is poorly developed and rarely explicit. Several studies in both the Old and New Worlds have recognized the manufacture of expedient tools and utilized flakes on local raw materials (Bamforth 1990; Nelson 1991; Parry and Kelly 1987; Straus 1991a). The meaning of this assemblage pattern is clarified by examining the organization of nonlocal raw materials in the same assemblage.

Resource maximization and time minimization relationships have usefully characterized aspects of hunter-gatherer ecology (Bamforth and Bleed 1997; Winterhalder 2001). Torrence (1994) appropriately warns that knappers produce foremost a tool adequate for its purpose and only indirectly consider transport costs or conservation of raw materials. These functional constraints are occasionally evident within tool assemblage attributes and can be supported by experimental and use-wear studies.

After hypothesis formulation regarding functional use, analysis should incorporate models of resource maximization and time minimization in procurement and processing of stone raw material. For example, if a local quartz is suitable for making thick scrapers, and no other raw materials have lower opportunity costs, then thick scrapers should be made from local raw materials.

Mobility Mirages: Exotic Raw Materials and Specialized Tool Forms

The role of labor-intensive, highly formalized tools has dominated most middle-range articulation of the organization of technology with mobility strategies (Carr 1994; Goodyear 1989; Hofman 1991; Larick 1987). Often these discussions reach an impasse when acknowledging the possibilities of exchange and trade for exotic materials (Meltzer 1989). More significantly, these arguments can suffer from spuriousness when concluding that a formal tool type or technology is linked to increased mobility. For example, if Paleoindians preferred (for whatever reason, functional or aesthetic) to fashion lanceolate points on high-quality raw materials occurring at only a single source on the landscape, it is tautological to argue that Paleoindian technological organization reflects high mobility, using distance to sources as supporting evidence. Lithic analysis must include an assessment of the use of local raw materials for manufacturing informal tools and must document the relationship of high-quality and more likely exotic raw materials with the formalized tool elements (Holdaway et al. 1996;

Brantingham 2003). Research in this direction is most notably represented by several Northern European Mesolithic studies in the Old World (Myers 1989; Jochim 1998; Vierra 1995), Paleoindian research in North America (Amick 1999; Ellis and Deller 2000; Sellet 2004), and studies on later Prehistoric assemblages in New Mexico (Walsh 1998).

Assemblage Variability and Regional Context

A scatter of stone artifacts may represent the location of specific human activities associated with certain artifact classes, a discard area (transformation), the impact of geological postdepositional processes, or—even worse for the archaeologist—all of the above (Hayden 1998; Isaac 1986; Stern 1993). Lithic technological organization may vary within the same cultural system or even individual behavior, depending on geographical location, specific activity, and social setting (Hayden et al. 1996; Phillipson 1980). Refitting studies and indirect analogs (such as nodule-type methods) are promising methodologies for exploring variability in this realm (Almeida 2000; Larson and Kornfeld 1997; Roebroeks and Hennekens 1990; Sellet 1999). Nevertheless, a holistic and representative sample of lithic technological variability from a hunter-gatherer system is essential for any modeling of mobility (Henry 1989; Thacker 1996; Williams 2000). Numerous lithic studies have demonstrated that a regional approach is the best solution to this sampling and theoretical issue (Demars 1982; Dibble 1991; Ebert and Camilli 1993; Floss 1994; Thacker 2000).

Regional approaches are a critical source of data. Only through an integration of land use, site types, activity variation, and their effects on lithic organization may complex models of raw material exploitation be linked to settlement and subsistence and hence mobility strategies. A regional approach does not assume recovery of an articulated settlement system or ethnographically meaningful reconstruction of past ranges and territories. Regional analysis seeks organizational structure and strategies of systems, rather than historically real movements and boundaries. Multiple site assemblages strengthen hypotheses of lithic organizational significance and avoid the potential idiosyncrasy of interpretations based on one activity area or assemblage (Blankholm 1991; Eriksen 1991; Feblot-Augustins 1997). Diachronic approaches are useful for evaluating the theoretical assumptions of a model as systemic change occurs (Bernaldo de Quiros and Cabrera Valdes 1996; Montes Barquín and Sanguino-González 1998).

THE IMPORTANCE OF LOCAL RAW MATERIAL EXPLOITATION
FOR RECONSTRUCTING HUNTER-GATHERER MOBILITY AT
THE PLEISTOCENE/HOLOCENE BOUNDARY: AN ARCHAEOLOGICAL
APPLICATION FROM PORTUGAL

Seven assemblages from the Terminal Magdalenian and Epipaleolithic of Portugal illustrate the importance of on-site or near-site resources for understanding the interface between technological organization and settlement strategies. All seven assemblages are from open-air sites in the Rio Maior vicinity and six of them yielded charcoal for absolute dating (Table 11.1). These large assemblages were selected from more than a dozen Late Pleistocene/Early Holocene sites in the region due to functional interpretation: all seven sites are currently considered campsites (Bicho 2000, Marks et al. 1994; Marks and Mishoe 1997; Zilhão 1997a), based on the presence of stone-lined hearth features and site size. Regional Terminal Paleolithic land-use patterns appear identical to those of the Epipaleolithic, suggesting a continuity in overall site location strategy within the site sample (Thacker 1996b).

The paleoenvironment of the Rio Maior vicinity from 11,000 BP until about 8500 BP was rather temperate by European standards, with the exception of a mild cold period during the end of Dryas III, as Nuno Bicho (1994) documents. Vegetational communities at the end of the Pleistocene were a mix of Atlantic and Mediterranean species, with pine, oak, and birch species present in the region throughout the Terminal Paleolithic. Faunal communities contain rabbit, hare, red deer, roe deer, horse, aurochs, and wild boar, as well as some colder-adapted species such as goat (*Capra*) and chamois (*Rupicapra*) (Bicho et al. 2000; Hockett and Bicho 2000; Haws 2000). These two latter species disappeared by the Early Holocene, and oak gradually replaced pine in many areas. Most reconstructions depict the Early Holocene paleoenvironment of central Estremadura as essentially equivalent to modern conditions (Bicho 1993, 1994; Marks and Mishoe 1997; Zilhão 1997b).

Table 11.1. Radiocarbon Dates for the Late Upper Paleolithic and Epipaleolithic Assemblages of the Rio Maior Vicinity

Cabeço do Porto Marinho I-U	12,220 ±110
Cabeço do Porto Marinho III S-U	11,810 ±110
Carneira-Pinhal	10,880 ±90
Cabeço do Porto Marinho V	9,100 ±160
Areeiro III	8,860 ±80; 8,850±50; 8,570±130
Fonte Pinheiro	8,450 ±190

Faunal assemblages from archaeological sites during this period are rare and come mainly from caves exhibiting better preservation conditions. Limited comparative data on human diet across the Pleistocene/Holocene boundary preliminarily indicate no major changes except of degree, with probable parallels to the Cantabrian pattern of subsistence intensification through both specialization and diversification, culminating during the Mesolithic (Straus 1992, 1999; Bicho 1994).

The continuity between the Magdalenian and Final Epipaleolithic lithic technology has been demonstrated by several studies (Bicho 2000; Zilhão 1997a), with chronological change being limited to an apparent (but poorly documented) increasing frequency of geometrics produced using a microburin technique as the Holocene progressed. Limited and mostly conjectural settlement system reconstruction has focused on two observations: the apparent increase in the number of sites in Portugal during the Epipaleolithic and the proposal that inland-coastal movement "increased" and took on a more logistical organizational character due to the presence of marine resources in Estremaduran cave sites (Bicho 1994, 1997, 1998, 2000).

THE DISTRIBUTION OF LITHIC RAW MATERIALS IN THE RIO MAIOR VICINITY

Using a lithic resource model requires detailing the occurrence of knappable raw materials in a region. There are two significant sources of lithic raw material in the Rio Maior vicinity, as discovered through a total coverage survey conducted from 1990 to 1993 (Thacker 2002). High-quality chert cobbles occur in secondary position, within gravels supporting the ridge separating the Rio Maior and Penegral drainage. These chert cobbles are exposed as intermittent and ephemeral perched streams incise the sands and gravels of the ridge during wet seasons. The second raw material source occurs as gravels of variable-quality quartz and quartzite that structurally support most of the landforms in the Rio Maior drainage. These gravel cobbles are exposed by stream and river erosion and occasionally by wind erosion/deflation of sand dunes or beds. More importantly, such gravels occur within a few minutes' walk of all the sites included in this chapter and, in fact, of any point in the drainage except on the limestone uplands. In addition to quartz and quartzites, these gravels contain small and variable frequencies of sandstones, siltstones, basalt, limestone, and rock crystal. Thus human groups in the valley during the Late Pleistocene and Early Holocene were selecting raw materials either from an essentially on- or near-site context or from the chert deposits, still well within a daily foraging radius.

Detecting the Shift to a Specialized, Diversified Subsistence/Settlement Strategy Using Stone Tool Assemblages

Current hypotheses concerning change in subsistence and settlement in Portuguese Estremadura during the Epipaleolithic propose the following:

(1) The increased use of specialized tool kits, especially geometrics, throughout the Epipaleolithic, culminating in the Mesolithic (Bicho 1998, 2000).

(2) An increasing logistical strategy of mobility, evidenced by faunal assemblages from inland Epipaleolithic cave sites (Straus 1992; Straus et al. 2000; Vierra 1995; Bicho 1994).

(3) An expedient use of local raw materials. Bicho (1997, 1998) proposed that, rather than being embedded in other activities, the "necessity" of chert procurement for manufacturing bladelet and geometric barbs impacted settlement systems, resulting in site location (albeit logistical camps) near known chert sources.

Diachronic tool type richness-diversity measures are a productive way to control for sample sizes and assess assemblage variability in tool form (Jones et al. 1989; Simek and Price 1990). Published tool typologies from the six sites are directly comparable and have minimal bias originating in methodological or investigator effects. All but one were analyzed by A. E. Marks and Bicho; Fonte Pinheiro was analyzed by Thacker and Bicho. The specific types included in the lists are based on the Upper Paleolithic typological scheme of Denise de Sonneville-Bordes, adapted for Portugal by Marks and João Zilhão (Marks et al. 1994; Zilhão 1997a). No retouched tool types are chronologically sensitive across the time range spanned by the assemblages (a necessary prerequisite for such diversity measures). The entire type list was chosen for the richness scale, because compressing individual types into tool classes does not, in this case, alter results.

As reported in Table 11.2, tool diversity at each site is predominantly a function of sample size. While the difference is not statistically significant, assemblages from the later periods contain slightly fewer tool types than the Terminal Magdalenian ones. Likewise, the percentage of geometrics and microburins in the tool assemblage shows no significant variability until Fonte Pinheiro, which is dated to the Final Epipaleolithic/Earliest Mesolithic (8450 years BP). In sum, using these assemblages from seemingly similar functional contexts (campsites), it is difficult to argue for major or even minor directional change in formal tool-kit design during the Terminal Pleistocene and Epipaleolithic. The only change in tool-kit diversity occurs at the Epipaleolithic/ Mesolithic transition.

Table 11.2. Lithic Tool Assemblage Diversity

Site	Number of tools	Number of tool types	% of geometrics
Cabeço do Porto Marinho I-U	1,481	72	0.4
Cabeçço do Porto Marinho III S-U	372	55	1.3
Carneira-Pinhal	200	42	0
Cabeço do Porto Marinho V	157	39	3.1
Carneira II	171	42	8.2
Areeiro III	554	45	0.2
Fonte Pinheiro	211	40	15.4

If Epipaleolithic hunter-gatherers were increasingly logistically organized, what structural changes should occur within large campsite lithic assemblages? Base camps or residential campsites are occupied for longer durations in logistical strategies, as the movement focus involves bringing "resources to people" (Kuhn 1992, 1995). Lithic assemblages from campsites occupied for longer durations will more likely contain several organizational strategies across different functional activities and possibly across raw material categories, which may be discernible through activity area differentiation (Torrence 2001). Conversely, the archaeological patterns produced by a highly specialized, logistical subsistence and settlement system will rarely exhibit extremely flexible, minimally differentiated reduction and use of tools. Unfortunately, site maintenance activities, such as surface sweeping and secondary discard activities, are also more likely to occur at longer-occupied locations. As emphasized above, equivalent functional context is critical for building these hypotheses using lithic assemblages.

Table 11.2 displays the frequencies of tools and cores in each Portuguese assemblage. The greatest variability between assemblages occurs during the Late Upper Paleolithic (CPM I-U and CPM III-S) and again at the Epipaleolithic/Mesolithic boundary. No significant change occurs through the Epipaleolithic in the frequency of tools or cores in the campsite assemblages. This observation corroborates earlier doubts as to a major transformation or trend in the organization of technology between the Late Upper Paleolithic and Epipaleolithic. While the percentage of on-site/near-site quartz and quartzite cores does not vary with time, the use of such local raw materials for tools does slightly increase at the Epipaleolithic/Mesolithic transition.

Table 11.3 demonstrates that in the Portuguese case, over 96% of informal tools throughout the Epipaleolithic were produced on chert rather than on quartz or quartzite. Again, a noticeable shift to more informal tools on local raw materials occurs at the Epipaleolithic/Mesolithic transition. Within the

Table 11.3. Informal Tool Production on On-Site or Near-Site Raw Materials

Site	% of informal tools on quartz and quartzite	% of quartz and quartzite tools that are informal
Cabeço do Porto Marinho I-U	0.03	0.33
Cabeço do Porto Marinho III S-U	0.04	0.36
Carneira-Pinhal	0.02	0.33
Cabeço do Porto Marinho V	0	0
Carneira II	0.02	0.50
Areeiro III	0.03	0.40
Fonte Pinheiro	0.29	0.50

quartz and quartzite tool assemblages from all periods, tools on local materials were more likely to be informal after about 9000 BP. This limited use of local materials for informal tools coupled with the lack of change in broader technological organization makes it unlikely that these Portuguese assemblages demonstrate a shift to longer-duration base camp settlement before about 9000 BP, and possibly not until around 8500 BP.

In conclusion, context-focused artifact analysis revises interpretation of lithic assemblages from Epipaleolithic Portugal. In central Estremadura there is no evidence for a significant change in technological organization or raw material exploitation until the end of the Epipaleolithic. If settlement strategies changed during the Epipaleolithic, they did so without fundamentally altering the technology or organization of lithic assemblages. Residential mobility between the coastal areas and inland Portugal, rather than logistical strategies, may explain the presence of coastal marine resources near Rio Maior. Transporting a previously unexploited (or untransported) food resource during a residential move, while a change in subsistence strategy, does not necessarily require a change in mobility. Lithic organizational patterns from the Rio Maior region are consistent with the hypothesis that the number of residential moves per year and distance traveled may not have changed significantly between the Late Upper Paleolithic and the Epipaleolithic.

Conclusions

Stone artifacts may reflect broad hunter-gatherer organizational attributes, such as mobility; but numerous other processes influence raw material use within lithic economies. Documenting the transport of an artifact differs fundamentally from identifying group movement or understanding the structural organization of mobility. Any middle-range theory linking the archaeological record of stone artifact assemblages to the dynamics of settlement system structure must emphasize the context of raw material exploitation within a technology. Prehistoric lithic economies can be contextualized by combining an ecological model of raw material availability with a regional approach integrating assemblage data across raw materials. As the Portuguese Epipaleolithic case study illustrates, focusing on the role of local raw materials and informal tool manufacturing *lessens* the risk of modeling spurious relationships between technological organization and hunter-gatherer mobility.

ACKNOWLEDGMENTS

I am grateful to Frédéric Sellet, Russell Greaves, and Pei-Lin Yu for the opportunity to participate in their symposium and in this volume. I also appreciate helpful comments provided by Nuno Bicho, Phil Carr, Jonathan Haws, and T. Douglas Price.

REFERENCES

Almeida, F.
2000   The Terminal Gravettian of Portuguese Estremadura: Ecological Variability of the Lithic Industries. Ph.D. dissertation. Southern Methodist University, Dallas.
Amick, D. S.
1999   Raw Material Variation in Folsom Stone Tool Assemblages and the Division of Labor in Hunter-Gatherer Societies. In *Folsom Lithic Technology: Explorations in Structure and Variation,* edited by D. Amick, pp. 169–187. Archaeological Series No. 12. International Monographs in Prehistory, Ann Arbor.
Andrefsky, W.
1994   Raw Material Availability and the Organization of Technology. *American Antiquity* 59: 21–35.
1998   *Lithics: Macroscopic Approaches to Analysis.* Cambridge University Press, Cambridge.
Bamforth, D. B.
1990   Settlement, Raw Material, and Lithic Procurement in the Central Mohave Desert. *Journal of Anthropological Archaeology* 9: 70–104.
Bamforth, D. B., and P. Bleed
1997   Technology, Flaked Stone Technology, and Risk. In *Rediscovering Darwin: Evolutionary Theory and Archeological Explanation*, edited by C. M. Barton and G. A. Clark, pp. 109–140. Archaeological Papers. American Anthropological Association, Washington, D.C.
Bernaldo de Quiros, F., and V. Cabrera Valdés
1996   Raw Material in the Paleolithic of Cueva del Castillo and in the Cantabrian Re-

gion. In *Non-Flint Stone Tools and the Palaeolithic Occupation of the Iberian Peninsula*, edited by N. Moloney et al., pp. 21–32. BAR, Oxford.

Bettinger, R. L.
1991   *Hunter-Gatherers: Archaeological and Evolutionary Theory*. Plenum Press, New York.

Bettinger, R. L., R. Malhi, and H. McCarthy
1997   Central Place Models of Acorn and Mussel Processing. *Journal of Archaeological Science* 24: 887–899.

Bicho, N.
1993   Late Glacial Prehistory of Central and Southern Portugal. *Antiquity* 67: 761–775.

1994   The End of the Paleolithic and the Mesolithic of Portugal. *Current Anthropology* 35: 664–673.

1997   Spatial, Technological, and Economic Organization after the Last Glacial Maximum in Portuguese Prehistory. In *El Món Mediterrani després del Pleniglacial (18,000–12,000 B.P.)*, edited by J. M. Fullola and N. Soler, pp. 213–223. Museu d'Árqueologia de Catalunya, Girona.

1998   The Pleistocene-Holocene Transition in Portuguese Prehistory: A Technological Perspective. In *The Organization of Lithic Technology in Late Glacial and Early Postglacial Europe*, edited by S. Milliken, pp. 39–62. BAR, Oxford.

2000   *Technological Change in the Final Upper Paleolithic of Rio Maior*. Arkeos, Tomar.

Bicho, N., B. Hockett, J. Haws, and W. Belcher
2000   Hunter-Gatherer Subsistence at the End of the Pleistocene: Preliminary Results from Picareiro Cave, Central Portugal. *Antiquity* 74: 500–506.

Binford, L. R.
1979   Organization and Formation Processes: Looking at Curated Technologies. *Journal of Anthropological Research* 35: 255–273.

1980   Willow Smoke and Dogs' Tails: Hunter-Gatherer Settlement Systems and Archaeological Site Formation. *American Antiquity* 45: 4–20.

2001   *Constructing Frames of Reference*. University of California Press, Berkeley.

Binford, L. R., and J. F. O'Connell
1984   An Alyawara Day: The Stone Quarry. *Journal of Anthropological Research* 40: 406–432.

Bird, D. W.
1997   Behavioral Ecology and the Archaeological Consequences of Central Place Foraging among the Meriam. In *Rediscovering Darwin: Evolutionary Theory and Archeological Explanation*, edited by C. M. Barton and G. A. Clark, pp. 291–306. Archaeological Papers. American Anthropological Association, Washington, D.C.

Blankholm, H. P.
1991   *Intrasite Spatial Analysis in Theory and Practice*. Aarhus University Press, Aarhus.

Bonzani, R. M.

1997   Plant Diversity in the Archaeological Record: A Means toward Defining Hunter-Gatherer Mobility Strategies. *Journal of Archaeological Science* 24: 1129–1139.

Brantingham, P. J.

2003   A Neutral Model of Stone Raw Material Procurement. *American Antiquity* 68(3): 487–509.

Carr, P. J.

1994   Technological Organization and Prehistoric Hunter-Gatherer Mobility: Examination of the Hayes Site. In *The Organization of North American Prehistoric Chipped Stone Tool Technologies*, edited by P. J. Carr, pp. 35–44. Archaeological Series No. 7. International Monographs in Prehistory, Ann Arbor.

Church, T.

1994   *Lithic Resource Studies: A Sourcebook for Archaeologists*. Lithic Technology Special Publication 3. Department of Anthropology, University of Tulsa, Tulsa.

1995   Comment on "Neutron Activation Analysis of Stone from the Chadron Formation and a Clovis Site on the Great Plains" by Hoard et al. (1992). *Journal of Archaeological Science* 22: 1–5.

Clark, G. A.

1994   Migration as an Explanatory Concept in Paleolithic Archaeology. *Journal of Archaeological Method and Theory* 1: 305–343.

Clark, J. E.

1999   Comments on "Stone Tools." *Lithic Technology* 24: 126–135.

Close, A.

1996   Carry That Weight: The Use and Transportation of Stone Tools. *Current Anthropology* 37: 545–553.

2000   Reconstructing Movement in Prehistory. *Journal of Archaeological Method and Theory* 7: 49–78.

Demars, P.

1982   *L'utilisation du silex au Paléolithique Supérieur: Choix, approvisionnement, circulation—L'Exemple du bassin de Brive*. CNRS, Paris.

Dibble, H.

1991   Mousterian Assemblage Variability on an Interregional Scale. *Journal of Anthropological Research* 47: 239–258.

Ebert, J., and E. Camilli

1993   Lithic Distributions and Their Analytical Potential: An Example. *Lithic Technology* 18: 95–105.

Edmonds, M.

1987   Rocks and Risk: Problems with Lithic Procurement Strategies. In *Lithic Analysis and Later British Prehistory*, edited by A. G. Brown and M. R. Edmonds, pp. 155–180. BAR, Oxford.

Ellis, C., and D. B. Deller

2000   *An Early Paleo-Indian Site near Parkhill, Ontario.* Canadian Museum of Civiliza-
        tion, Quebec.

Elston, R. G.

1990   A Cost-Benefit Model of Lithic Assemblage Variability. In *The Archaeology of
        James Creek Shelter,* edited by R. G. Elston and E. E. Budy, pp. 153–163. Anthro-
        pological Paper 115. University of Utah Press, Salt Lake City.

Eriksen, B. V.

1991   *Change and Continuity in a Prehistoric Hunter-Gatherer Society.* Archaeologica
        Venatoria, Tübingen.

Feblot-Augustins, J.

1997   *La circulation des matières premières au Paléolithique.* 2 vols. ERAUL, Liège.

Floss, H.

1994   *Rohmaterialversorgung im Paläolithikum des Mittelrheingebietes.* RGZM Mono-
        graphs No. 21. Habelt, Bonn.

Gamble, C. S., and W. A. Boismier

1991   *Ethnoarchaeological Approaches to Mobile Campsites.* International Monographs in
        Prehistory, Ann Arbor.

Goodyear, A. C.

1989   A Hypothesis for the Use of Cryptocrystalline Raw Materials among Paleoindian
        Groups of North America. In *Eastern Paleoindian Lithic Resource Use,* edited by
        C. Ellis and J. Lothrop, pp. 1–9. Westview Press, Boulder.

Gould, R. A., and S. Saggers

1985   Lithic Procurement in Central Australia: A Closer Look at Binford's Idea of Em-
        beddedness in Archaeology. *American Antiquity* 50: 117–136.

Grayson, D. K., and F. Delpech

1998   Changing Diet-Breadth in the Early Upper Paleolithic of Southwestern France.
        Journal of Archaeological Science 25: 1119–1129.

Haury, C. E.

1994   Defining Lithic Procurement Terminology. In *Lithic Resource Studies: A Source-
        book for Archaeologists,* edited by T. Church, pp. 26–31. Lithic Technology Special
        Publication 3. Department of Anthropology, University of Tulsa, Tulsa.

Haws, J.

2000   Tardiglacial Subsistence and Settlement in Central Portugal. In *Paleolítico da
        Península Ibérica: Actas do 3 Congresso de Arqueologia Peninsular,* vol. 2, edited by
        V. O. Jorge, pp. 403–413. ADECAP, Porto.

Hayden, B.

1998   Practical and Prestige Technologies: The Evolution of Material Systems. *Journal
        of Archaeological Method and Theory* 5: 1–59.

Hayden, B., E. Bakewell, and R. Gargett

1996   The World's Longest Lived Corporate Group: Lithic Analysis Reveals Prehistoric

Social Organization near Lillooet, British Columbia. *American Antiquity* 61: 341–356.

Hayden, B., N. Franco, and J. Spafford

1996 Evaluating Lithic Strategies and Design Criteria. In *Stone Tools: Theoretical Insights into Human Prehistory*, edited by G. Odell, pp. 9–47. Plenum Press, New York.

Henry, D.

1989 Correlations between Reduction Strategies and Settlement Patterns. In *Alternative Approaches to Lithic Analysis*, edited by D. Henry and G. Odell, pp. 139–156. Archaeological Papers. American Anthropological Association, Washington, D.C.

Hill, A.

1994 Early Hominid Behavioural Ecology: A Personal Postscript. *Journal of Human Evolution* 27: 321–328.

Hockett, B. S., and N. Bicho

2000 Small Mammal Hunting during the LUP of Central Portugal. In *Paleolítico da Península Ibérica: Actas do 3 Congresso de Arqueologia Peninsular*, vol. 2, edited by V. O. Jorge, pp. 415–424. ADECAP, Porto.

Hofman, J. L.

1991 Folsom Land Use: Projectile Point Variability as a Key to Mobility. In *Raw Material Economies among Prehistoric Hunter-Gatherers*, edited by A. Montet-White and S. Holen, pp. 335–356. Publications in Anthropology 19. University of Kansas, Lawrence.

Holdaway, S., S. McPherron, and B. Roth

1996 Notched Tool Reuse and Raw Material Availability in French Middle Paleolithic Sites. *American Antiquity* 61(2): 377–387.

Holt, J. Z.

1996 Beyond Optimization: Alternative Ways of Examining Animal Exploitation. *World Archaeology* 28: 89–109.

Ingbar, E. E.

1992 The Hanson Site and Folsom on the Northwestern Plains. In *Ice Age Hunters of the Rockies*, edited by D. Stanford and J. Day, pp. 169–192. Denver Museum of Natural History, Denver.

1994 Lithic Material Selection and Technological Organization. In *The Organization of North American Prehistoric Chipped Stone Tool Technologies*, edited by P. J. Carr, pp. 45–56. Archaeological Series No. 7. International Monographs in Prehistory, Ann Arbor.

Isaac, G. L.

1986 Foundation Stones: Early Artefacts as Indicators of Activities and Abilities. In *Stone Age Prehistory: Studies in Memory of Charles McBurney*, edited by G. N. Bailey and P. Callow, pp. 221–241. Cambridge University Press, Cambridge.

Jochim, M. A.

1989   Optimization and Stone Tool Studies: Problems and Potential. In *Time, Energy, and Stone Tools*, edited by R. Torrence, pp. 106–111. Cambridge University Press, Cambridge.

1990   The Ecosystem Concept in Archaeology. In *The Ecosystem Approach in Anthropology: From Concept to Practice*, edited by E. F. Moran, pp. 75–90. University of Michigan Press, Ann Arbor.

1998   *A Hunter-Gatherer Landscape: Southwest Germany in the Late Paleolithic and Mesolithic*. Plenum Press, New York.

Jones, G. T., C. Beck, and D. K. Grayson

1989   Measures of Diversity and Expedient Lithic Technology. In *Quantifying Diversity in Archaeology*, edited by R. Leonard and G. T. Jones, pp. 69–78. Cambridge University Press, Cambridge.

Kelly, R.

1992   Mobility/Sedentism: Concepts, Archaeological Measures, and Effects. *Annual Review of Anthropology* 21: 43–66.

1994   Some Thoughts on Future Directions in the Study of Stone Tool Technological Organization. In *The Organization of North American Prehistoric Chipped Stone Tool Technologies*, edited by P. J. Carr, pp. 132–136. Archaeological Series No. 7. International Monographs in Prehistory, Ann Arbor.

1995   *The Foraging Spectrum: Diversity in Hunter-Gatherer Lifeways*. Smithsonian Institution Press, Washington, D.C.

Kuhn, S.

1989   Hunter-Gatherer Foraging Organization and Strategies of Artifact Replacement and Discard. In *Experiments in Lithic Technology*, edited by D. Amick and R. Mauldin, pp. 33–45. BAR, Oxford.

1992   On Planning and Curated Technologies in the Middle Paleolithic. *Journal of Anthropological Research* 48: 185–214.

1994   A Formal Approach to the Design and Assembly of Transported Toolkits. *American Antiquity* 59: 426–442.

1995   *Mousterian Lithic Technology: An Ecological Perspective*. Princeton, Princeton University Press.

Larick, R.

1987   Perigord Cherts: An Analytical Frame for Investigating the Movement of Paleolithic Hunter-Gatherers and Their Resources. In *The Scientific Study of Flint and Chert*, edited by G. Sieveking and M. Hart, pp. 111–120. Cambridge University Press, Cambridge.

Larson, M. L., and M. Kornfeld

1997   Chipped Stone Nodules: Theory, Method, and Examples. *Lithic Technology* 22: 4–18.

Madsen, D., and D. Schmitt
1998   Mass Collecting and the Diet-Breadth Model: A Great Basin Example. *Journal of Archaeological Science* 25: 445–455.

Marks, A. E., N. Bicho, J. Zilhão, and C. R. Ferring
1994   Upper Pleistocene Prehistory in Portuguese Estremadura: Results of Preliminary Research. *Journal of Field Archaeology* 21: 53–68.

Marks, A. E., and M.-B. Mishoe
1997   The Magdalenian of Portuguese Estremadura. In *El Món Mediterrani després del Pleniglacial (18,000–12,000 B.P.)*, edited by J. M. Fullola and N. Soler, pp. 225–232. Museu d'Árqueologia de Catalunya, Girona.

Meltzer, D.
1989   Was Stone Exchanged among Eastern North American Paleoindians? In *Eastern Paleo-Indian Lithic Resource Use*, edited by C. Ellis and J. Lothrop, pp. 11–39. Westview Press, Boulder.

Mithen, S. J.
1989   Modeling Hunter-Gatherer Decision Making: Complementing Optimal Foraging Theory. *Human Ecology* 17: 59–83.
1990   *Thoughtful Foragers: A Study in Prehistoric Decision Making*. Cambridge University Press, Cambridge.

Montes Barqúin, R., and J. Sanguino-González
1998   Diferencias en las estrategias de adquisición de recursos líticos entre el Paleolítico Inferior y Medio en el centro de la Región Cantábrica: Implicaciones económicas y territoriales. In *Los recursos abióticos en la prehistoria*, edited by J. Bernabeu et al., pp. 55–71. Universitat de València, Valencia.

Montet-White, A., and S. Holen
1991   *Raw Material Economies among Prehistoric Hunter-Gatherers*. Publications in Anthropology 19. University of Kansas, Lawrence.

Myers, A.
1989   Reliable and Maintainable Technological Strategies in the Mesolithic of Mainland Britain. In *Time, Energy, and Stone Tools*, edited by R. Torrence, pp. 78–90. Cambridge University Press, Cambridge.

Nash, S.
1996   Is Curation a Useful Heuristic? In *Stone Tools: Theoretical Insights into Human Prehistory*, edited by G. Odell, pp. 81–99. Plenum Press, New York.

Nelson, M. C.
1991   The Study of Technological Organization. *Archaeological Method and Theory* 3: 57–100.

Odell, G.
1984   Chert Resource Availability in the Lower Illinois Valley: A Transect Sample. In *Prehistoric Chert Exploitation: Studies from the Midcontinent*, edited by B. Butler and E. May, pp. 45–67. Center for Archaeological Investigations Occasional Papers 2. Southern Illinois University, Carbondale.

1994   Assessing Hunter-Gatherer Mobility in the Illinois Valley: Exploring Ambiguous Results. In *The Organization of North American Prehistoric Chipped Stone Tool Technologies*, edited by P. J. Carr, pp. 70–86. Archaeological Series No. 7. International Monographs in Prehistory, Ann Arbor.

1996   Economizing Behavior and the Concept of Curation. In *Stone Tools: Theoretical Insights into Human Prehistory*, edited by G. Odell, pp. 51–79. Plenum Press, New York.

2000   Stone Tool Research at the End of the Millennium: Procurement and Technology. *Journal of Archaeological Research* 8(4): 269–331.

Parry, W. J., and R. L. Kelly

1987   Expedient Core Technology and Sedentism. In *The Organization of Core Technology*, edited by J. Johnson and C. Morrow, pp. 285–304. Westview Press, Boulder.

Phillipson, D. W.

1980   Technological Disparity or the Contemporaneity of Diverse Industries. In *Proceedings of the 8th Panafrican Congress of Prehistory and Quaternary Studies*, edited by R. Leakey et al., pp. 15–16. ILLMIAP, Nairobi.

Reid, K. C.

1997   Gravels and Travels: A Comment on Andrefsky's "Cascade Phase Lithic Technology." *North American Archaeologist* 18(1): 67–81.

Roebroeks, W., and P. Hennekens

1990   Transport of Lithics in the Middle Paleolithic: Conjoining Evidence from Maastricht-Belvedere (NL). In *The Big Puzzle*, edited by E. Cziesla et al., pp. 283–295. Holos, Bonn.

Sellet, F.

1999   A Dynamic View of Paleoindian Assemblages at the Hell Gap Site, Wyoming: Reconstructing Lithic Technological Systems. Ph.D. dissertation. Department of Anthropology, Southern Methodist University. University Microfilms, Ann Arbor.

2004   Beyond the Point: Projectile Manufacture and Behavioral Influence. *Journal of Archaeological Science* 31: 1553–1566.

Shackley, M. S.

1998   Gamma Rays, X-Rays, and Stone Tools: Some Recent Advances in Archaeological Geochemistry. *Journal of Archaeological Science* 25: 259–270.

Shott, M. J.

1996   An Exegesis of the Curation Concept. *Journal of Anthropological Research* 52(3): 259–280.

Simek, J., and H. Price

1990   Chronological Change in Perigord Lithic Assemblage Diversity. In *The Emergence of Modern Humans*, edited by P. Mellars, pp. 243–261. Cornell University Press, Ithaca.

Smith, E. A.

1987   Optimization Theory in Anthropology: Application and Critiques. In *The Latest on the Best: Essays on Evolution and Optimality*, edited by J. Dupre, pp. 201–249. Cambridge University Press, Cambridge.

Smith, E. A., and B. Winterhalder (editors)

1992   *Evolutionary Ecology and Human Behavior*. Aldine de Gruyter, New York.

Stern, N.

1993   The Structure of the Lower Pleistocene Archaeological Record—A Case Study from the Koobi Fora Formation. *Current Anthropology* 34(3): 201–226.

Straus, L. G.

1991a   The Role of Raw Materials in Upper Paleolithic and Mesolithic Stone Artifact Assemblage Variability in Southwest Europe. In *Raw Material Economies among Prehistoric Hunter-Gatherers*, edited by A. Montet-White and S. Holen, pp. 169–186. Publications in Anthropology 19. University of Kansas, Lawrence.

1991b   Southwestern Europe at the Last Glacial Maximum. *Current Anthropology* 32: 189–199.

1992   *Iberia before the Iberians*. University of New Mexico Press, Albuquerque.

1999   High Resolution Archeofaunal Records across the Pleistocene Holocene Transition on a Transect between 43 and 51 Degrees North Latitude in Western Europe. In *Zooarchaeology of the Pleistocene/Holocene Boundary*, edited by J. Driver, pp. 21–29. BAR, Oxford.

Straus, L., N. Bicho, and A. Winegardner

2000   Mapping the Upper Paleolithic regions of Iberia. *Journal of Iberian Archaeology* 2: 7–42.

Thacker, P.

1996a   Hunter-Gatherer Lithic Economy and Settlement Systems: Understanding Regional Assemblage Variability in the Upper Paleolithic of Portuguese Estremadura. In *Stone Tools: Theoretical Insights into Human Prehistory*, edited by G. Odell, pp. 101–126. Plenum Press, New York.

1996b   A Landscape Perspective on Upper Paleolithic Settlement in Portuguese Estremadura. Ph.D. dissertation. Southern Methodist University, Dallas.

2000   The Relevance of Regional Analysis for Upper Paleolithic Archaeology: A Case Study from Portugal. In *Regional Approaches to Adaptation in Late Pleistocene Western Europe*, edited by G. Peterkin and H. Price, pp. 25–45. BAR. Archeopress, Oxford.

2002   Residential Mobility and Lithic Economizing Behavior: Explaining Technological Change in the Portuguese Upper Paleolithic. In *Lithic Raw Material Economies in Late Glacial and Early Postglacial Europe*, edited by L. Fisher and B. Eriksen, pp. 147–159. BAR. Archeopress, Oxford.

Torrence, R.

1989   *Time, Energy, and Stone Tools*. Cambridge University Press, Cambridge.

1994   Strategies for Moving On in Lithic Studies. In *The Organization of North Ameri-*

can Prehistoric Chipped Stone Tool Technologies, edited by P. J. Carr, pp. 123–131. Archaeological Series No. 7. International Monographs in Prehistory, Ann Arbor.

2001   Hunter-Gatherer Technology: Macro- and Micro-Scale Approaches. In *Hunter-Gatherers: An Interdisciplinary Perspective*, edited by C. Panter-Brick, R. Layton, and P. Rowley-Conwy, pp. 73–98. Cambridge University Press, Cambridge.

Torrence, R., J. Specht, R. Fullagar, and G. R. Summerhayes

1996   Which Obsidian Is Worth It?: A View from the West New Britain Sources. In *Oceanic Culture History*, edited by J. M. Davidson et al., pp. 211–224. Special Publication of the *New Zealand Journal of Archaeology*.

Vierra, B. J.

1995   *Subsistence and Stone Tool Technology: An Old World Perspective*. Anthropological Papers 47. Arizona State University, Tempe.

Walsh, M. R.

1998   Lines in the Sand: Competition and Stone Selection on the Pajarito Plateau, New Mexico. *American Antiquity* 63: 573–593.

Williams, J. K.

2000   Land Use and Technological Trends in the Levantine Upper Paleolithic. *Journal of the Israel Prehistoric Society* 30: 33–47.

Williams, N. M., and E. S. Hunn (editors)

1982   *Resource Managers: North American and Australian Hunter-Gatherers*. AAAS Selected Symposium 67. Westview Press, Boulder.

Winterhalder, B.

2001   The Behavioral Ecology of Hunter-Gatherers. In *Hunter-Gatherers: An Interdisciplinary Perspective*, edited by C. Panter-Brick, R. Layton, and P. Rowley-Conwy, pp. 12–38. Cambridge University Press, Cambridge.

Zilhão, J.

1997a   *O Paleolítico Superior da Estremadura Portuguesa*. Edições Colibri, Lisbon.

1997b   The Paleolithic Settlement of Portuguese Estremadura after the Last Glacial Maximum. In *El Món Mediterrani després del Pleniglacial (18,000–12,000 B.P.)*, edited by J. M. Fullola and N. Soler, pp. 233–242. Museu d'Árqueologia de Catalunya, Girona.

12

Cycles of Aridity and Human Mobility

Risk Minimization among Late Pleistocene Foragers of the Western Desert, Australia

PETER VETH

Australia is the most arid continent to have been colonized by modern humans and the Western Desert arguably the most marginal landscape within it. This is largely due to its high temperatures, lack of coordinated drainage, paucity of permanent waters, and patchy and unpredictable resources (cf. Mabbutt 1977; Veth 1993b). These factors have historically created conditions of high risk and stress for arid-land foragers (Yellen 1976). Remarkably, we now have firm dates for occupation of the Western Desert in excess of 30,000 BP (Smith et al. 1997; Thorley 1998b) and increasing evidence for early and widespread use of most landforms. There is also mounting evidence that major changes in settlement and mobility patterns occurred with the onset of hyperarid conditions associated with the last glacial maximum (O'Connor et al. 1999; Veth n.d.b).

In this chapter I discuss the central role of mobility in the risk-minimizing strategies employed by Western Desert foragers and examine assemblages from three sites dating to the first phase of occupation from 32,000 BP to 22,000 BP (Smith et al. 1997; Thorley 1998b; Veth n.d.b). I examine eight different lines of archaeological evidence that may be used to infer residential mobility from these Pleistocene assemblages.

RISK MINIMIZATION AND THE ROLE OF MOBILITY IN THE AUSTRALIAN DESERTS

There is increasing evidence from the Western Desert that conditions were more favorable for humans during the first phase of occupation and then worsened significantly toward the height of the last glacial maximum by approximately 18,000 years ago (Smith 1996; Thorley 1998b; Veth n.d.b). This period saw the likely loss of a considerable number of permanent water sources and food staples (Smith 1989: 102). Increasing aridity associated with the last glacial maximum probably represented the first test of human adjustment to truly arid conditions,

Table 12.1. Behavioral and Archaeological Correlates of Resource Structure Model

Behavioral Correlates	Resource Structure—unpredictable and scarce
Residential mobility	Very high, opportunistic
Territorial strategy	Undefended—very permeable
Information exchange	High
Group size	Very small
Population density	Very low
Diet breadth	Very high

Archaeological Correlates	Resource Structure—unpredictable and scarce
Occupation site intensity	Very low at home base
Macro-regional assemblage variability	High stylistic uniformity
Raw material sources	Local and distant exotics
Intrasite spatial organization	Poorly structured
Faunal and floral diversity	Very high, mostly plant

with a majority view that humans were unable to cope with the ensuing hyper-arid conditions and restricted their territory in response (Hiscock 1988; Thorley 1998a; Veth 1995).

I propose that during the Pleistocene Western Desert foragers engaged in a pattern of settlement dominated by high residential mobility to low logistical mobility (after Binford 1980), coupled with a "territorial" mobility strategy whereby adjacent localities were abandoned for varying periods (after Kelly 1992, 1995; Varien 1999). Mobility configurations likely had great dynamism, given the ethnographically observed patterns for Western Desert Aborigines, whereby strategy switching occurred in response to drought. Strategy switching describes how groups responded by either withdrawing into their "home" territory or, alternatively, avoiding areas altogether through temporary abandonment (see Gould 1991). This extreme flexibility in mobility is seen to underpin a raft of risk-minimizing strategies in the Western Desert (see Tonkinson 1991).

The nature of Pleistocene arid-zone adaptations may be examined by using a resource structure model such as that proposed by Stanley Ambrose and Karl Lorenz (1990: 10), where the archaeological correlates of hunter-gather behaviors in an unpredictable and scarce resource regime are defined (Table 12.1). The marginality of the Western Desert for humans is largely a function of poor rainfall reliability and generally impoverished species diversity in comparison to other desert regions of the world (Gould 1991: 17). The archaeological predictions of the resource structure model are examined in the following sections.

PLEISTOCENE SITES OF THE WESTERN DESERT

The sites of Serpent's Glen (O'Connor et al. 1998), Puritjarra (Smith 1989, 1996), and Kulpi Mara (Thorley 1998a, 1998b) all have occupational deposits dating to before 23,000 BP (Figure 12.1). While this sample is not large, it does provide early sites from a range of different settings located within and on the margins of the Western Desert. All dates discussed here are uncalibrated.

Serpent's Glen Rockshelter is located within quartz sandstone ranges among an expanse of low-relief dunes. The uplands contain several springs that provide reliable and potable water (O'Connor et al. 1998). Excavation revealed three major stratigraphic layers; the lower layer has a minimum date of 23,550 BP and contains a sparse assemblage of stone artifacts, ochre, charcoal, and bone. The middle layer is culturally sterile, while the upper layer dates from 4710 BP to the modern period and contains over 98% of the site's lithic materials in addition to significant quantities of ochre, charcoal, and bone and minor quantities of emu eggshell, bird eggshell, and macrofloral remains.

Puritjarra Rockshelter is located within a western outlier of the Central Australian Ranges and has a major permanent water source nearby. Puritjarra has a Pleistocene layer that contains a sparse flaked stone, charcoal, ochre, and bone assemblage dated to as early as 32,000 BP (Smith et al. 1998). While M. A. Smith (1989) has argued that occupation at Puritjarra spans the subsequent period of 22,000 BP to 13,000 BP, coinciding with peak glacial aridity, this has been challenged by Peter Hiscock (1988) and P. B. Thorley (1998a, 1998b) on both theoretical and empirical grounds. Sparse assemblages are certainly registered by 12,000 BP; and by approximately 6500 BP the quantity of flaked stone increases substantially, with an efflorescence in both quantity and breadth of cultural materials dating to the last millennium.

Kulpi Mara is a large sandstone rockshelter on the escarpment of the Central Australian Ranges and, in contrast to the two previous sites, has no permanent water source nearby (Thorley 1998a, 1998b). The excavation revealed three layers, the lower dating to approximately 24,000 to 30,000 BP. Cultural assemblages in this unit are typically sparse. A chronological hiatus is then recorded with the middle layer, dating from approximately 12,000 BP, when an increase in site use is registered. The middle of the upper layer is dated to 2500 BP, with the upper spits containing abundant lithics, charcoal, and also ochre, wooden, and resin artifacts and spun fibers.

At all three sites the passage of time coinciding with peak glacial aridity is registered as either culturally sterile sediments or very slow rates of sediment and artifact accumulation, suggesting abandonment or, at most, several minor episodes of ephemeral occupation (cf. Thorley 1998a: 41). This pattern is evident

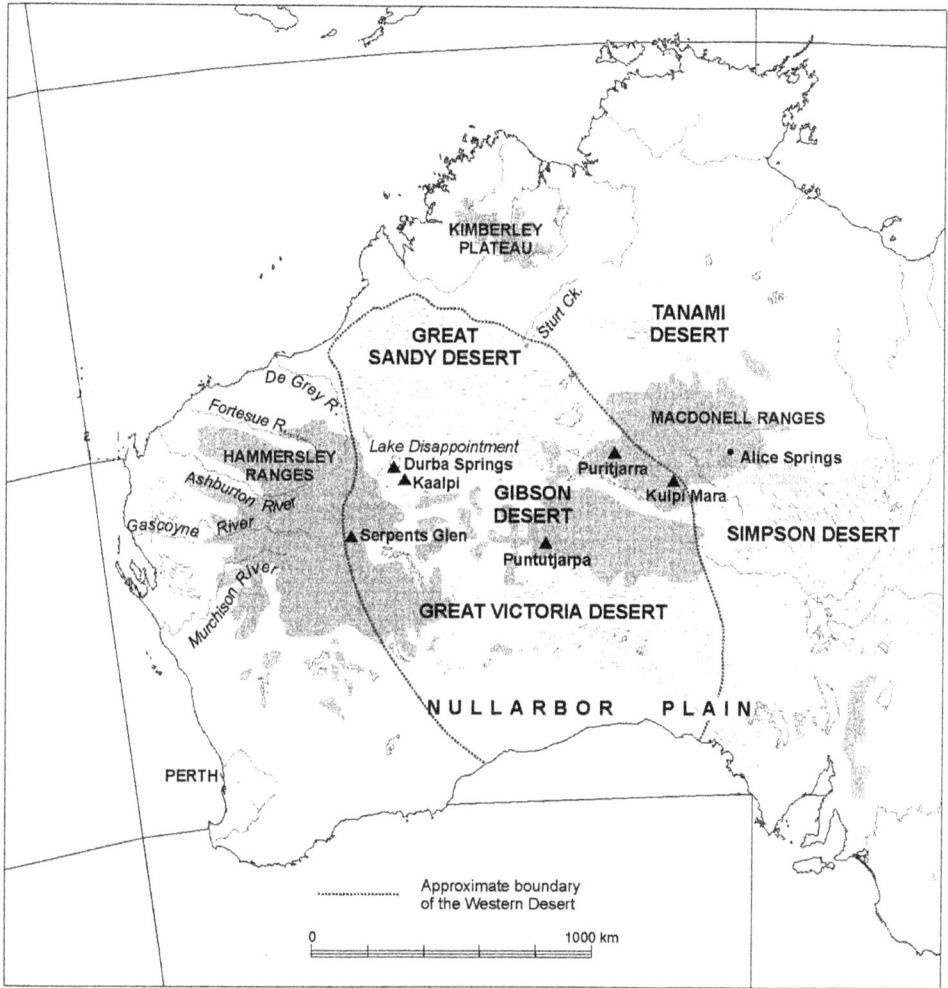

Figure 12.1. Location of key Pleistocene sites in the Western Desert.

in other desert lowland sites surrounding the Western Desert, such as at Noala Cave, which dates to 27,000 BP, and Mandu Mandu Rockshelter, dating from 34,000 BP (Morse 1994; O'Connor et al. 1999; Veth 1993a, 1995; see Figure 12.1).

As a first step it is necessary to establish that some level of resource stress was experienced at these sites, and ideally this should be independent from other categories of archaeological data. That is required to support assumptions that patchy and unpredictable resource structure in the terminal Pleistocene would have resulted in stress to Western Desert foragers.

AN INDEPENDENT INDEX OF STRESS

It is possible to calculate an independent index of relative protein stress from selected Western Desert and Central Australian sites. On the basis of ethnographic accounts that Western Desert Aborigines pulverized whole game and extracted marrow to maximize protein yield, high levels of bone reduction in arid zone sites are seen to be an indicator of protein stress. Where these patterns can be reliably associated with processing and pulverizing of game, as opposed to taphonomic factors, it is possible to consider relative levels of stress (Gould 1978, 1996; Haley 1999; O'Connor et al. 1998).

Indices of fragmentation (IoF) can be calculated for combined fractions of economic fauna from all spits (total bone weight/number of fragments) for the sites of Kaalpi and Serpent's Glen and the Holocene sites of Puntutjarpa and Intitjikula (see Figure 12.1). Although Holocene data are included here, the sites show long-term consistencies in the degree of bone comminution (Gould 1996: 74, 81). A related index of protein stress is found in the ratio of broken to whole macropod teeth. Breakage of these robust elements has been documented as a by-product of the crushing and pulverizing of crania and mandibles (Gould 1996; O'Connor et al. 1998: 16).

Paleoenvironmental models for the Australian arid zone would predict greater resource scarcity and unpredictability, and hence stress, in the lowlands of the Western Desert and less so on the periphery and within the Central Australian Ranges (for example, Gould 1991; Ross et al. 1992; Veth 1993b). It follows that the degree of bone reduction should be greater within the lowlands, and this should be registered in lower values for index of fragmentation and higher ratios of broken to whole macropod teeth. Some consideration must be made of relative species distributions between sites, although this is problematic because the levels of taxonomic assignment are low, ranging from 8.8% to 12%. There is some evidence that the proportion of larger macropods present on the lowland sites is lower than on the ranges, which would likely serve to inflate the differences in IoFs between these sites (cf. Gould 1996).

These predictions of relative protein stress are generally met when the faunal assemblages are analyzed. Serpent's Glen and the nearby site of Kaalpi, both located within expansive dunefields, register IoFs of 0.063 gm/fragment and 0.104 gm/fragment, respectively (Haley 1999; O'Connor et al. 1998). Over 98% of macropod teeth are broken. The site of Puntutjarpa to the east registers an average IoF of 0.530 gm/fragment, with 100% of macropod teeth being fractured. In contrast, the site of Intirtekwerle within the Central Australian Ranges registers an IoF of 0.500 gm/fragment (similar to Puntutjarpa), but only 60% of macropod teeth have been broken (see Table 12.2).

Table 12.2. Values for Index of Fragmentation and Macropod Tooth Breakage from Serpent's Glen, Kaalpi, Puntutjarpa, and Intirtekwerle

Site name	Index of fragmentation	% macropod teeth broken
Serpent's Glen	0.063	100
Kaalpi	0.104	> 98
Puntutjarpa	0.530	100
Intirtekwerle	0.500	60

Source: Gould 1996; Haley 1999; O'Connor et al. 1998.

A number of conclusions can be drawn from this comparison. First, conditions of protein stress have obtained throughout the history of occupation at all of these sites and are predicted also to have been present at Puritjarra and Kulpi Mara. Second, the cline in increasing protein stress away from the ranges lends support to models of marginality that predict that the highest levels of residential mobility will occur on the desert lowlands (cf. Smith 1989).

ARCHAEOLOGICAL INDICES OF HIGH RESIDENTIAL MOBILITY

The Intensity of Stone Reduction

On the assumption that occupation was by small groups and for a short duration, it is expected that the intensity of reduction of locally available lithics will be relatively low.

Locally available and abundant stone materials account for 90–100% of the Pleistocene assemblages. The intensity of utilization of these materials is seen to be a product of occupation intensity and to reflect group mobility. Assuming that assemblage variability can be strongly determined by occupation intensity (after Rolland and Dibble 1990), this factor is examined by considering changes in artifact discard rates (Smith 1988), degree of artifact curation (cf. Shott 1989: 26), and the presence of economizing strategies, such as bipolar knapping (after Hiscock 1996).

The number of artifacts clearly ascribed to the preglacial layers of the Western Desert sites is small and in two of the three cases contrasts markedly with numbers of artifacts from the Holocene units. Estimates of artifact discard rates for the three sites are provided in Table 12.3. Differences of this high order in themselves must indicate that different settlement patterns have operated through time.

With reference to curation at Serpent's Glen, the Pleistocene layers contained no retouched/utilized artifacts (O'Connor et al. 1998: 13). At Puritjarra there are only amorphous retouched artifacts in Layer 2, and these are substantially

Table 12.3. Estimated Artifact Discard Rates from Pre–Last Glacial Maximum and Holocene Layers from Pleistocene Western Desert Sites

Site	Pre-LGM	Late Holocene
Serpent's Glen	5	1,600
Puritjarra	6	2,087
Kulpi Mara	140	250

Note: Number of artifacts per kyr.

Table 12.4. Mean Weight of Amorphous Retouched Artifacts in Grams, Puritjarra

Pit	N9	N10	QR9	Z10	Total No.
Layer 1a	6.9	8.0	37.7	14.1	51
Layer 1b	18.2	8.0	10.8	8.0	33
Layer 2	17.0	75.5	691.7	104.1	9

Source: After Smith 1988: 119.

Table 12.5. Mean Weights of Total Artifacts in Grams, Puritjarra, Pit N19

Pit	Local lithics	Exotic lithics
Layer 1a	2.3	0.7
Layer 1b	3.5	1.1
Layer 2	24.0	15.3

Source: After Smith 1988: Table 4.10.

larger than those in the upper Holocene layer (Smith 1988: 119; see Table 12.4). Size reduction is evident in both local (silcrete/silicified sandstone) and exotic (chert/chalcedony) raw materials (see Table 12.5).

Only three retouched/utilized artifacts have been found in the pre–Last Glacial Maximum (LGM) layer from Kulpi Mara (Square C). Thorley (1998b: 235) concludes: "Use of local materials appears to have been predominantly opportunistic and the production of whole silcrete/quartzite flakes . . . independent of [later] backed artifact manufacture and discard."

The combination of low numbers of artifacts and their low degree of modification attests to a low intensity of site occupation, with expedient use of locally available raw materials. It is possible that tools brought into these sites as part of a high residential mobility pattern were made from higher-quality "exotics" that were extensively curated and transported until a preferred resupply zone was encountered, thereby increasing the expedient use of local lithics.

Available data from Puritjarra and Kulpi Mara do not presently allow an assessment of reduction stages. The small assemblage from Serpent's Glen, however, is typical of other assemblages in the Western Desert dominated by early stages of reduction of local raw materials (see Veth 1993b, 1996). None of the site reports indicate any bipolar reduction in the Pleistocene, argued to be an indicator of lower residential mobility from northern Australia (Hiscock 1996).

The Diversity of Artifacts

A number of studies have illustrated that there is an inverse relationship between assemblage diversity and level of residential mobility (cf. Andrefsky 1998: 204; Shott 1986: 25). This relationship is shown in Figure 12.2. Assemblage diversity equates with the number of artifact types (tools) recovered from an assemblage. A more sensitive index of technological diversity may be defined by measuring the number of tool classes used during daily activities (Shott 1986: 23). Michael Shott's analysis was based on artifact data and mobility information from a dozen ethnographic studies of hunter-gatherer groups. Base camps (after Binford 1980) where a more logistical strategy was practiced showed a positive correlation between artifact diversity and length of stay.

William Andrefsky (1998: 206) contrasts this pattern with the low diversity expected at task-specific sites and states:

> Although no ethnographic data on artifact diversity were available for special-task-oriented or field camps, it would be logical to assume that special-task-oriented camps, such as hunt camps, plant collecting stations, or butchering sites, would have a relatively low diversity of artifacts. In other words, if a narrow range of activity took place at a particular location, one would expect to find a relatively low number of artifact types.

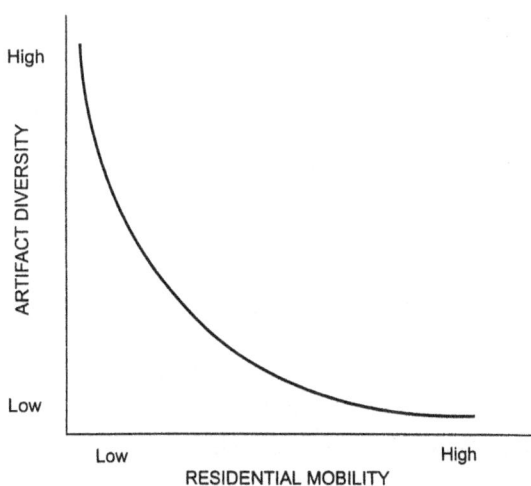

Figure 12.2. Proposed relationship between artifact diversity and residential mobility (Andrefsky 1998: Figure 8.5, adapted from Shott 1986: 25).

Table 12.6. Artifact Diversity from Western Desert Sites

Site type	No. sites	Artifact diversity
Residential base camp	9	8
Ephemeral residence	34	4
Task-specific	15	2

Source: Veth 1993b, n.d.b.

Such data are available, however, from ethnoarchaeological work that I carried out from 1986 to 1996 with Martu Aborigines of the Western Desert (see Veth 1987, 1989, 1993b, 1996, unpublished notes). I analyzed a range of artifact sites that had been used by Martu before contact with Europeans from the 1920s to the 1960s (cf. Tonkinson 1991). Because of these historical associations, localities could be reliably ascribed, at least for the historic period, as (a) major residential base camps that were used for aggregations, (b) smaller residential camps used predominantly in an ephemeral mode, or (c) camps that were clearly task-specific or field camps. A combination of high water permanency and high diversity of economic plant species predicts greater permanency of occupation at sites and often occupation by larger groups of people. Localities near ephemeral water sources and less diverse plant patches witnessed the highest residential mobility.

Assumed aggregation sites located near permanent waters had the highest values for artifact diversity, contained the highest proportion of exotic materials, evidenced a greater proportion of artifacts in later reduction stages, contained more tools exhibiting high levels of curation and recycling, and had significantly more basal grinding stones (Veth 1993b, 1996). The sites used by Martu custodians as ephemeral residences and what they referred to as "passing-through places" generally reflected the opposite pattern.

I am not suggesting here that the late Holocene assemblages are directly comparable to those of the Pleistocene. Rather, I argue that the material outcomes of different mobility configurations in the Western Desert are tangible and do not support previous models arguing for the homogeneity of residential sites (Hayden 1976). Tools identified in sites used by Martu in the historic period included retouched/utilized flakes, amorphous grindstones, millstones, edge-ground axes, backed blades, backed geometrics, tula adzes, and burren adzes. Artifact diversity values for three classes of sites are shown in Table 12.6.

It was expected that the Pleistocene assemblages of the three Western Desert sites would have diversity values closer to those of the ephemeral and task-specific sites in contrast to the base camps. It must be noted, however, that the introduction of new standardized tools from the mid-Holocene, such as the adzes and backed pieces, will inflate the diversity values for the Holocene sites (cf. Hiscock

Table 12.7. Artifact Diversity Values for Pleistocene Assemblages from Western Desert Sites

Site	Layer	Artifact diversity
Serpent's Glen	3	1
Puritjarra	2	2
Kulpi Mara	3	2

and Veth 1991). Diversity values for the Pleistocene-aged sites are provided in Table 12.7. The tools located from these three sites may be incorporated into the category of retouched/utilized flakes (referred to as amorphous retouched by Smith 1988: 126), where clear evidence of modification to a blank is visible through retouch/use-wear.

These artifacts are made on a wide variety of blank morphologies and are consistent with the expedient tools recorded by Brian Hayden (1979) from the Western Desert. Another category is also recognized here: primary flakes that may have microscopic evidence for use-wear. This is simply a cautionary category that serves to inflate the artifact diversity values in lieu of use-wear analyses, which are obviously in order.

These values are clearly very low and stand in stark contrast to those recorded from the major aggregation sites, as shown in Table 12.6. It should be stressed that sample size effect and the known increase in formal tools by the mid-Holocene will serve to exaggerate this difference. It is reasonable to conclude on the basis of the available data, however, that there is a low diversity of tools in the Pleistocene assemblages.

The Presence of Local and Exotic Lithics

It is expected that large foraging ranges associated with a high residential mobility strategy will result in the curation, maintenance, and occasional discard of lithics that may come from distant sources and hence be termed exotic, especially where suitable alternative sources are not readily available. The presence of abundant though highly variable quality silcretes and quartzites in the vicinity of all three Western Desert sites will have provided the range of blanks generally required for expedient maintenance tools in the Western Desert. The majority of artifacts from the three sites are from local materials (90–100%), with the presence of exotics likely a reflection of the scale of territorial range (Binford and Stone 1985 vs. Gould 1985).

Increases in the proportion of exotics are associated with the later appearance of standardized hafted tools, such as tula adzes and backed blades, by the mid-Holocene (for example, Hiscock 1994; Hiscock and Veth 1991; Thorley 1998b:

235) and particularly during the subsequent emergence in the late Holocene of what are argued to be intensifying long-distance exchange and ritual networks (Thorley 1998b). Major aggregation sites that served as centers of "production" for such regional networks evidence a significant increase in the proportion of exotics. This increase in exotics is argued to be a reflection of longer-distance exchange networks and the higher number of source areas from which participating groups traveled (for example, Cane 1984; Thorley and Gunn 1996; Veth 1993b).

The Quantity of Grinding Material

Pleistocene assemblages from arid Australia rarely contain seed grinders (for example, Gorecki et al. 1997). This may be linked to a range of factors, including the function and permanency of site use, assemblage size effect, and a general lack of intensive seed/food processing (see, however, Edwards and O'Connell 1995). Grindstones conforming to a wide range of morphologies are commonly found in stratified contexts from the arid and semiarid zone from after approximately 3500 BP (Smith 1986). This has been argued to represent one of the major transitions in the economy and settlement of arid land peoples as they shifted toward the exploitation of unpredictable yet densely clustered resources, with consequent changes in mobility strategy (Thorley 1998b).

At the site of Serpent's Glen, Sue O'Connor and I (Veth and O'Connor 1996) recorded hundreds of whole and fragmented formal millstones, mullers, and amorphous grindstones on the surface and surrounding sand plains, which are assumed to date to the late Holocene. The Pleistocene layer of Serpent's Glen, in contrast, contains no ground artifacts. Equally, there are no grinding implements before the mid-Holocene at Puritjarra and Kulpi Mara (Thorley 1998b: 323). Abundant ethnohistorical evidence (see Smith 1986) indicates that seed-grinding stations were established at major aggregation sites and that these helped to underpin larger gatherings of people for extended periods. The mobility configuration of such sites was clearly oriented more toward a logistical strategy. The lack of any seed-grinding materials in the Pleistocene layers of the

Table 12.8. Mean Number of Whole Grindstones Located at Sites Adjacent to Permanent, Semipermanent, and Ephemeral Water Sources

Water permanency	Number of sites	Mean number of grindstones
Permanent	8	16.625
Semipermanent	23	4.739
Ephemeral	16	1.188

Source: Veth 1993b.

three Western Desert sites is consistent with their use as ephemeral occupation sites.

Data collected from occupation sites adjacent to water sources of varying permanency (Veth 1993b) show a clear correlation between water permanency and number of intact grindstones (see Table 12.8). Martu Aborigines noted that the sites adjacent to permanent waters had acted as major aggregation venues and that groups from many linguistic affiliations had come to such places, sometimes traveling hundreds of kilometers in small family groups.

Stylistic Uniformity of Rock Art

Patterning in art has been used by a range of researchers to indicate aggregation and dispersion patterns, resting largely on interpretations of inside/outside (exclusive/inclusive) access to information networks (Conkey 1980; Galt-Smith 1997; Gould 1980). The high residential mobility scenario of the Pleistocene would predict high stylistic uniformity and the predominant use of outside symbols.

Both Serpent's Glen and the sites of Kaalpi and Durba Springs to the north in the Little Sandy Desert have extensive rock art galleries, with both petroglyphs and pictographs. Although dating of the rock art is in the early stages, the majority of pictographs are thought to date to the late Holocene, with ochre specimens from the site of Kaalpi dating from 1300 BP and essentially absent before this time (Haley 1999: 61). Recent dates from a shelter with paintings at Durba Springs also support a relatively recent origin for the paintings (Veth n.d.a).

In contrast, a large number of the petroglyphs occur on quartz sandstone panels that have been significantly weathered, chemically altered, and in some cases covered in very thick coatings of natural varnish and other crusts. It is quite likely that many of these engraved motifs date to the Pleistocene.

Preliminary recording of motifs at these three sites suggests significant differences in the style, spatial patterning, and frequency of different classes of motifs between the engravings and the paintings. These may reflect a change in site function from the Pleistocene to the late Holocene (Veth n.d.b). Ephemeral residential sites in the Pleistocene may have become major aggregation sites in the Late Holocene. The engravings include a range of tracks, circles, concentric circles, and arcs in addition to a range of naturalistic motifs, including macropods, marsupials, and various birds. Noticeable are some very large anthropomorphs with extensive infilling and the so-called archaic faces: disembodied faces distributed through the arid zone from the coastal Pilbara to the Cleland Hills of central Australia, where Puritjarra is located (Dix 1977; Veth n.d.b). The geometric motifs, archaic faces, and large anthropomorphs can be found in differing proportions at Kaalpi, Serpent's Glen, and Durba. The majority of the motifs appear to refer to foraging themes, with the geometric motifs typical of those that have been documented as acting as mnemonic devices to map both water-hole and ancestral routes at the regional level (after Gould 1980). This is assumed to reflect an outside or inclusive symbolic configuration (Figures 12.3–12.6).

Left: Figure 12.3. Martu custodian at engraving site showing a highly patinated anthropomorph with complex pecked infill near the central Canning Stock Route (photograph by Peter Veth).

Below: Figure 12.4. Trail of engraved "bush-turkey" tracks—intaglio Panaramitee style—as commonly found throughout the Western Desert (photograph by Jo McDonald).

Figure 12.5. Diminutive pigment anthropomorph with exaggerated headdress and arranged parallel boomerangs (photograph by Peter Veth).

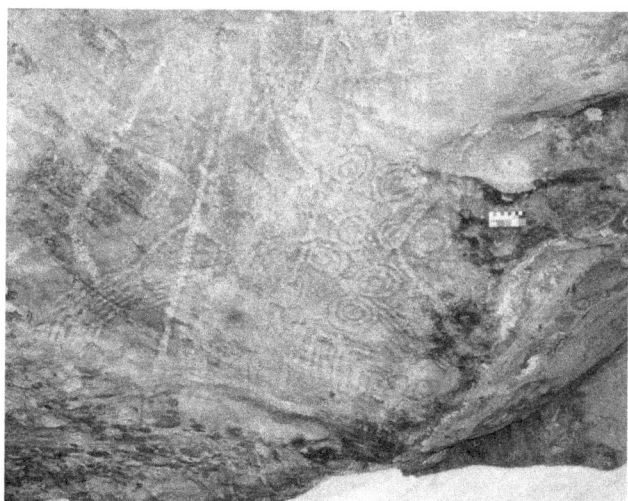

Figure 12.6. Complex superimposition involving multiple phases of art production with concentric circles, figurative "bush-turkey" figures, and attenuated anthropomorphs with distinctive body adornments and appendages—specific to this locality (photograph by Jo McDonald).

In contrast the paintings display greater assemblage diversity and contain a higher proportion of complex designs, although many of the fundamental geometric motifs are also present. Several Martu with whom I have worked had participated in ceremonies and gatherings at Kaalpi and Durba before contact with Europeans and were able to provide some mythological referents for the figurative paintings. Very detailed anthropomorphs are depicted with expanded headdresses, only recorded from this specific locality of the Western Desert. As such they may be signifiers of corporate identity. The use of inside/outside symbols is consistent with the increasing role of these sites for aggregation rather than dispersion (cf. Galt-Smith 1997).

Changes in Sedimentary Records

A shift from ephemeral occupation of a site under a high residential mobility regime to greater permanency may be registered in the matrix of a deposit. Such a process has been identified in the Pleistocene-aged occupation horizons at Puritjarra. Smith (1989: 99) concluded that "a major change in the use of the shelter is clear from changes in the character of the sediments, with large amounts of finely divided charcoal and fine rock fragments and a more open fabric."

Short-duration and low-intensity occupation episodes are more likely to leave intact living surfaces, due to longer periods of site abandonment and ongoing sedimentation processes and the decreased likelihood of disturbance during subsequent occupations (O'Connor et al. 1993). Shorter site visits also result in lower indices for artifact breakage due to treadage, as has been demonstrated by Hiscock (1988) for the Colless Creek sites.

Changes in Ochre Provenance

The study of ochre distributions from Puritjarra dating to the terminal Pleistocene demonstrates changing provenance (Smith et al. 1998). From 32,000 BP to 13,000 BP the red ochre source is located 125 km away, across major dunefields of the Western Desert. The occupants of Puritjarra either were moving over large portions of the desert lowlands to obtain ochre or were in contact with groups that had access to this quarry. The pattern changes significantly after 13,000 BP and particularly between 7500 BP and 1000 BP, with the ochre supply shifting to a much closer source in the Central Australian Ranges. More intensive use of the rockshelter and increased quantities of ochre are registered after 7500 BP. Smith et al. (1998: 287) synthesize both the changes in ochre source and other archaeological signatures of territoriality:

> If we see the production of pigment art . . . at the site as reflecting an increasing need to assert corporate rights and relationships to the site, and the more restricted catchment reflected in the ochres as reflecting an actual

reduction in residential mobility, we get a picture of a shift from an open spatially extensive pattern of land use to a more closed system with smaller group territories after 13,000 BP.

Continuity of Occupation at Sites

The three Western Desert sites all evidence a decrease, if not an absence, of occupational material during the last glacial maximum (O'Connor et al. 1998: 16; Smith 1988: Table 4; Thorley 1998b: 239). At least two of the sites are located adjacent to permanent water sources, a rare commodity in the Western Desert at any time during its human occupation. The expectation for groups continuing to occupy a region as conditions became harsher and semipermanent water sources were lost would be greater tethering at permanent waters. This is not the case for any of the three sites. As Thorley (1998b: 323) concludes:

> As a site near permanent water, Puritjarra would be expected to show indications of more intensive occupation around the time ephemerally-watered sites such as Kulpi Mara ceased to be occupied. This would be consistent with a model of populations falling back on more reliable "refuge" sites. Yet like Kulpi Mara, Puritjarra also shows a slowing of sediment and artifact deposition rates during the glacial maximum.

The glacial record from all three sites more comfortably fits groups who have restricted their territorial range rather than those who have continued to occupy a landscape employing a high residential mobility strategy.

IMPLICATIONS OF MOBILITY INDICATORS

In this chapter I have examined eight different lines of archaeological evidence that may be used to infer residential mobility from Pleistocene assemblages in the Western Desert. Such assemblages are relatively intractable when compared to similar-aged assemblages from tropical northern Australia. Multiple lines of evidence are clearly preferable and can act as potential cross-checks. The Pleistocene assemblages conform to the behavioral and archaeological predictions of Ambrose and Lorenz (1990: 10), where the archaeological correlates of hunter-gatherer behaviors in an unpredictable and scarce resource regime are defined.

I have argued that the identification of changes in configurations of residential mobility is central to understanding variability in the assemblages of these three Western Desert sites. The assumption that early occupation was by small and highly mobile groups finds empirical support in this analysis. Independent lines of evidence for resource stress, such as the index of fragmentation, help to calibrate differences in resource scarcity and hence stress from different regions. Detailed analyses of the intensity of stone reduction, the provenance of stone artifacts and ochre, the diversity of tools and art, and sedimentary records all have the potential to shed light on mobility strategies. It is clear from this review that flexibility in mobility patterns, including territorial abandonment and strategy switching, has always been a necessary part of risk minimization in Western Desert foraging behavior.

AUTHOR'S NOTE

A version of this chapter appeared previously under the title "Cycles of Aridity and Human Mobility: Risk-Minimization amongst Late Pleistocene Foragers of the Western Desert, Australia," in *Desert Peoples: Archaeological Perspectives*, edited by P. Veth, M. A. Smith, and P. Hiscock (Oxford: Blackwell Publishing, 2005).

REFERENCES

Ambrose, S. H., and K. G. Lorenz
1990   Social and Ecological Models of the Middle Stone Age in Southern Africa. In *The Emergence of Modern Humans*, edited by P. Mellars, pp. 3–33. Edinburgh University Press, Edinburgh.

Andrefsky, W.
1998   *Lithics: Macroscopic Approaches to Analysis*. Cambridge University Press, Cambridge.

Binford, L. R.
1980   Willow Smoke and Dogs' Tails: Hunter-Gatherer Settlement Systems and Archaeological Site Formation. *American Antiquity* 45: 4–20.

Binford, L. R., and N. M. Stone
1985   "Righteous Rocks" and Richard Gould: Some Observations on Misguided Debate. *American Antiquity* 50(1): 151–153.

Cane, S.
1984   Desert Camps: A Case Study of Stone Artifacts and Aboriginal Behaviour in the Western Desert. Ph.D. dissertation. Department of Prehistory, Australian National University, Canberra.

Conkey, M. W.
1980   The Identification of Hunter-Gatherer Aggregation Sites: The Case of Altamira. *Current Anthropology* 21: 609–630.

Dix, W.
1977   Facial Representations in Pilbara Rock Engravings. In *Form in Indigenous Art*, edited by P. J. Ucko, pp. 277–285. Prehistory and Material Culture Series No. 13. Australian Institute of Aboriginal Studies. Duckworth, London.

Edwards, D. A., and J. F. O'Connell

1995　Broad Spectrum Diets in Arid Australia. *Antiquity* 69: 769–783.

Galt-Smith, B.

1997　Motives for Motifs: Identifying Aggregation and Dispersion Settlement Patterns in the Rock Art Assemblages of Central Australia. Honors thesis. Department of Archaeology and Paleoanthropology, University of New England, Armidale.

Gorecki, P., M. Grant, S. O'Connor, and P. Veth

1997　The Morphology, Function and Antiquity of Grinding Implements in Northern Australia. *Archaeology in Oceania* 32: 141–150.

Gould, R. A.

1978　Archaeological Signatures of Stress in the Australian Desert. Paper presented at the 1978 Annual Meeting of the American Anthropological Association, Los Angeles.

1980　*Living Archaeology*. Cambridge University Press, Cambridge.

1985　The Empiricist Strikes Back: Reply to Binford. *American Antiquity* 50(3): 638–644.

1991　Arid-Land Foraging as Seen from Australia: Adaptive Models and Behavioral Realities. *Oceania* 62: 12–33.

1996　Faunal Reduction at Puntutjarpa Rockshelter, Warburton Ranges, Western Australia. *Archaeology in Oceania* 31(2): 72–86.

Haley, M.

1999　Kaalpi: Investigation of Archaeological Assemblages from the Calvert Ranges Western Desert, Western Australia. Honors thesis. Department of Archaeology, James Cook University, Townsville.

Hayden, B.

1976　Australian Western Desert Lithic Technology: An Ethno- archaeological Study of Variability in Material Culture. Ph.D. dissertation. Department of Anthropology, University of Toronto.

1979　*Paleolithic Reflections*. Australian Institute of Aboriginal Studies, Canberra.

Hiscock, P.

1988　Prehistoric Settlement Patterns and Artifact Manufacture at Lawn Hill, Northwest Queensland. Ph.D. dissertation. University of Queensland, Brisbane.

1994　Technological Responses to Risk in Holocene Australia. *Journal of World Prehistory* 8(3): 267–292.

1996　Mobility and Technology in the Kakadu Coastal Wetlands. *Indo-Pacific Prehistory Association Bulletin* 15(2): 151–157.

Hiscock, P., and P. Veth

1991　Change in the Australian Desert Culture: A Reanalysis of Tulas from Puntutjarpa. *World Archaeology* 22: 332–345.

Kelly, R. L.

1992　Mobility/Sedentism: Concepts, Archaeological Measures, and Effects. *Annual Review of Anthropology* 21: 43–66.

1995   *The Foraging Spectrum*. Smithsonian Institution Press, Washington, D.C.

Mabbutt, J. A.

1977   *Desert Landforms*. Vol. 2. Australian National University Press, Canberra.

Morse, K.

1994   West Side Story: Towards a Prehistory of the Cape Range Peninsula. Ph.D. dissertation. Centre for Archaeology, University of Western Australia.

O'Connor, S., P. Veth, and A. Barham

1999   Cultural versus Natural Explanations for Lacunae in Aboriginal Occupation Deposits in Northern Australia. *Quaternary International* 59: 61–70.

O'Connor, S., P. Veth, and C. Campbell

1998   Serpent's Glen Rockshelter: Report of the First Pleistocene-Aged Occupation Sequence from the Western Desert. *Australian Archaeology* 46: 12–21.

O'Connor, S., P. Veth, and N. Hubbard

1993   Changing Interpretations of Postglacial Human Subsistence and Demography in Sahul. In *Sahul in Review: The Archaeology of Australia, New Guinea and Island Melanesia*, edited by M. A. Smith, M. Spriggs, and B. Fankhauser, pp. 95–105. Australian National University Press, Canberra.

Rolland, N., and H. L. Dibble

1990   A New Synthesis of Middle Paleolithic Variability. *American Antiquity* 55(3): 480–499.

Ross, A., T. Donnelly, and R. Wasson

1992   The Peopling of the Arid Zone: Human-Environment Interactions. In *The Naïve Lands*, edited by J. Dodson, pp. 76–114. Longman-Cheshire, Melbourne.

Shott, M. J.

1986   Technological Organization and Settlement Mobility: An Ethnographic Examination. *Journal of Anthropological Research* 42: 15–51.

1989   On Tool-Class Use Lives and the Formation of Archaeological Assemblages. *American Antiquity* 54(1): 9–30.

Smith, M. A.

1986   The Antiquity of Seed Grinding in Central Australia. *Archaeology in Oceania* 21: 29–39.

1988   The Pattern and Timing of Prehistoric Settlement in Central Australia. Ph.D. dissertation. University of New England, Armidale.

1989   The Case for a Resident Human Population in the Central Australian Ranges during Full Glacial Aridity. *Archaeology in Oceania* 24: 93–105.

1996   Prehistory and Human Ecology in Central Australia: An Archaeological Perspective. In *Exploring Central Australia: Society, Environment and the 1894 Horn Expedition*, edited by S. R. Morton and D. J. Mulvaney, pp. 61–73. Surrey Beatty and Sons, Chipping Norton.

Smith, M. A., B. Fankhauser, and M. Jercher

1998   The Changing Provenance of Red Ochre at Puritjarra Rockshelter Central Australia: Late Pleistocene to Present. *Proceedings of the Prehistoric Society* 64: 275–292.

Smith, M. A., J. R. Prescott, and M. J. Head
1997   Comparison of 14C and Luminescence Chronologies at Puritjarra Rock Shelter, Central Australia. *Quaternary Geochronology (Quaternary Science Reviews)* 16: 1–22.

Thorley, P. B.
1998a   Pleistocene Settlement in the Australian Arid Zone: Occupation of an Inland Riverine Landscape in the Central Australian Ranges. *Antiquity* 72: 34–45.
1998b   Shifting Location, Shifting Scale: A Regional Landscape Approach to the Prehistoric Archaeology of the Palmer River Catchment, Central Australia. Ph.D. dissertation. School of Southeast Asian and Australian Studies, Northern Territory University, Darwin.

Thorley, P., and B. Gunn
1996   Archaeological Research from the Eastern Border Lands of the Western Desert. Paper presented at the Western Desert Origins Workshop. Australian Linguistic Institute, Canberra.

Tonkinson, R.
1991   *The Mardu Aborigines: Living the Dream in Australia's Desert.* Holt, Rinehart and Winston, Fort Worth.

Varien, M. D.
1999   *Sedentism and Mobility in a Social Landscape.* University of Arizona Press, Tucson.

Veth, P.
1987   Martujarra Prehistory: Variation in Arid Zone Adaptations. *Australian Archaeology* 25: 102–111.
1989   Islands in the Interior: A Model for the Colonization of Australia's Arid Zone. *Archaeology in Oceania* 24: 81–92.
1993a   The Aboriginal Occupation of the Montebello Islands, Northwest Australia. *Australian Aboriginal Studies* 2: 39–50.
1993b   *Islands in the Interior: The Dynamics of Prehistoric Adaptations within the Arid Zone of Australia.* Archaeological Series No. 3. International Monographs in Prehistory, Ann Arbor.
1995   Aridity and Settlement in North West Australia. *Antiquity* 69: 733–746.
1996   Current Archaeological Evidence from the Little and Great Sandy Deserts. In *Archaeology of Northern Australia: Regional Perspectives*, edited by P. Veth and P. Hiscock, pp. 50–65. Tempus, Archaeology and Material Culture Studies in Anthropology No. 4. Anthropology Museum, University of Queensland, St. Lucia.
2005   Conclusion: Major Themes and Future Research Directions. In *Desert Peoples: Archaeological Perspectives*, edited by P. Veth, M. A. Smith, and P. Hiscock, pp. 293–300. Blackwell Publishing, Oxford.

n.d.a   Kaalpi: The Archaeology of a Sandstone Outlier in the Western Desert. In preparation.

n.d.b   Origins of the Western Desert Language: Convergence in Linguistic and Archaeological Space and Time Models. *Archaeology in Oceania*. In press.

Veth, P., and S. O'Connor
1996   A Preliminary Analysis of Basal Grindstones from the Carnarvon Range, Little Sandy Desert. *Australian Archaeology* 43: 20–22.

Yellen, J. E.
1976   Long-term Hunter-Gatherer Adaptation to Desert Environments: A Biogeographical Perspective. *World Archaeology* 8(3): 262–274.

Contributors

Michael S. Alvard, associate professor, Department of Anthropology, Texas A&M University, College Station, Texas.

Lewis R. Binford, professor emeritus, Department of Anthropology, Southern Methodist University, Dallas, Texas.

Claudia Chang, professor, Department of Anthropology, Sweet Briar College, Sweet Briar, Virginia.

Nathan Craig, lecturer, Department of Anthropology, University of California, Santa Barbara, California.

Napoleon A. Chagnon, professor emeritus, Department of Anthropology, University of California, Santa Barbara, California.

Russell D. Greaves, research associate, Peabody Museum of Archaeology and Ethnology, Harvard University, Cambridge, Massachusetts and adjunct associate professor, Department of Anthropology, University of Utah, Salt Lake City, Utah.

Robert L. Kelly, professor, Department of Anthropology, University of Wyoming, Laramie, Wyoming.

Marsha D. Ogilvie, adjunct assistant professor, Department of Anthropology, University of New Mexico, Albuquerque, New Mexico.

Gustavo G. Politis, professor, Universidad Nacional de La Plata, La Plata, Argentina.

Lin Poyer, professor, Department of Anthropology, University of Wyoming, Laramie, Wyoming.

Frédéric Sellet, associate professor, Department of Anthropology, University of Kansas, Lawrence, Kansas.

Paul T. Thacker, assistant professor, Department of Anthropology, Wake Forest University, Winston-Salem, North Carolina.

Bram Tucker, assistant professor, Department of Anthropology, University of Georgia, Athens, Georgia.

Peter Veth, director of research, Australian Institute of Aboriginal and Torres Strait Islander Studies, Canberra, Australia.

Pei-Lin Yu, Power Office archaeologist, U.S. Bureau of Reclamation, Boise, Idaho.

Index

Page numbers in *italics* refer to illustrations.

www.ingramcontent.com/pod-product-compliance
Lightning Source LLC
Chambersburg PA
CBHW020829270326
41928CB00006B/472